Microwave Indices from Active and Passive Sensors for Remote Sensing Applications

Microwave Indices from Active and Passive Sensors for Remote Sensing Applications

Special Issue Editors

Emanuele Santi
Simonetta Paloscia

MDPI • Basel • Beijing • Wuhan • Barcelona • Belgrade

Special Issue Editors
Emanuele Santi
Consiglio Nazionale delle Ricerche
Italy

Simonetta Paloscia
Consiglio Nazionale delle Ricerche
Italy

Editorial Office
MDPI
St. Alban-Anlage 66
4052 Basel, Switzerland

This is a reprint of articles from the Special Issue published online in the open access journal *Remote Sensing* (ISSN 2072-4292) from 2017 to 2019 (available at: https://www.mdpi.com/journal/remotesensing/special_issues/microwave_RS)

For citation purposes, cite each article independently as indicated on the article page online and as indicated below:

LastName, A.A.; LastName, B.B.; LastName, C.C. Article Title. *Journal Name* **Year**, *Article Number*, Page Range.

ISBN 978-3-03897-820-6 (Pbk)
ISBN 978-3-03897-821-3 (PDF)

Contents

About the Special Issue Editors

Emanuele Santi Emanuele Santi received his Master's degree in Electronic Engineering from the University of Florence and his PhD in Earth's Remote Sensing Techniques from the University of Basilicata. Since 1998, he has been part of the Microwave Remote Sensing Group at the Institute of Applied Physics of the National Research Council (IFAC–CNR), and in 2009, he obtained a permanent staff position. His research deals with the study of microwave emission and scattering from sea and land surfaces and their relationships with snow, soil, and vegetation parameters, using data from passive and active microwave sensors operating from ground-based platforms, aircraft, and satellites. The main topic of his research activity is the development and validation of forward electromagnetic models and statistical inversion algorithms for estimating the geophysical parameters of soil, sea, snow, and vegetation. Recently, he focused on the application of machine learning methods and statistical approaches to the retrieval of geophysical parameters from microwave satellite data. The main outcomes of this activity were prototypal algorithms based on artificial neural networks (ANNs) for estimating soil moisture, vegetation biomass, and snow depth/water equivalent at the local and global scale using multifrequency data from active and passive microwave sensors. He also developed a procedure for merging experimental data and EM model simulations for training the ANNbased retrieval algorithms and a disaggregation method for improving the spatial resolution of satellite radiometers, which is currently implemented in the LPRM AMSR2 Downscaled Soil Moisture product hosted by NASA. He was and is currently involved in many international programs funded by the European Community, the European and Italian Space Agencies (ESA and ASI), and the Japanese Space Agency (JAXA), acting as a team leader, Work Package leader, or co-investigator. He has authored or coauthored 128 papers, including ISI books, journals, and proceedings of international conferences (source: Scopus; h-index = 18), and since 2003 has reviewed 178 papers for 16 international journals, receiving the 'Top Reviewers Award for Sentinels of Science: Earth and Planetary Sciences' in 2016 (source: Publons). He has also been a referee of the International Geoscience and Remote Sensing Symposium (IGARSS) since 2010. He has received the 2017 J-STARS best paper prize award for the paper "Vegetation Water Content Retrieval by Means of Multifrequency Microwave Acquisitions from AMSR2" (DOI: 10.1109/JSTARS.2017.2703629). He is acting as an Associate Editor for the MDPI journal *Remote Sensing*, and has been a Guest Editor for the Special Issue *"Microwave Indices from Active and Passive Sensors for Remote Sensing Applications"* of the same journal. He is a member of the Institute of Electric and Electronic Engineers (IEEE) and of the Microwave Remote Sensing Center (CETEM). He is the conference Chair of SPIE Europe Remote Sensing and Vice-Chair of the GRS29 Central–North Italy chapter of GRSS.

Simonetta Paloscia joined the National Research Council (CNR), and since 1982, when she won a national competition, has been a staff scientist there. She started her work on the remote sensing of land surfaces by studying experimental relationships between microwave emission and vegetation features. Since 1987, she has been with the Institute of Applied Physics (IFAC–CNR), where she has continued her research in the remote sensing of soil moisture and vegetation by using data from both passive and active microwave sensors and studying relationships between emission and scattering from vegetation-covered soils. Since 1993, she has extended her interest to the remote sensing of forests, and in the late 1990s, to snow-covered surfaces. Her current research interest concerns the study of microwave emission and scattering of land surfaces for the retrieval of soil moisture, vegetation biomass, and snow water equivalent.Since 2001, she has been a Senior Scientist at IFAC–CNR, and since January 2004, she has been the leader of the IFAC Microwave Remote Sensing Group. In 2010, she was nominated Head of Research of CNR.She had a temporary teaching contract of "Microwave Remote Sensing Applications" for the Professional Master "Geomatics and Natural Resources Evaluation" at the "Istituto Agronomico per l'Oltremare" of the Ministry of Foreign Affairs in Florence from 1994 to 2010.Simonetta Paloscia has been a coinvestigator in several international microwave remote sensing airborne campaigns since the first SAR-580 in 1981, and the following AGRISAR86, AGRISCATT'87, AGRISCATT'88, MAC'91, EMISAR, and EMIRAD. In particular, she was deeply involved in the NASA/DLR/ASI SIR-C/X-SAR project, where she coordinated the activity on the Supersite for Hydrology in Tuscany, Italy. She also organized and coordinated several field campaigns on different sites in Italy with ground-based and airborne microwave radiometers designed, realized, and operated by the IFAC Microwave Remote Sensing Group. Simonetta Paloscia participated as a PI or CI for IFAC–CNR in many international projects of the European Commission (NOPEX-Forest Dynamo, ReSeDA, ENVISNOW, FLAUBERT, FLOODMAN, ERANET), European Space Agency (ESA Soil Moisture, LEIMON, GRASS, SMOS, Sentinel-1), ESA-AO (ERS-1/2 and ENVISAT/ASAR), and NASDA/JAXA-AO (JERS-1, AQUA/AMSR-E, GCOM-W/AMSR2). She is a member of the JAXA/Advanced Earth Observing Satellite Science Team for the use of ADEOS II-AMSR/GCOM-AMSR2 microwave data in algorithms for measuring soil moisture and vegetation biomass. She is part of the Science Team in the NASA/SMAP (Soil Moisture Active and Passive) project. Simonetta Paloscia was chair of the local organizing committee of the IEEE International Geoscience and Remote Sensing Symposium IGARSS'95 in Florence, Italy, and since then has frequently been a member of the organizing/technical committee of IGARSS. She was the organizer and General Co-Chair of the Specialist Meeting on Microwave Radiometry and Remote Sensing Applications, MICRORAD (sponsored by IEEE–GRSS), held in Florence, Italy, in 1999 and 2008, and is a permanent member of the MICRORAD Steering Committee. She was the organizer and General Chair of the URSI comm-F meeting (Spectral Signatures) held in Florence, Italy, in 2010.

Editorial

Editorial for the Special Issue "Microwave Indices from Active and Passive Sensors for Remote Sensing Applications"

Simonetta Paloscia and Emanuele Santi *

National Research Council, Institute of Applied Physics, 50019 Florence, Italy; s.paloscia@ifac.cnr.it
* Correspondence: e.santi@ifac.cnr.it; Tel.: +39-055-5226431

Received: 16 January 2019; Accepted: 17 January 2019; Published: 7 March 2019

Keywords: Microwave Radiometry; Microwave Indices; Soil Moisture Content; Vegetation Biomass; Snow Depth and Snow Water Equivalent

---

Since the early 1980s, the capabilities of satellite sensors operating at microwaves for the remote sensing of Earth's surface have been widely assessed in a number of studies (e.g., [1]). Due to the high sensitivity of microwave emission and scattering to water content in the observed surfaces, microwave sensors, both active (scatterometers, SAR) and passive (radiometers), can provide useful information on the hydrological and carbon cycles, enabling the retrieval of the main soil, snow and vegetation parameters (e.g., [2,3]).

Further research demonstrated that a significant accuracy improvement could be obtained by combining data collected at different frequencies and polarizations in appropriate indices, with respect to the one achievable with single frequency/polarization observations. In particular, Microwave Indices have been successfully related to the main geophysical parameters associated to land hydrological cycle, such as soil moisture content, plant water content and snow depth or snow water equivalent (e.g., [4–6]).

The research conducted on microwave radiometry pointed out that given combinations of data acquired by satellite sensors such as SSM/I, AMSR-E and AMSR2 can be successfully related to snow and vegetation parameters. Among these indices, the Frequency Index, which is the difference between the brightness temperatures at two frequencies (i.e., Ka and Ku bands), showed a marked sensitivity to snow cover [7], while the Spectral Polarization Indices, which are obtained combining different polarizations and frequencies, were able to correctly identify relevant values of snow depths [8–11]. Moreover, the Polarization Indices, obtained at lower frequencies (i.e., X- and C-bands) as the difference between the two polarizations (H and V), showed a significant correlation to the vegetation biomass, allowing the correction for the effects of vegetation in the retrievals of soil moisture and the retrieval of herbaceous and agricultural vegetation biomass [11].

Similar results have been obtained in the case of active sensors (i.e., SAR and scatterometers) operating at C- and X- bands; their derived indices, as the Radar Vegetation Index (RVI) [12], were proven to be highly related to vegetation structure and vegetation biomass. These indices have been largely used for correcting the effect of vegetation cover in soil moisture retrievals and in estimating the vegetation biomass [13,14].

This special issue was aimed at providing an overview of the capabilities of Microwave Indices for remote sensing applications. Besides demonstrating their sensitivity to the main parameters of soil, vegetation and snow, the special issue was focused on the use of Microwave Indices in retrieval algorithms devoted to the estimate of soil moisture, vegetation biomass and snow depth/water equivalent, at both the local and global scale.

The obtained results sufficiently demonstrated that the synergy between observations at different frequencies and polarizations can significantly improve the capabilities of microwave sensors in observing and retrieving parameters related to the hydrological and the carbon cycles.

In particular, the studies belonging to this special issue demonstrated the capabilities of microwave derived indices for a number of applications that span from the retrieval of soil moisture based on Sentinel-1 data in wetland ecosystems [15] to the assessment of RVI as microwave metric of vegetation cover [16]. Also, the potential of Sentinel-1 cross ratio VH/VV in monitoring crop conditions was successfully evaluated [17] and the use of SAR derived vegetation descriptors for improving the soil moisture retrieval was exploited in [18], while the fusion of scatterometric and SAR data, aimed again at the soil moisture retrieval, is proposed in [19]. Finally, the synergy of Sentinel-1 and 2 in detecting grassland phenology in mountain regions is exploited in [20]. As applications devoted to snow monitoring, the co-pol. and dual-frequency ratios at Ku-, X- and C-bands were demonstrated to be adequately sensitive to variations in snow thickness for the Antarctic first year ice [21]. In [22], the combination of active and passive microwave acquisitions allowed characterizing the seasonal behavior of snow properties and retrieving the effective correlation length and the snow water equivalent. Finally, an overview of the potential of microwave indices obtained from multi-frequency/polarization radiometry in monitoring soil moisture, snow depth and vegetation biomass was presented in [23].

Author Contributions: The two authors contributed equally to all aspects of this editorial.

Acknowledgments: The authors would like to thank the authors who contributed to this Special Issue and to the reviewers who dedicated their time for providing the authors with valuable and constructive recommendations.

Conflicts of Interest: The authors declare no conflict of interest.

References

1. Hollinger, J.; Lo, R.; Poe, G.; Savage, R.; Pierce, J. *Special Sensor Microwave/Imager User's Guide*; Naval Research Laboratory: Washington, DC, USA, 1987.
2. Ulaby, F.T.; Moore, R.K.; Fung, A.K. *Microwave Remote Sensing: Active and Passive, Vol. II*; Addison-Wesley: Reading, MA, USA, 1982.
3. Ulaby, F.T.; Moore, R.K.; Fung, A.K. *Microwave Remote Sensing: Active and Passive, Vol. III*; Artech House: Nordwood, MA, USA, 1986.
4. Njoku, E.G.; Jackson, T.J.; Lakshmi, V.; Chan, T.K.; Nghiem, S.V. Soil moisture retrieval from AMSR-E. *IEEE Trans. Geosci. Remote Sens.* **2003**, *41*, 215–229. [CrossRef]
5. Chukhlantsev, A.A. *Microwave Radiometry of Vegetation Canopies*; Springer: Dordrecht, The Netherlands, 2006.
6. Chang, T.C.; Foster, J.L.; Hall, D.K.; Rango, A.; Hartline, B.K. Snow water equivalent estimation by microwave radiometry. *Cold Reg. Sci. Technol.* **1982**, *5*, 259–267. [CrossRef]
7. Kelly, R.; Chang, A.T.C. Development of a passive microwave global snow depth retrieval algorithm for Special Sensor Microwave Imager (SSM/I) and Advanced Microwave Scanning Radiometer-EOS (AMSR-E) data. *Radio Sci.* **2003**, *38*, 8040–8076. [CrossRef]
8. Foster, J.L.; Chang, A.T.C.; Hall, D.K. Comparison of snow mass estimates from a prototype passive microwave snow algorithm, a revised algorithm and a snow depth climatology. *Remote Sens. Environ.* **1997**, *62*, 132–142. [CrossRef]
9. Tsutsui, H.; Koike, T. Development of Snow Retrieval Algorithm Using AMSR-E for the BJ Ground-Based Station on Seasonally Frozen Ground at Low Altitude on the Tibetan Plateau. *J. Meteor. Soc. Jpn.* **2012**, *90*, 99–112. [CrossRef]
10. Paloscia, S.; Pampaloni, P. Microwave Polarization Index for Monitoring Vegetation Growth. *IEEE Trans. Geosci. Remote Sens.* **1988**, *26*, 617–621. [CrossRef]
11. Santi, E.; Pettinato, S.; Paloscia, S.; Pampaloni, P.; Macelloni, G.; Brogioni, M. An algorithm for generating soil moisture and snow depth maps from microwave spaceborne radiometers: HydroAlgo. *Hydrol. Earth Syst. Sci.* **2012**, *16*, 3659–3676. [CrossRef]

12. Kim, Y.; Jackson, T.; Bindlish, R.; Lee, H.; Hong, S. Radar Vegetation Index for Estimating the Vegetation Water Content of Rice and Soybean. *IEEE Geosci. Remote Sens. Lett.* **2012**, *9*, 564–568. [CrossRef]

13. Paloscia, S.; Pettinato, S.; Santi, E.; Notarnicola, C.; Pasolli, L.; Reppucci, A. Soil moisture mapping using Sentinel-1 images: Algorithm and preliminary validation. *Remote Sens. Environ.* **2013**, *134*, 234–248. [CrossRef]

14. Kim, Y.; Jackson, T.; Bindlish, R.; Hong, S.; Jung, G.; Lee, K. Retrieval of Wheat Growth Parameters With Radar Vegetation Indices. *IEEE Geosci. Remote Sens. Lett.* **2014**, *11*, 808–812. [CrossRef]

15. Dabrowska-Zielinska, K.; Musial, J.; Malinska, A.; Budzynska, M.; Gurdak, R.; Kiryla, W.; Bartold, M.; Grzybowski, P. Soil Moisture in the Biebrza Wetlands Retrieved from Sentinel-1 Imagery. *Remote Sens.* **2018**, *10*, 1979. [CrossRef]

16. Szigarski, C.; Jagdhuber, T.; Baur, M.; Thiel, C.; Parrens, M.; Wigneron, J.-P.; Piles, M.; Entekhabi, D. Analysis of the Radar Vegetation Index and Potential Improvements. *Remote Sens.* **2018**, *10*, 1776. [CrossRef]

17. Vreugdenhil, M.; Wagner, W.; Bauer-Marschallinger, B.; Pfeil, I.; Teubner, I.; Rüdiger, C.; Strauss, P. Sensitivity of Sentinel-1 Backscatter to Vegetation Dynamics: An Austrian Case Study. *Remote Sens.* **2018**, *10*, 1396. [CrossRef]

18. Li, J.; Wang, S. Using SAR-Derived Vegetation Descriptors in a Water Cloud Model to Improve Soil Moisture Retrieval. *Remote Sens.* **2018**, *10*, 1370. [CrossRef]

19. Bauer-Marschallinger, B.; Paulik, C.; Hochstöger, S.; Mistelbauer, T.; Modanesi, S.; Ciabatta, L.; Massari, C.; Brocca, L.; Wagner, W. Soil Moisture from Fusion of Scatterometer and SAR: Closing the Scale Gap with Temporal Filtering. *Remote Sens.* **2018**, *10*, 1030. [CrossRef]

20. Stendardi, L.; Karlsen, S.R.; Niedrist, G.; Gerdol, R.; Zebisch, M.; Rossi, M.; Claudia Notarnicola, C. Exploiting time series of Sentinel-1 and Sentinel-2 imagery to detect grassland phenology in mountain regions. *Remote Sens.* **2019**, *11*, 542. [CrossRef]

21. Nandan, V.; Geldsetzer, T.; Mahmud, M.; Yackel, J.; Ramjan, S. Ku-, X- and C-Band Microwave Backscatter Indices from Saline Snow Covers on Arctic First-Year Sea Ice. *Remote Sens.* **2017**, *9*, 757. [CrossRef]

22. Lemmetyinen, J.; Derksen, C.; Rott, H.; Macelloni, G.; King, J.; Schneebeli, M.; Wiesmann, A.; Leppänen, L.; Kontu, A.; Pulliainen, J. Retrieval of Effective Correlation Length and Snow Water Equivalent from Radar and Passive Microwave Measurements. *Remote Sens.* **2018**, *10*, 170. [CrossRef]

23. Paloscia, S.; Pampaloni, P.; Santi, E. Radiometric Microwave Indices for Remote Sensing of Land Surfaces. *Remote Sens.* **2018**, *10*, 1859. [CrossRef]

Article

Ku-, X- and C-Band Microwave Backscatter Indices from Saline Snow Covers on Arctic First-Year Sea Ice

Vishnu Nandan *, Torsten Geldsetzer, Mallik Mahmud, John Yackel and Saroat Ramjan

Cryosphere Climate Research Group, Department of Geography, University of Calgary, Calgary, AB T2N 1N4, Canada; geldsetz@ucalgary.ca (T.G.); msmahmud@ucalgary.ca (M.M.); yackel@ucalgary.ca (J.Y.); saroat.ramjan@ucalgary.ca (S.R.)
* Correspondence: vishnunandan.nandaku@ucalgary.ca

Academic Editors: Emanuele Santi, Simonetta Paloscia, Xiaofeng Li and Prasad S. Thenkabail
Received: 19 June 2017; Accepted: 19 July 2017; Published: 23 July 2017

Abstract: In this study, we inter-compared observed Ku-, X- and C-band microwave backscatter from saline 14 cm, 8 cm, and 4 cm snow covers on smooth first-year sea ice. A Ku-, X- and C-band surface-borne polarimetric microwave scatterometer system was used to measure fully-polarimetric backscatter from the three snow covers, near-coincident with corresponding in situ snow thermophysical measurements. The study investigated differences in co-polarized backscatter observations from the scatterometer system for all three frequencies, modeled penetration depths, utilized co-pol ratios, and introduced dual-frequency ratios to discriminate dominant polarization-dependent frequencies from these snow covers. Results demonstrate that the measured co-polarized backscatter magnitude increased with decreasing snow thickness for all three frequencies, owing to stronger gradients in snow salinity within thinner snow covers. The innovative dual-frequency ratios suggest greater sensitivity of Ku-band microwaves to snow grain size as snow thickness increases and X-band microwaves to snow salinity changes as snow thickness decreases. C-band demonstrated minimal sensitivity to changes in snow salinities. Our results demonstrate the influence of salinity associated dielectric loss, throughout all layers of the three snow covers, as the governing factor affecting microwave backscatter and penetration from all three frequencies. Our "plot-scale" observations using co-polarized backscatter, co-pol ratios and dual-frequency ratios suggest the future potential to up-scale our multi-frequency approach to a "satellite-scale" approach, towards effective development of snow geophysical and thermodynamic retrieval algorithms on smooth first-year sea ice.

Keywords: active microwaves; snow; sea ice; co-pol ratio; dual-frequency ratios

---

1. Introduction and Background

Arctic sea ice extent, age, volume, and thickness have undergone rapid decrease during the past three decades, with the Arctic Ocean on a path to a new climate regime influenced by a thinner sea ice cover and being more and more dominated by first-year ice (FYI) [1,2]. Widespread decline of multi-year ice (MYI) replaced by FYI, with associated decline in spring snow depth are well accepted and documented [1,3,4]. A warming Arctic triggers delayed sea ice "freeze-up" which could lead to thinner FYI, thereby decreasing adequate time for snow accretion on FYI. These thinner snow covers on FYI are likely to become more saline owing to the accentuated vapor and temperature gradients across the ocean–atmosphere interface [5]. Snow electro-thermophysical properties (through accumulation and redistribution) on FYI exhibit high spatiotemporal variability [6] from hourly to seasonal time scales throughout its annual cycle, and plays a central role in regulating sea ice growth and decay processes [7].

Active microwave remote sensing techniques employing space-borne scatterometry and Synthetic Aperture Radar (SAR) have proven to be effective tools to characterize the electrical and thermodynamic state of snow covered FYI, where snow cover plays a critical role in microwave interactions (propagation and scattering) within the snow/sea ice system [8–14]. With changes in snow thickness and its associated thermophysical properties such as snow temperature, snow salinity, snow density and snow grain size (grain radius or specific surface area) on FYI, microwaves exhibit characteristic variations within different snow cover types on FYI [15,16]. Snow cover can influence and modify microwave interactions on FYI (dependent on incidence angle (θ_{inc}), polarization and wavelength) in two ways. First, through thermodynamically controlled effects (e.g., snow wetness and brine volume) on snow dielectrics (dielectric permittivity and loss), and secondly due to microwave scattering (surface and volume scattering) owing to different snow thermophysical properties. Fluctuations in near-surface air temperature change the snow temperature, which in turn modifies the brine volume at/near the snow/sea ice interface and within-snow layers, following the eutectic phase distribution curve. Snow salinity controls the penetration depth, and it influences the partitioning between scattering at interfaces, i.e. surface scattering (air/snow, within-snow at density gradients, snow/ice) and volume scattering by modulating the absorption of both the incident and the reflected/scattered microwave radiation. This in turn alters dielectric and thermodynamic properties of the snow cover, which in turn could lead to uncertainties in snow thickness estimations on FYI.

Understanding complex microwave interactions utilizing a multi-frequency approach from different snow cover types on Arctic FYI requires further examination. Space-borne scatterometer systems such as the Advanced Scatterometer (ASCAT) (C-band; 5.2 GHz) and Quick Scatterometer (QuikSCAT) (Ku-band; 13.4 GHz), and SAR systems such as Sentinel-1, RADARSAT-2 (C-band; 5.5 GHz), TerraSAR-X and Constellation of Small Satellites for the Mediterranean basin Observation (Cosmo-SkyMed) (X-band; 9.6 GHz) operate over a wide range of varying spatiotemporal resolutions, coverage areas, and polarization combinations. Owing to high spatiotemporal variability of the snow cover on FYI, correlating a SAR pixel to the underlying snow thermophysical properties adds significant uncertainty for direct thermophysical interpretation. Additionally, all of the above-mentioned space-borne platforms operate over coarse temporal resolutions making it extremely difficult to quantify plot-scale variations in microwave backscatter due to the dynamic changes in snow thicknesses. Plot-scale studies are crucial to understand detailed high-resolution behavior of various thermophysical processes from different snow cover types, which dictate the microwave backscattering behavior at multiple frequencies. Using surface-based and air-borne multi-frequency and multi-polarization measurements, a significant amount of research has investigated microwave backscatter sensitivity to plot-scale changes in snow thermophysical properties (e.g., [14,17–20]). However, no previous studies have explored the potential of characterizing these plot-scale polarization-dependent multi-frequency microwave backscatter diversity from different snow cover types on Arctic FYI.

2. Research Objectives

This study presents surface-based fully-polarimetric microwave backscatter measurements acquired at Ku- (17.25 GHz), X- (9.65 GHz) and C-band (5.52 GHz) frequencies from saline snow covers (14 cm, 8 cm and 4 cm) on smooth FYI. These frequencies relate closely to the center frequencies of recent and currently operational space-borne scatterometer and SAR systems like ASCAT, QuikSCAT, TerraSAR-X, COSMO-SkyMed, and RADARSAT-2. Utilizing a surface-based multi-frequency polarimetric microwave scatterometer system, this study explores the potential of a multi-frequency observational approach to characterize the diversity of Ku-, X-, and C-band σ_{VV}^{0} and σ_{HH}^{0}, its derived co-polarization ratios (γ_{co}), and the innovative dual-frequency ratios ($\gamma_{DFR[PP]}$), from the three different snow cover cases. Here, σ^{0} or sigma-naught is the conventional normalized measure of the radar return per unit area, from a distributed target, and VV or HH denotes co-polarized backscatter in vertical and horizontal polarizations, respectively. To accomplish our research objectives within the study context, we address the following questions:

(a)   What are the observable differences in Ku-, X- and C-band σ^0_{VV} and σ^0_{HH} and modeled penetration depths, as a function of θ_{inc}, from the saline 14 cm, 8 cm and 4 cm snow covers on FYI?

(b)   How do various polarization-dependent dual-frequency ratios ($\gamma_{DFR[PP]}$) and co-pol ratios (γ_{co}) change with 14 cm, 8 cm and 4 cm snow covers on FYI?

(c)   Based on differences in σ^0_{VV}, σ^0_{HH}, γ_{co}, and $\gamma_{DFR[PP]}$ from the 14 cm, 8 cm and 4 cm snow covers answered by (a) above, which polarization and frequency exhibit the greatest sensitivity with respect to changes in snow thickness?

3. Methods

3.1. Study Area

The surface based Ku-, X- and C-band scatterometer data (σ^0_{VV}, σ^0_{HH}, σ^0_{HV} and σ^0_{VH}) were acquired on 19 May 2012, from homogenous and saline 14 cm, 8 cm and 4 cm snow covers. The scatterometer measurements were acquired at ~9:45 a.m. (for 14 cm), ~12:25 p.m. (for 4 cm) and ~9:30 p.m. (for 8 cm) local time, respectively. The study site is situated near Resolute Bay, Nunavut, Canada (74.70°N, 95.63°W) (Figure 1), and dominated by relatively smooth, slightly deformed land-fast first-year ice types. The snow covers selected for this study are representative of snow covers on smooth FYI in the Canadian Arctic Archipelago (CAA), and their thicknesses fall closely to the mean interannual snow thickness ranges prior to melt-onset, previously reported by [6,9]. These snow cover cases fall on the lower end of the snow thicknesses, and is important to investigate, since increasingly thinner snow covers on Arctic sea ice are observed during mid- to late-winter seasons [3]. The air temperature (T_a) (in °C) measured using an on-sea ice installed micro-meteorological station (measured at one minute intervals and resampled hourly), were found to be consistently cold close to -13.5 °C, throughout the sampling period (Figure 2). No significant precipitation events such as snow falls were reported on 19 May 2012.

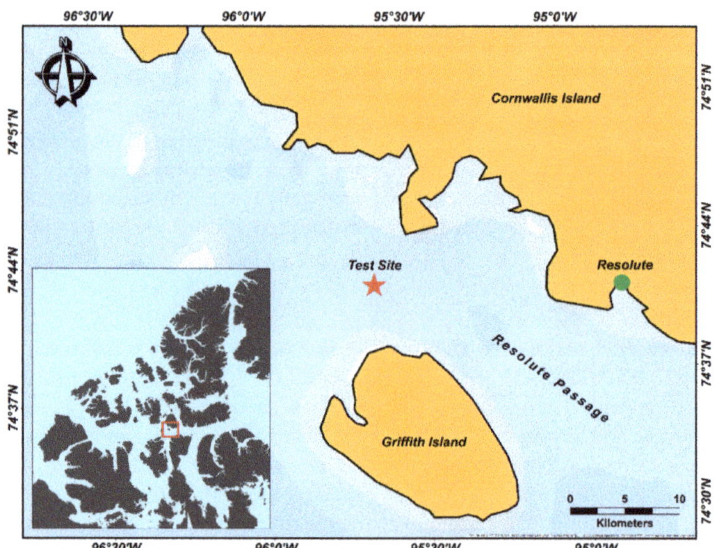

Figure 1. Map of Resolute Bay region (indicated in green dot) in Resolute Passage in the Canadian Arctic, Nunavut, Canada. Study site location is indicated in red star. Snow covered first-year ice accumulated areas are depicted in light blue, and land in brown. Note: A similar figure with different color scheme can be found in [14].

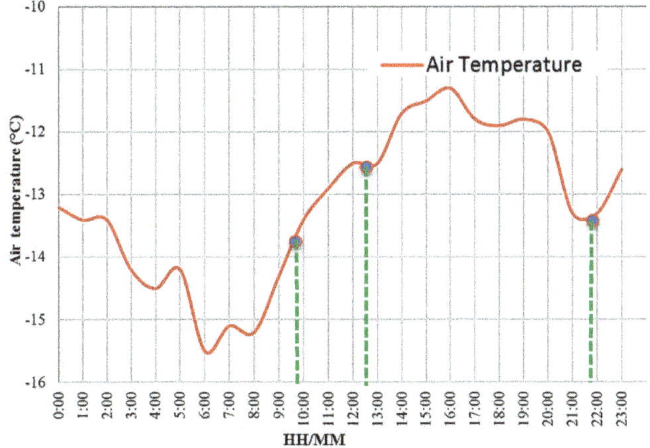

Figure 2. Hourly air temperature measured on 19 May 2012, from the on-ice micro-meteorological station. Colored dots represent times of in-situ snow property measurements at ~9:45 a.m. (for 14 cm), ~12:25 p.m. (for 4 cm) and ~9:30 p.m. (for 8 cm) local times. Green vertical lines denote the timing of the scatterometer measurements quasi-coincident with the in-situ snow thermophysical property measurements.

3.2. Data Collection

3.2.1. Ku-, X- and C-Band Multifrequency Polarimetric Microwave Scatterometer System

The Ku-, X- and C-band fully-polarimetric σ^0_{VV}, σ^0_{HH}, σ^0_{HV}, σ^0_{VH} measurements were acquired at a range resolution of ~30 cm using a surface-based multi-frequency scatterometer system (Figure 3). The Ku- (17.25 GHz) and X-band (9.65 GHz) UW-Scat scatterometer system operated concurrently with the C-band (5.52 GHz) scatterometer, with three overlapping replicate scans, completed within an hour. Detailed description of both scatterometer system specifications can be found under Table 1 in [14]; and details of calibration process, near-field correction and error determination documented in [21] (for Ku- and X-bands) and [18] (for C-band). In a 60° azimuth scan range, the scatterometer scan lines are averaged, as a function of system geometry and antenna beam-width, in order to obtain a minimum of 10 (Ku- and C-bands) and 15 (X-band) independent samples per scan line. The UW-Scat system acquired microwave backscatter at θ_{inc} between 21° and 81° at 2° increments, while the C-band scatterometer system acquired data between 15° and 75° at 3° increments. A rough illustration of scatterometer acquisition method can be found in Figure 3b in [14]. Only the co-polarized backscatter coefficients (σ^0_{VV} and σ^0_{HH}) are used within our current study context, as the magnitude of cross-polarized σ^0_{HV} is very low, owing to the high noise level (-50 dBm^2 for Ku- and X-band and -36 dBm^2 for C-band), restricting the usability of scatterometer measurements for upscaling to recently operating space-borne dual-polarized SAR systems.

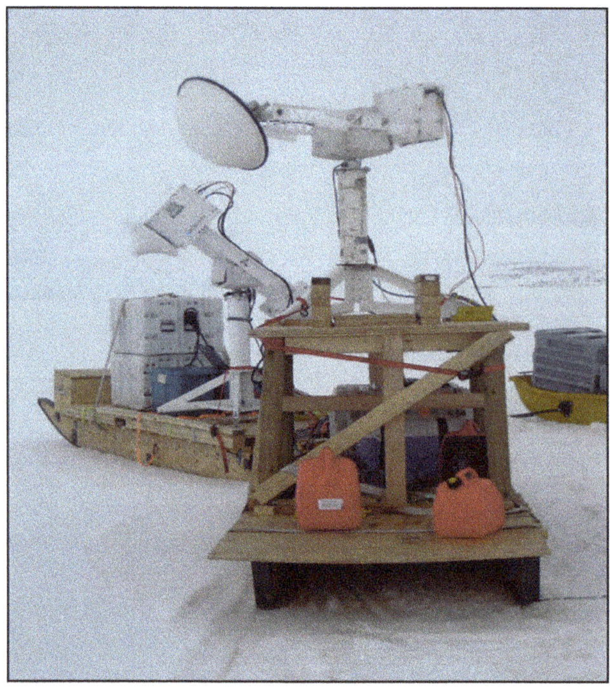

Figure 3. Surface-based multi-frequency polarimetric microwave scatterometer system: C-band scatterometer (foreground), and UW-Scat (Ku- and X-bands) (background).

3.2.2. Snow Thermophysical Property Observations

The 14 cm, 8 cm and 4 cm snow pits were located adjacent to the scatterometer scan area (~25 m × 25 m) at three different locations (located within ~100 m^2), and snow thermophysical measurements were sampled quasi-coincident with the Ku-, X- and C-band scatterometer observations. Snow density (ρ_s; sampled using a 66.35 cm^3 density cutter with an accuracy of ±0.01 g), snow temperature (T_s; measured using a 0.1 °C resolution Digi-Sense RTD thermometer probe at an accuracy of ±0.2 °C), snow salinity (S_s; measured using a WTW Cond 330i conductivity meter with an accuracy of ±0.5%) and snow grain radius (R_s; measured from disaggregated snow grains on a 2 mm grid crystal plate and classified following [22] were sampled from all three snow covers, sampled every 2 cm in vertical profile. Post-measurement destructive snow sampling (including snow thickness and thermophysical measurements) of the scatterometer scan area revealed that the snow cover was consistent with the adjacent snow pit used for sampling snow thermophysical measurements. However, it has to be noted that stochastic variability in snow thermophysical properties from a single snow cover from different snow pits located within small spatial scales may exist [16]. This study also assumes the air/snow interface and within-snow interfaces to be radar smooth [14]. This is important to be noted as the both the scatterometer systems are not co-located and scan a relatively wide area at 60° azimuth scan range; and therefore the snow covered area covering the scatterometer scan area is assumed to be isotropic in nature.

The 8 cm snow cover (Figure 4b) exhibited similar snow vertical structure as the 14 cm snow cover (Figure 4a) (detailed schematic and description of the vertical stricture of the 14 cm snow cover based on snow grain types can be found in [14]. The 4 cm snow cover consisted of highly brine-wetted rounded depth hoar crystals (Figure 4c). All layers for all three snow covers contained significant amounts of brine. Field measurements from the three snow thicknesses show a highly saline bottom

4 cm basal layer (~6—~20 parts per thousand (ppt)) consisting of somewhat-rounded depth hoar crystals (Tables 1–3). High salinities in the basal snow layers are due to significant upward brine wicking [16,23,24]. The presence of brine throughout the snow cover during the late winter season are less commonly observed as snow covers are brine-wetted usually during freeze-up, and snow covers overlaying highly saline frost flowers. [16,23].

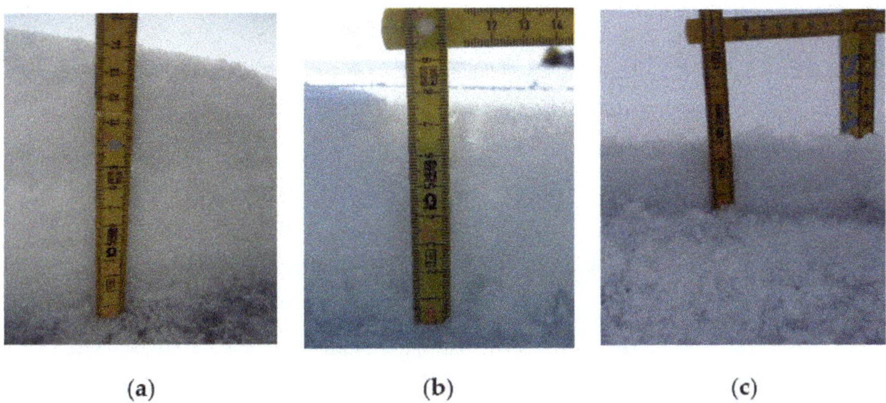

(a) (b) (c)

Figure 4. Sample snow cover on first-year sea ice (FYI) located adjacent to the scatterometer scan area: (a) 14 cm; (b) 8 cm; and (c) 4 cm.

Table 1. Snow cover and sea ice thermophysical properties for the 14 cm snow cover on smooth FYI for 19 May 2012. The total sea ice thickness was 1.3 m, and "sea ice" in the table represent the topmost 2 cm frazil ice core section. Note that snow and sea ice thermophysical properties for the 14 cm snow cover used here are adopted from Table 2 in [14].

Layer Number	Thickness (m)	Density ρ_s (g/cm^3)	Temperature T_s (°C)	Salinity S_s (ppt)	Grain Radius R_s (mm)
14–12	0.02	0.41	−9.1	3.2	0.7
12–10	0.02	0.42	−9.3	2.6	0.75
10–8	0.02	0.42	−9.4	2.7	0.8
8–6	0.02	0.44	−9.3	4.2	0.9
6–4	0.02	0.42	−9.1	7.1	0.9
4–2	0.02	0.27	−8.5	12.9	1.0
2–0	0.02	0.24	−8.5	12.9	1.25
Sea ice	0.02	0.85	−8.3	14.2	0.78

Table 2. Snow cover and sea ice thermophysical properties for the 8 cm snow cover on smooth FYI for 19 May 2012. The total sea ice thickness was 1.3 m, and "sea ice" in the table represent the topmost 2 cm frazil ice core section.

Layer Number	Thickness (m)	Density ρ_s (g/cm^3)	Temperature T_s (°C)	Salinity S_s (ppt)	Grain Radius R_s (mm)
8–6	0.02	0.33	−7.1	3.6	0.8
6–4	0.02	0.45	−7.0	5.7	1.0
4–2	0.02	0.25	−7.4	11.7	1.0
2–0	0.02	0.26	−7.5	12.4	1.5
Sea ice	0.02	0.85	−7.5	14.7	-

Table 3. Snow cover and sea ice thermophysical properties for the 4 cm snow cover on smooth FYI for 19 May 2012. The total sea ice thickness was 1.3 m, and "sea ice" in the table represent the topmost 2 cm frazil ice core section.

Layer Number	Thickness (m)	Density ρ_s (g/cm^3)	Temperature T_s (°C)	Salinity S_s (ppt)	Grain Radius R_s (mm)
4–2	0.02	0.22	−7.5	6.9	1.5
2–0	0.02	0.26	−7.6	8.6	1.5
Sea ice	0.02	0.85	−7.3	19.8	-

3.3. Brine Wetted Snow Dielectric Modeling and Penetration Depth

The dielectric permittivity, ε' and dielectric loss, ε'' for all three brine-wetted snow covers are calculated for each frequency (Tables 2 and 3). The first step involves calculating layer-wise brine volume φ_{bs} following [23].

$$\varphi_{bs} = \left[\frac{\varphi_{bsi}\rho_b}{(1 - \varphi_{bsi})\rho_i + \varphi_{bsi}\rho_b} \right] \left[\frac{\rho_s}{\rho_b} \right] \tag{1}$$

where ρ_b is the density of brine in g/cm^3, ρ_i is the temperature-dependent density of pure ice in g/cm^3, ρ_s is the snow density in g/cm^3, and φ_{bsi} is the temperature-dependent brine volume fraction of sea ice. From the layer-wise φ_{bs} and ρ_s, we calculate the ε' and ε'' for each snow layer for all three snow covers. For fresh dry snow on FYI, the dielectric permittivity ε' equals the permittivity of dry snow ε'_{ds}, which is frequency independent [25]:

$$\varepsilon'_{ds} = 1 + 2.55\rho_{ds} \tag{2}$$

where ρ_{ds} is the dry snow density in g/cm^3. For snow brine-wetted snow layers, ε' becomes the permittivity of brine-wetted snow ε'_{bs}, calculated using a dielectric mixture model following [25].

$$\varepsilon'_{bs} = \varepsilon'_{ds} + S\varphi_{bs}\varepsilon'_b \tag{3}$$

where S is a saturation-dependent dielectric depolarization factor, set to 1.33 [26], and ε'_b is the frequency- and temperature-dependent permittivity of brine [26].

For fresh dry snow on FYI, the dielectric loss ε'' equals the dielectric loss of dry snow (ε''_{ds}), which is <0.01 according to [27]; here ε'' is set at 0.001. For brine-wetted snow layers, ε'' becomes the dielectric loss of brine-wetted snow (ε''_{bs}), also calculated using a dielectric mixture model according to [25].

Employing the modeled ε' and ε'', the penetration depth δ_P into the snow cover, ignoring scattering losses is derived following [28]:

$$\delta_P = \frac{\lambda_0}{4\pi} \left\{ \frac{\varepsilon'}{2} \left[\left(1 + \left(\frac{\varepsilon''}{\varepsilon'} \right)^2 \right)^{1/2} - 1 \right] \right\}^{-1/2} \tag{4}$$

where λ_0 is the free space sensor wavelength. The two-way loss for a snow layer is given by

$$L = exp\left(\frac{-2K_e\tau}{\cos\theta} \right) \tag{5}$$

where the extinction coefficient K_e is the inverse of δ_P, θ is the incidence angle within the snow layer based on the degree of refraction, and τ is the snow layer thickness (0.02 m). The maximum δ_P is attained using the equation $P(d)/P(0_+) = 1/e$, where $P(0_+)$ is the power at the air/snow interface and $P(d)$ is the power at depth d [14,29]. Hence, the deepest layer to which microwaves penetrate corresponds to approximately one third of the initial power that enters the snow layer [14,30].

3.4. Microwave Co-Polarization Ratio (γ_{co}) and Dual-Frequency Ratios ($\gamma_{DFR[PP]}$)

The transmissivity and reflectivity influence the observed Ku-, X- and C-band σ^0_{VV} and σ^0_{HH} depending on polarization and θ_{inc}, with the VV-polarized waves exhibiting greater transmissivity than HH-polarized waves, and HH-polarized waves exhibiting greater reflectivity. The co-polarization ratio or co-pol ratio (γ_{co}) (in dB) is used to quantify this polarization-based difference in linear backscatter at the same frequency. For e.g., Ku-band γ_{co} is given by

$$\gamma_{co} = \frac{\sigma^0_{VV(Ku)}}{\sigma^0_{HH(Ku)}} \tag{6}$$

With contrast in dielectric mismatch between the air/snow, within-snow and snow/sea ice interfaces, transmissivity and reflectivity of VV- and HH-polarized waves differ, modeled after Fresnel reflection coefficients [31]. Moreover, salinity gradients within all three snow covers are high and different, which in turn can modify the snow dielectrics at Ku-, X- and C-bands. This introduces substantial frequency-dependent polarization diversity between σ^0_{VV} and σ^0_{HH}, which in turn modifies γ_{co}, from all three snow cover cases. Previous studies used γ_{co} employing a single-frequency approach (C-band), for discriminating snow covered FYI properties based on dielectric effects [10,12,15,32].

The dual-frequency ratio ($\gamma_{DFR[PP]}$) (in dB) is the innovative parameter introduced in this study, and is the difference in the radar backscatter between any two frequencies at the same polarization. For example,

$$\gamma_{DFR[VV](Ku,X)} = \frac{\sigma^0_{VV}(Ku)}{\sigma^0_{VV}(X)} \tag{7}$$

$$\gamma_{DFR[HH](Ku,X)} = \frac{\sigma^0_{HH}(Ku)}{\sigma^0_{HH}(X)}$$

For X- and C-bands, ($\gamma_{DFR[VV](X,C)}$ and $\gamma_{DFR[HH](X,C)}$); and for Ku- and C-bands, ($\gamma_{DFR[VV](Ku,C)}$ and $\gamma_{DFR[HH](Ku,C)}$) are also calculated. $\gamma_{DFR[PP]}$ is used in addition to γ_{co}, in order to investigate frequency-sensitive differences in σ^0_{VV} and σ^0_{HH} and also to understand the dominant behavior of a particular frequency with changes in snow electro-thermophysical properties for the three different snow cover cases.

3.5. Multilayer Snow and Ice Backscatter (MSIB) Model

A first-order multilayer snow and ice backscatter (MSIB) model is used to calculate surface scattering [32] and volume scattering [15] contributions of/within each snow layer, from within all three snow covers, also accounting for reflection, refraction and attenuation. The MSIB model utilizes snow layer thickness (m; 0.02 m in our case), S_s (ppt), T_s (°C), ρ_s (g/cm^3), R_s (mm) and surface roughness parameters (root-mean square interface roughness (m); 0.005 m and correlation length (m), 0.03 m) as model inputs. The MSIB model does not include backscatter contributions from large scale deformed FYI features such as ridges, as the study area in this research falls under relatively smooth FYI. Previous studies [10,11,13,14,30] have used the MSIB model (modified version of methods originally formulated by [28,33] for various snow cover on sea ice related studies.

The MSIB model simulates the surface scattering contribution under the scalar estimates of the Kirchhoff physical optics method, for relatively smooth surfaces described by a Gaussian distribution function. The volume scattering is modeled based on the number density of ice/brine inclusions within each snow layer and their backscatter cross-sections [33]. The structure and radii of snow brine inclusions are unknown; therefore, we assume that they have a structure similar to that of water inclusions. The radii of snow brine inclusions used in this study fall within the Rayleigh scattering region, sensitive to volume scattering at C-, X- and Ku-band frequencies. Although snow grain radius ≥ 1 mm causes Ku-band microwave scattering from these snow grains to fall in the Mie scattering region, it is anticipated that the dominant volume scattering mechanism originate from/within the brine inclusions, given their significantly higher ε' and ε''. At C- and X-band, both the brine inclusions and the snow grains fall within the Rayleigh scattering region. Therefore, a Rayleigh volume scattering model integrated into the MSIB model is justified in this study. Detailed description of the MSIB algorithm can be found in [30].

3.6. Analysis Structure

The measured Ku-, X- and C-band σ^0_{VV} and σ^0_{HH} and associated γ_{co}, and $\gamma_{DFR[PP]}$ (hereby jointly referred altogether as "microwave parameters") are compared between the 14 cm, 8 cm and 4 cm snow cover cases, at near-($21° \leq \theta_{inc} \leq 30°$) (NR), mid-($33° \leq \theta_{inc} \leq 42°$) (MR) and far-range ($45° \leq \theta_{inc} \leq 60°$) (FR) incidence angles. Snow dielectrics for the three different snow covers are calculated for each frequency, and used in conjunction with the in situ measured snow salinities. This is to provide an enhanced thermophysical perspective on how changes in these parameters at different snow thicknesses, affect changes in Ku-, X- and C-band microwave parameters. The modeled penetration depths provide theoretical insight into the potential propagation capability of all three microwave frequencies and polarizations for different snow thicknesses. MSIB modeled surface and volume scattering contributions, as a function of frequency and polarization, are simulated for each of three snow covers to support explanations of the variations in the individual scattering mechanisms at varying snow thermophysical conditions from the three different snow covers. These measures could contribute to a better understanding of the interaction between changing snow thermophysical properties and corresponding dielectric properties and surface-based Ku-, X- and C-band microwave backscatter.

4. Results

This section individually illustrates and analyzes the Ku-, X- and C-band microwave parameters, at NR, MR and FR, against the three different snow cover cases. Snow thermophysical properties, modeled snow dielectrics and penetration depths are utilized to support the frequency- and polarization-diversities with change in snow thickness.

4.1. Ku-, X- and C-Band σ^0_{VV} and σ^0_{HH} and Modeled Penetration Depths

The observed Ku-, X-band σ^0_{VV} and σ^0_{HH} measured at NR for the 4 cm snow cover are ~5 dB and ~8 dB greater than for the 8 cm and 14 cm snow cover cases. (Figure 5a,b). Throughout all θ_{inc}, Ku-band exhibits slightly greater σ^0_{VV} than σ^0_{HH} (~0.5 dB to 1.25 dB), for all three snow cover cases, when compared to X- and C-bands (Figure 5a). C-band consistently exhibits strong reflective behavior, with $\sigma^0_{HH} > \sigma^0_{VV}$ by ~2.5 dB for all three snow cover cases (Figure 5c) [30,31], owing to its greater ε' mismatch at the air/snow interface [10–12,14,14,33,34].

Modeled penetration depths show C-band exhibiting the maximum penetration depth of top 6 cm for the 14 cm snow cover, 4 cm (for the 8 cm snow cover) and 3 cm (for the 4 cm snow cover case) (Table 5). X-band penetrate only to the top 4 cm of the 14 cm snow cover, 3 cm in case of the 8 cm snow cover and 2 cm for the 4 cm snow cover case. On the other hand, Ku-band penetrate only to the top 1 cm and 2 cm, respectively, for all three snow covers (Table 5). All three frequencies are absorbed within the topmost snow layers due to the cumulative drop in microwave power while propagating through the lossy high brine volumes of the snow covers.

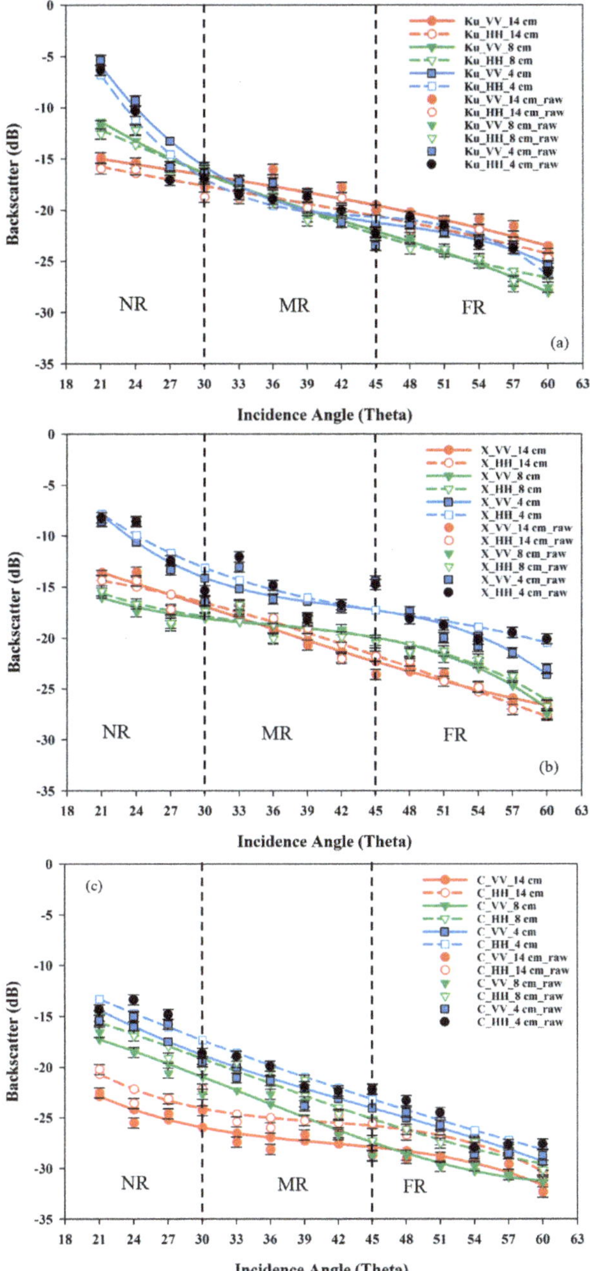

Figure 5. Observed Ku-, X- and C-band backscatter (σ_{VV}^0 and σ_{HH}^0), from 14 cm, 8 cm and 4 cm snow covers on FYI acquired on 19 May 2012: (**a**) Ku-band; (**b**) X-band; and (**c**) C-band. Scatterometer backscatter trend lines are cubic fits. Colored points represent measurement points with error bars indicating min-max deviation. Vertical black dotted lines partition near-range (NR), mid-range (MR) and far-range (FR) incidence angles.

Table 4. Modeled multi-layer snow dielectric permittivity (ε') for the 14 cm, 8 cm and 4 cm snow covers on smooth FYI, for Ku-, X- and C-bands.

Layer Number	Dielectric Permittivity (ε')								
	14 cm			8 cm			4 cm		
	Ku	X	C	Ku	X	C	Ku	X	C
14–12	2.18	2.31	2.44						
12–10	2.25	2.37	2.49						
10–8	2.24	2.36	2.47						
8–6	2.39	2.58	2.78	2.06	2.23	2.40			
6–4	2.52	2.84	3.18	2.62	2.99	3.37			
4–2	2.11	2.47	2.86	2.14	2.53	2.93	1.82	2.02	2.23
0	1.97	2.29	2.62	2.20	2.63	3.06	1.96	2.22	2.50
Sea ice	4.57	5.72	6.92	4.80	6.15	7.55	5.44	7.32	9.26

Table 5. Modeled Ku-, X- and C-band penetration depths (from air/snow interface) (Equation (4)) from the 14 cm, 8 cm and 4 cm snow covers on FYI, acquired on 19 May 2012.

Snow Thickness	Penetration Depth		
	Ku-Band	X-Band	C-Band
14 cm	2 cm	4 cm	6 cm
8 cm	2 cm	3 cm	4 cm
4 cm	1 cm	2 cm	3 cm

4.2. Ku-, X- and C-Band Co-Pol Ratio (γ_{co})

The Ku-, X- and C-band γ_{co} demonstrate frequency separation, with prominent variability of γ_{co} with θ_{inc} from all three different snow cover cases (Figure 6). For the 14 cm and 8 cm snow covers, Ku-band γ_{co} clearly shows greater σ^0_{VV} than σ^0_{HH}, especially in NR and MR, suggesting dominant volume scattering from the top most decomposed and fragmented precipitation particles (R_s ~0.8 mm; Tables 1 and 2) (Figure 6a,b). X-band γ_{co} exhibits increasing reflective behavior with decreasing snow thickness, suggesting greater sensitivity of X-band HH-polarized waves to the increase in snow salinities and snow dielectrics (Tables 4 and 6) (Figure 6a–c). C-band γ_{co} clearly exhibits greater σ^0_{HH} than σ^0_{VV} from all three snow covers, justifying strong reflective behavior of C-band HH-polarized microwaves (Figure 6a–c). Nevertheless, C-band microwaves do not exhibit characteristic variability of γ_{co} with θ_{inc}, when compared to Ku- and X-bands, especially with the thicker 14 cm and 8 cm snow covers. The mechanisms responsible for this behavior are clear, as both 14 cm and 8 cm snow covers exhibit almost similar thermophysical (Tables 1–3) and dielectric properties (Tables 4 and 6).

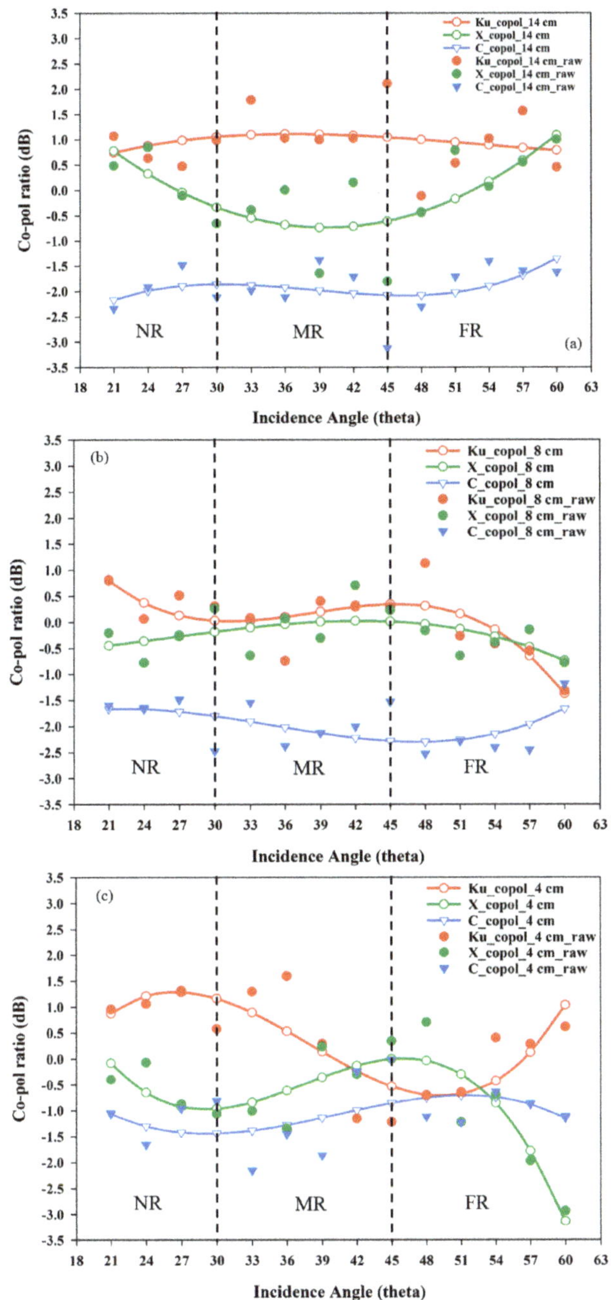

Figure 6. Calculated Ku-, X- and C-band co-pol ratios (γ_{co}) from 14 cm, 8 cm and 4 cm snow covers on FYI: (**a**) Ku-band; (**b**) X-band; and (**c**) C-band. Co-pol ratio trend lines are cubic fits. Colored points represent measurement points. Vertical black dotted lines partition near-range (NR), mid-range (MR) and far-range (FR) incidence angles.

Table 6. Modeled multi-layer snow dielectric loss (ε'') for the 14 cm, 8 cm and 4 cm snow covers on smooth FYI, for Ku-, X- and C-bands.

Layer Number	Dielectric Loss (ε'')								
	14 cm			8 cm			4 cm		
	Ku	X	C	Ku	X	C	Ku	X	C
14–12	0.21	0.25	0.23						
12–10	0.20	0.24	0.22						
10–8	0.19	0.23	0.21						
8–6	0.31	0.38	0.35	0.28	0.34	0.30			
6–4	0.53	0.65	0.60	0.61	0.74	0.67			
4–2	0.60	0.74	0.67	0.65	0.78	0.71	0.33	0.40	0.36
2–0	0.52	0.64	0.58	0.70	0.85	0.77	0.44	0.54	0.48
Sea ice	2.53	3.44	4.07	2.92	3.94	4.59	4.06	5.46	6.33

4.3. Ku-, X- and C-Band Dual-Frequency Ratios ($\gamma_{DFR[PP]}$)

For the 14 cm snow cover case, the σ^0_{HH} and σ^0_{VV} increases with θ_{inc} at Ku-band than at X-band by ~4 dB to 5 dB, clearly depicted by $\gamma_{DFR[VV](Ku,X)}$ and $\gamma_{DFR[HH](Ku,X)}$ (Figure 7a,b). This suggests the greater sensitivity of Ku-band microwaves to snow grain size, especially from the topmost decomposed and fragmented layers. This is clearly observed from MR to FR, where Ku-band separates X-band by crossing each other and exhibits asymptotic increase in Ku-band backscatter at these θ_{inc}. However, in the case of 8 cm and 4 cm snow covers, $\gamma_{DFR[VV](Ku,X)}$ and $\gamma_{DFR[HH](Ku,X)}$ shows a steep decrease by ~7 dB, especially at NR, suggesting lesser sensitivity of Ku-band VV- and HH-polarized waves to thermophysical changes in thinner snow covers, when compared to X-band microwaves. Greater snow salinity gradients in the upper layers of the 8 cm and 4 cm snow covers (Tables 2 and 3) lead to greater X-band ε' and ε'' (Tables 4 and 6). This results in increased X-band backscatter, which can be considered to be the dominant factor causing this frequency shift from thicker to thinner saline snow cover cases.

Notable changes are also observed in $\gamma_{DFR[VV](Ku,X)}$ and $\gamma_{DFR[HH](Ku,X)}$ at FR for the 4 cm snow cover, where X-band σ^0_{HH} increases by ~4 dB, while Ku-band σ^0_{VV} illustrates an increasing trend, suggesting the importance of snow grain size affecting both Ku- and X-band total backscatter (Figure 7a,b). $\gamma_{DFR[HH](Ku,X)}$ for the 8 cm and the 4 cm snow cover cases exhibit strong divergence towards Ku-band, thereby demonstrating the dominance of snow grain size affecting microwave backscatter especially at higher frequencies for thicker snow covers (Figure 7b).

The X- and C-band $\gamma_{DFR[VV](X,C)}$ and $\gamma_{DFR[HH](X,C)}$ shows strong sensitivity of X-band over C-band microwaves, throughout all θ_{inc}, however fluctuating for different snow covers (Figure 7c,d). For the 14 cm snow cover, both $\gamma_{DFR[VV](X,C)}$ and $\gamma_{DFR[HH](X,C)}$ exhibit greater X-band magnitude at NR, when compared to C-band microwaves. This may be caused by X-band's greater sensitivity to scale-dependent variations in surface roughness. It could also be due to the fact that, spatial variations in ε'' reduce the mean absorption, allowing greater X-band surface scattering from within-upper snow layer interfaces, when compared to C-band. This characteristic has been previously reported in several studies using the 14 cm snow cover case (e.g., [14,33,34]). As snow thickness decreases, polarization-dependent frequency separation reduces at MR, and both frequencies exhibit minimal frequency diversity. For example, as the snow thickness changes from 14 cm to 8 cm to 4 cm, separation between X- and C-band σ^0_{HH} and σ^0_{VV} reduces from ~9.5 dB for 14 cm to ~6.5 dB for 8 cm, to ~2.5 dB for 4 cm (Figure 7c,d). As θ_{inc} increases from MR to FR, $\gamma_{DFR[VV](X,C)}$ and $\gamma_{DFR[HH](X,C)}$ from the 4 cm snow cover show a notable steep increase, suggesting possible X-band volume scattering contributions from the upper snow layer snow grains, along with surface scattering from the air/brine-wetted snow interface (Figure 7c,d). However, $\gamma_{DFR[VV](X,C)}$ and $\gamma_{DFR[HH](X,C)}$ drop off at the end of FR and continue to decrease, tending towards minimal frequency diversity.

The Ku- and C-band $\gamma_{DFR[VV](Ku,C)}$ and $\gamma_{DFR[HH](Ku,C)}$ show dominant Ku-band σ^0_{VV} sensitivity over C-band σ^0_{VV} throughout all θ_{inc}, especially for the 14 cm thick snow cover case (Figure 7e,f). This suggests the predominance of higher frequency Ku-band microwaves and its strong sensitivity to snow grain size over lower frequency C-band microwaves. The effect of snow grain size is also visible for 8 cm; however, snow salinities in the upper layers of the snow cover (Table 1) contributes to strong surface scattering, thereby causing a decrease in the frequency separation between Ku- and C-bands. Snow salinities show its strongest effect for the 4 cm snow cover, where $\gamma_{DFR[VV](Ku,C)}$ and $\gamma_{DFR[HH](Ku,C)}$ clearly show a strong angular drop off in the NR, separating C-band from Ku-band. This drop-off is steeper in $\gamma_{DFR[HH](Ku,C)}$, justifying the greater sensitivity of C-band HH-polarized waves to changes in snow salinities, when compared to Ku-band HH-polarized waves (Figure 7f). As θ_{inc} increases from NR to MR, volume scattering contributions from the depth hoar rounded grains dominates causing both $\gamma_{DFR[VV](Ku,C)}$ and $\gamma_{DFR[HH](Ku,C)}$ to gradually increase. However, at FR, $\gamma_{DFR[HH](Ku,C)}$ shows a decreasing trend, suggesting greater sensitivity of HH-polarized waves to snow salinities over snow grain microstructure. However, further study is required in this aspect.

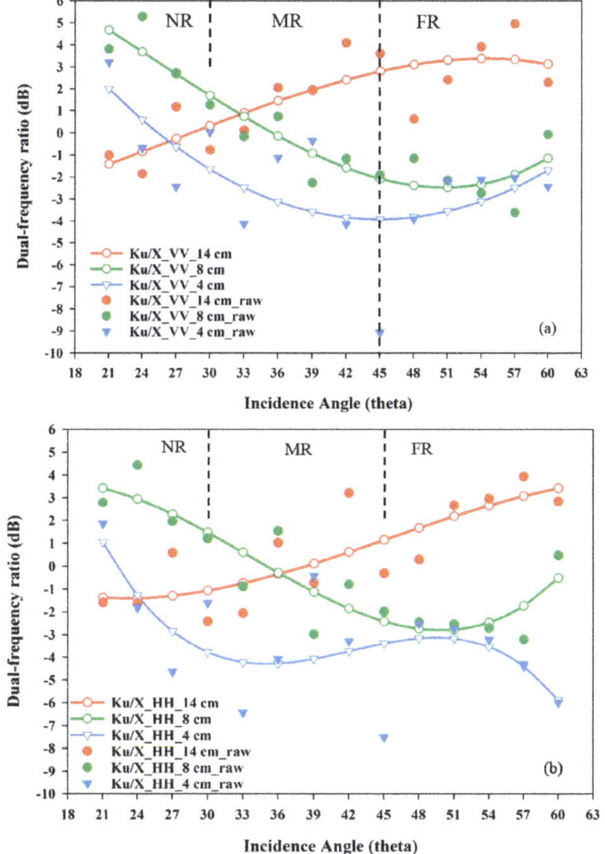

Figure 7. *Cont.*

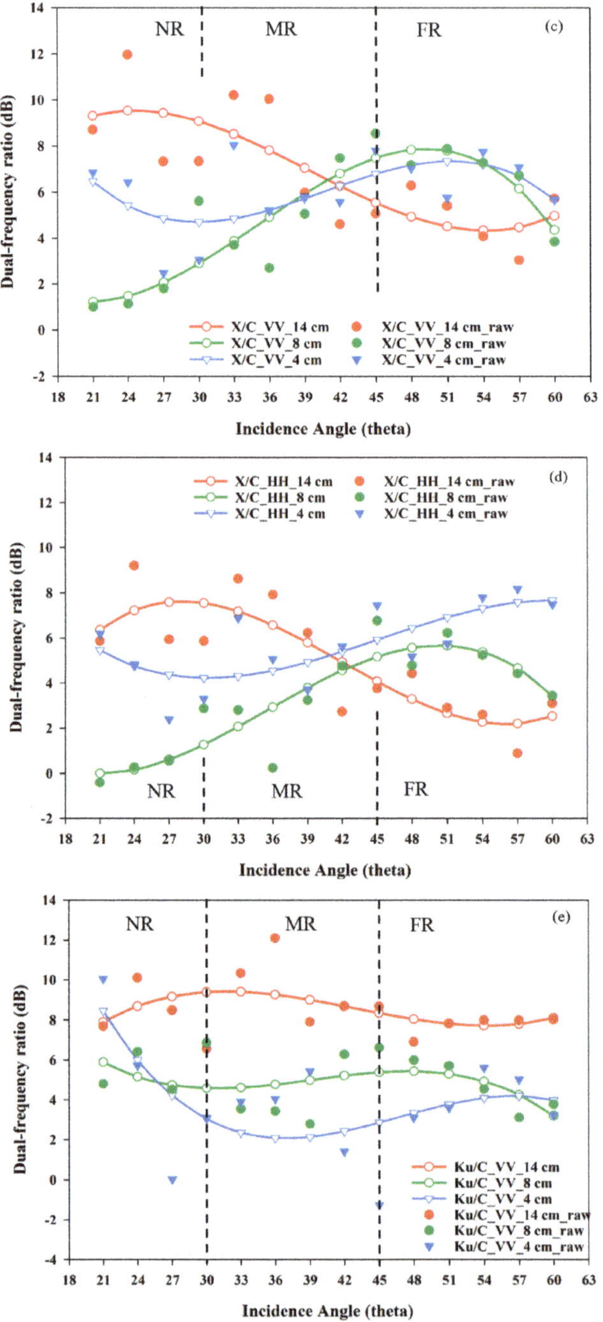

Figure 7. *Cont.*

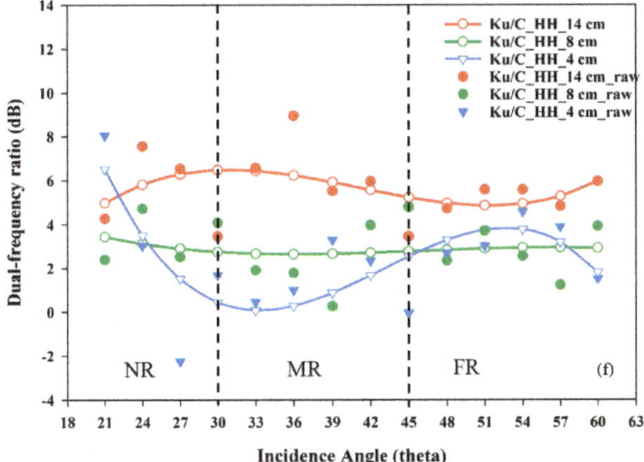

Figure 7. Derived (Ku-, X-), (X-, C-) and (Ku-, C-) VV and HH dual-frequency ratios (γ_{DFR}) from 14 cm, 8 cm and 4 cm snow covers on FYI: (**a,b**) Ku- and X-band VV and HH γ_{DFR}; (**c,d**) X- and C-band VV and HH γ_{DFR}; and (**e,f**) Ku- and C-band VV and HH γ_{DFR}. Dual-frequency ratio trend lines are cubic fits. Colored points represent dual-frequency ratios from measurement points. Vertical black dotted lines partition near-range (NR), mid-range (MR) and far-range (FR) incidence angles.

5. Discussions

5.1. Ku-, X- and C-Band σ^0_{VV} and σ^0_{HH} and Modeled Penetration Depths

Ku-band exhibits the highest σ^0_{VV} of all three frequencies from all three snow cover cases. This is attributed to greater sensitivity of Ku-band VV-polarized waves to volume scattering [14] originating from the uppermost dense brine-wetted decomposed and fragmented precipitation particles of the 14 cm and 8 cm snow covers (Figure 5a), and also from rounded depth hoar crystals in the 4 cm snow cover. These snow grains act as effective scattering centers at Ku-band, leading to greater σ^0_{VV}. Moreover, at higher microwave frequencies such as Ku-band, the influence of snow grains on microwave backscatter increases, leading to greater potential volume scattering estimates [14,15]. Moreover, MSIB model employing first-order Rayleigh scattering assumptions has shown its inability to simulate accurate microwave backscatter estimates at higher frequencies such as Ku-band [14,30], and warrants further research. Detailed description of sensitivity of higher microwave frequencies to snow microstructure can be found in [30].

Greater salinities in the topmost layer of the 4 cm snow cover (~6.9 ppt) (Table 3) results in the dielectric loss ε'' to be the strongest (Table 6), when compared to 3.6 ppt and 3 ppt observed for the topmost layer of the 8 cm and the 14 cm snow covers, respectively (Tables 1 and 2). This results in the strongest surface scattering at the air/snow interface for the 4 cm snow cover, resulting in greater σ^0_{VV} and σ^0_{HH} for all three frequencies, followed by its relative from 8 cm and 14 cm snow covers. MSIB model shows ~80% to ~90% of the total backscatter to be surface scattering for all three frequencies.

Overall, the dominant factor controlling the penetration depth from all three brine-wetted snow covers is the dielectric loss (Table 6) associated with high salinity throughout the snow covers, with greatest effect on σ^0_{HH} from the 4 cm snow cover, followed by 8 cm and 14 cm snow covers, respectively. Greater effect of snow salinity on HH-polarized waves is attributed to the greater surface scattering contributions originating from the dielectrically mismatched air/snow interface and/or within-snow interfaces, resulting in greater reflectivity of HH-polarized waves

5.2. Ku-, X- and C-Band Co-Pol Ratio (γ_{co})

In the case of the 4 cm snow cover, Ku-band microwaves exhibit strong characteristic variability of γ_{co} with θ_{inc} from NR to FR, when compared to the 14 cm and 8 cm snow covers. (Figure 6c). For the 4 cm snow cover, Ku-band exhibits greater volume scattering from the larger depth hoar rounded grains (R_s ~1.5 mm; Table 3), overpowering the surface scattering contributions from the air/snow interface. This is attributed to the greater sensitivity of higher frequency microwaves such as Ku-band to snow grain size. Previous studies and theory has demonstrated that Ku-band backscatter cross-section from larger snow grains and/or brine inclusions in snow depend on the 6th power of the grain radius, and is inversely proportional to the 4th power of microwave wavelength [15].

X-band microwaves exhibit a "U-shaped" signature from the 14 cm snow cover, from NR to FR, suggesting a combined effect of surface and volume scattering contributions from the topmost snow layers (Figure 6a). However, as snow salinity increases in 8 cm and 4 cm, the characteristic variability of γ_{co} with θ_{inc} drastically increases (Figure 6b,c), overpowered by greater σ^0_{HH} due to stronger S_s (Tables 2 and 3), with correspondingly higher ε' and ε'' gradients in the 4 cm snow cover (Tables 4 and 6), when compared to the 14 cm and 8 cm snow cover cases. σ^0_{HH} increases substantially in the case of 4 cm snow cover, and fluctuates especially from MR to FR (Figure 6b). This suggests X-band's greater sensitivity to micro-scale surface roughness variations, when compared to Ku-band [32].

However, C-band γ_{co} exhibits slight difference between the 14 cm and 8 cm snow covers at NR, with the 8 cm snow cover exhibiting slightly greater σ^0_{HH} than σ^0_{VV} (~1 dB), when compared to the 14 cm snow cover, indicating the σ^0_{HH} from the 8 cm snow cover to be more isotropic at steeper θ_{inc}. Even though C-band γ_{co} exhibits dominant σ^0_{HH}, as snow thickness decreases, the magnitude of σ^0_{HH} decreases slightly by 0.5 dB at MR and FR. This could result from volume scattering contributions from the depth hoar snow grains, especially from the 4 cm snow cover (Table 3) which has entirely different thermophysical structure when compared to the 14 cm and 8 cm snow covers (Tables 1 and 2).

Overall, the salinity of the upper snow layers from all three snow covers is the primary source of greater sensitivity of X- and C-band HH-polarized waves over VV-polarized waves with change in snow thickness. When compared to C-band, X-band σ^0_{HH} shows greater sensitivity to changes in snow salinity from the upper layers. Ku-band microwaves show fluctuating polarization diversity as snow thickness decreases, however the magnitude of fluctuations are less, when compared to X-band microwaves. This could be due to greater sensitivity of Ku-band microwaves to snow grain size and/or presence of brine inclusions in snow.

5.3. Ku-, X- and C-Band Dual-Frequency Ratios ($\gamma_{DFR[PP]}$)

This study also investigates how different $\gamma_{DFR[PP]}$ combinations behave for a single snow cover. In the case of 14 cm snow cover, $\gamma_{DFR[VV](Ku,C)}$ and $\gamma_{DFR[HH](Ku,C)}$ shows the strongest frequency diversity (Figure 7e,f), followed by $\gamma_{DFR[VV](X,C)}$ and $\gamma_{DFR[HH](X,C)}$ (Figure 7c,d); and $\gamma_{DFR[VV](Ku,X)}$ and $\gamma_{DFR[HH](Ku,X)}$ (Figure 7a,b). Snow grain size shows a stronger effect from the relatively thick 14 cm snow cover, with greater sensitivity from Ku-band microwaves. In the case of 8 cm snow cover, the Ku- and C-band $\gamma_{DFR[PP]}$ shows consistent greater Ku-band sensitivity to C-band (~5.5 dB) throughout all θ_{inc}. X- and C-band $\gamma_{DFR[PP]}$ and Ku- and X-band $\gamma_{DFR[PP]}$ shows increasing and decreasing trends, respectively, with increasing θ_{inc}, suggesting strong X-band sensitivity, followed by Ku- and C-band. In the case of 4 cm snow cover, all three γ_{DFR} combinations show clear separability between frequencies, with the strongest consistent separation (~7 dB) between Ku- and X-band $\gamma_{DFR[PP]}$ and Ku- and C-band $\gamma_{DFR[PP]}$, throughout all θ_{inc}. X- and C-band $\gamma_{DFR[PP]}$ falls in between the other two $\gamma_{DFR[PP]}$ combinations.

Overall, the different $\gamma_{DFR[PP]}$ combinations from the three different snow covers demonstrate its utility to be classified as a new "polarimetric parameter" to provide new information, on the sensitivity of polarization- and frequency-dependent microwave backscatter, to changes in snow thickness with corresponding fluctuations in snow thermophysical properties. The $\gamma_{DFR[PP]}$ (based on multi-frequency approach) also provides additional information, when compared to γ_{co} (based on single-frequency approach). Therefore, $\gamma_{DFR[PP]}$ helps in separating dominant polarization-dependent frequencies, sensitive to changes in snow thermophysical properties

6. Conclusions

The research evaluated an observational multi-frequency polarimetric microwave dataset to investigate Ku-, X- and C-band microwave co-polarized backscatter and modeled microwave penetration depths, acquired from three different saline snow covers (14 cm, 8 cm and 4 cm) overlying smooth land-fast first-year sea ice. We compared and investigated differences in co-polarized backscatter observations (σ^0_{VV} and σ^0_{HH}) and co-pol ratios γ_{co}, as a function of incidence angle (θ_{inc}), for all three frequencies from all three different snow cover cases. The newly-introduced polarization-dependent dual-frequency ratios $\gamma_{DFR[PP]}$ illustrated distinctive separability of frequencies and polarizations from all three snow cover cases.

Scatterometer observations, supported by in-situ snow thermophysical parameters, modeled snow dielectrics and penetration depths, demonstrate differences in Ku-, X- and C-band σ^0_{VV} and σ^0_{HH} for all three snow cover cases. As expected, the completely brine-wetted 4 cm snow cover (bulk snow salinity of 7.5 ppt) demonstrated increased σ^0_{VV} and σ^0_{HH}, when compared to its relative backscatter from 8 cm and 14 cm snow cover cases. C-band achieved the maximum penetration with 6 cm penetration for the 14 cm snow cover, while 4 cm for the 8 cm snow cover, and 3 cm for 4 cm snow cover (Table 5). Overall, the dominant factor controlling the penetration depth from all three brine-wetted snow covers is the dielectric loss associated with high salinity throughout the snow covers, with greatest effect on HH-polarized waves from the 4 cm snow cover, followed by 8 cm and 14 cm snow covers, respectively.

The calculated co-pol ratios (γ_{co}) clearly showed fluctuations as a function of frequency and polarizations from all three snow covers. X-band γ_{co} showed the greatest sensitivity to changes in snow thickness with greater observable polarization-dependent backscatter separation for the three different snow covers, as a function of incidence angle. This could possibly be due to greater sensitivity of X-band microwaves to plot-scale surface roughness variations and dielectric loss. Snow salinity measurements at higher resolution, to quantify micro-scale variability of snow dielectrics, are recommended in this regard. Ku-band γ_{co} showed minimal polarization diversity for a single snow cover, however differs with variation in VV-backscatter magnitude with change in snow thickness. Variability in snow grain microstructure variability is a factor for fluctuations within the dominant VV polarization. Even though the topmost snow layers showed substantial changes in snow salinities, C-band γ_{co} showed almost negligible polarization diversity between the three snow cover cases. C-band HH-polarized waves showed a comparatively greater backscatter than VV-polarized waves, justifying the strong reflective behavior of C-band microwaves, due to strong salinities throughout the snow covers, resulting in dielectric mismatches between air/brine-wetted snow interface and/or within-upper snow layer interfaces. This could suggest a lower sensitivity of C-band to changes in snow salinities. Further study is recommended in this regard.

The newly introduced dual-frequency ratios ($\gamma_{DFR[PP]}$) provided a combinational metric approach to provide enhanced understanding on how multiple microwave frequencies interact with varying snow thicknesses on FYI. In general, all three dual-frequency ratios show strong sensitivity to changes in snow thickness, dependent on the incidence angles. Ku- and X-band γ_{DFR} show strong Ku-band sensitivity to snow grain microstructure for thicker snow covers at near-range incidence angles, while X-band microwaves dominated as snow thickness reduced from 14 cm to 4 cm, possibly owing to X-band's strong sensitivity to changes in snow salinity. Interestingly, HH-polarized waves generally showed greater sensitivity over VV-polarized waves at far-range incidence angles, with increasing snow thicknesses, reminding the importance of snow grain microstructure affecting the backscatter, even under saline snow conditions. X- and C-band γ_{DFR} showed notable frequency separation between the three snow covers, at near-range incidence angles, possibly suggesting strong surface scattering effects with increase in snow salinities as snow thicknesses decreased. The Ku- and C-band γ_{DFR}

showed strong sensitivity between Ku- and C-bands at 14 cm and 8 cm snow covers, with differences indicating changes in snow grain microstructure as a dominant factor affecting backscatter. Overall, the innovative dual-frequency ratio demonstrated its ability to determine dominant polarizations and frequencies, sensitive to changes in snow thicknesses with corresponding changes in snow thermophysical properties. From an application point of view, the dual-frequency ratio can added as a new polarimetric parameter to currently existing polarimetric parameters, when microwave backscatter from multiple polarizations and frequencies are available from any operational SAR system.

Our results using co-pol ratios and dual-frequency ratios suggest the potential to utilize the multi-frequency microwave approach to characterize frequency-, incidence angle- and polarization-dependent changes in microwave backscatter from different snow cover types on smooth FYI. Further research should investigate exploiting the dual-frequency ratios on a satellite-scale approach over basin- and regional-scales to investigate snow covers on different FYI types. Moreover, our approach should also be tested under colder atmospheric and snow geophysical conditions for different snow covers, in order to investigate variations in penetration depths between frequencies, frequency and polarization diversity under changing atmospheric and snow geophysical conditions. Results from these plot-scale studies using the multi-frequency approach can be further upscaled to regional and hemispherical scale "snow on sea ice" applications such as estimation of snow thickness and snow water equivalent (SWE), using the recent, currently and upcoming space-borne SAR missions TerraSAR-X, RADARSAT Constellation Mission and NISAR (C-band); and ALOS-2 PALSAR-2 (L-band).

Acknowledgments: The authors would like to thank ProSensing, Inc. for providing UW-Scat system to obtain scatterometer data; Jack Landy, Randall Scharien, and Megan Shields at the Center for Earth Observation Science (CEOS), University of Manitoba for their logistical and technical assistance; and NSERC Discovery, NRS Research and PCSP grants for project funding.

Author Contributions: Vishnu Nandan, Torsten Geldsetzer and John Yackel designed the experiment. Vishnu Nandan formulated the research methodology and wrote the manuscript. Mallik Mahmud provided all high-resolution figures and graphs, along with proofreading. Saroat Ramjan contributed with additional inputs during proofreading.

Conflicts of Interest: The authors declare no conflict of interest.

References

1. Stroeve, J.C.; Serreze, M.C.; Holland, M.M.; Holland, M.M.; Kay, J.E.; Malanik, J.; Barett, A.P. The Arctic's rapidly shrinking sea ice cover: A research synthesis. *Clim. Chang.* **2012**, *110*, 1005–1027. [CrossRef]
2. Maslanik, J.A.; Fowler, C.; Stroeve, J.; Drobot, S.; Zwally, S.; Yi, D.; Emery, W. A younger, thinner Arctic ice cover: Increased potential for rapid, extensive sea-ice loss. *Geophys. Res. Lett.* **2007**, *34*, L24501. [CrossRef]
3. Webster, M.A.; Rigor, I.G.; Nghiem, S.V.; Kurtz, N.T.; Farrell, S.L.; Perovich, D.K.; Strum, M. Interdecadal changes in snow depth on Arctic sea ice. *J. Geophys. Res. Oceans* **2014**, *119*, 5395–5406. [CrossRef]
4. Comiso, J.C. Large Decadal Decline of the Arctic Multiyear Ice Cover. *J. Clim.* **2012**, *25*, 1176–1193. [CrossRef]
5. Blanchard-Wrigglesworth, E.; Farrell, S.L.; Newman, T.; Bitz, C.M. Snow cover on Arctic sea ice in observations and an Earth System Model. *Geophys. Res. Lett.* **2015**, *42*, 10342–10348. [CrossRef]
6. Iacozza, J.; Barber, D.G. An examination of snow redistribution over smooth land-fast sea ice. *Hydrol. Process.* **2010**, *24*, 850–865. [CrossRef]
7. Curry, J.A.; Schramm, J.L.; Ebert, E.E. Sea Ice-Albedo Climate Feedback Mechanism. *J. Clim.* **1995**, *8*, 240–247. [CrossRef]
8. Barber, D.G.; Fung, A.K.; Grenfell, T.C.; Nghiem, S.V.; Onstott, R.G.; Lytle, V.L.; Perovich, D.K.; Gow, A.J. The Role of Snow on Microwave Emission and Scattering over First-Year Sea Ice. *IEEE Trans. Geosci. Remote Sens.* **1998**, *36*, 1750–1763. [CrossRef]
9. Yackel, J.J.; Barber, D.G.; Papakyriakou, T.N.; Breneman, C. First-year sea ice spring melt transitions in the Canadian Arctic Archipelago from time-series synthetic aperture radar data, 1992–2002. *Hyrdol. Process.* **2007**, *21*, 253–265. [CrossRef]
10. Scharien, R.K.; Geldsetzer, T.; Barber, D.G.; Yackel, J.J. Physical, dielectric, and C band microwave scattering properties of first-year sea ice during advanced melt. *J. Geophys. Res.* **2010**, *115*, 1–16. [CrossRef]

11. Fuller, M.C.; Geldsetzer, T.; Gill, J.P.S.; Yackel, J.J.; Derksen, C. C-band backscatter from a complexly-layered snow cover on first-year sea ice. *Hydrol. Process.* **2014**, *28*, 4614–4625. [CrossRef]
12. Gill, J.P.S.; Yackel, J.J.; Geldsetzer, T.; Fuller, M.C. Sensitivity of C-band synthetic aperture radar polarimetric parameters to snow thickness over landfast smooth first-year sea ice. *Remote Sens. Environ.* **2015**, *166*, 34–49. [CrossRef]
13. Fuller, M.C.; Geldsetzer, T.; Yackel, J.J.; Gill, J.P.S. Comparison of a coupled snow thermodynamic and radiative transfer model with in situ active microwave signatures of snow-covered smooth first-year sea ice. *Cryosphere* **2015**, *9*, 2149–2161. [CrossRef]
14. Nandan, V.; Geldsetzer, T.; Islam, T.; Yackel, J.J.; Gill, J.P.S.; Fuller, M.C.; Gunn, G.; Duguay, C. Ku-, X- and C-band measured and modeled microwave backscatter from a highly saline snow cover on first-year sea ice. *Remote Sens. Environ.* **2016**, *187C*, 62–75. [CrossRef]
15. Drinkwater, M.R. LIMEX '87 ice surface characteristics: Implications for C-band SAR backscatter signatures. *IEEE Trans. Geosci. Remote Sens.* **1989**, *27*, 501–513. [CrossRef]
16. Barber, D.G.; Reddan, S.P.; LeDrew, E.F. Statistical characterization of the geophysical and electrical properties of snow on landfast first-year sea ice. *J. Geophys. Res.* **1995**, *100*, 2673–2686. [CrossRef]
17. Onstott, R.G.; Moore, R.K.; Weeks, W.F. Surface-Based Scatterometer Results of Arctic Sea Ice. *IEEE Trans. Geosci. Electron.* **1979**, *GE-17*, 78–85. [CrossRef]
18. Geldsetzer, T.; Mead, J.B.; Yackel, J.J.; Scharien, R.K.; Howell, S.E.L. Surface-Based Polarimetric C-Band Scatterometer for Field Measurements of Sea Ice. *IEEE Trans. Geosci. Remote Sens.* **2007**, *45*, 3405–3416. [CrossRef]
19. Yueh, S.; Dinardo, S.; Akgiray, A.; West, R.; Cline, D.; Elder, K. Airborne Ku-band polarimetric radar remote sensing of terrestrial snow cover. *IEEE Trans. Geosci. Remote Sens.* **2009**, *47*, 3347–3364. [CrossRef]
20. Werner, C.; Wiesmann, A.; Strozzi, T.; Schneebeli, M.; Mätzler, C. The Snowscat Ground-Based Polarimetric Scatterometer: Calibration and Initial Measurements from Davos, Switzerland. In Proceedings of the 2010 IEEE International Geoscience and Remote Sensing Symposium (IGARSS), Honolulu, HI, USA, 25–30 July 2010; pp. 2363–2366.
21. King, J.M.L.; Kelly, R.; Kasurak, A.; Duguay, D.; Gunn, G.; Mead, J. UW-Scat: A Ground-Based Dual-Frequency Scatterometer for Observation of Snow Properties. *IEEE Geosci. Remote Sens. Lett.* **2013**, *10*, 528–532. [CrossRef]
22. International Hydrological Programme of the United Nations Educational, Scientific and Cultural Organization (UNESCO-IHP). *The International Classification for Seasonal Snow on the Ground*; IHP-VII Technical Documents in Hydrology N°83, IACS Contribution N°1; International Hydrological Programme of the United Nations Educational, Scientific and Cultural Organization (UNESCO-IHP): Paris, France, 2009.
23. Drinkwater, M.R.; Crocker, G.B. Modeling changes in the dielectric and scattering properties of young snow covered sea ice at GHz frequencies. *J. Glaciol.* **1988**, *34*, 274–282. [CrossRef]
24. Barber, D.G.; Nghiem, S.V. The role of snow on thermal dependence of microwave backscatter over sea ice. *J. Geophys. Res.* **1999**, *104*, 25789–25803. [CrossRef]
25. Geldsetzer, T.; Langlois, A.; Yackel, J.J. Dielectric properties of brine-wetted snow on first-year sea ice. *Cold Reg. Sci. Technol.* **2009**, *58*, 47–56. [CrossRef]
26. Stogryn, A.; Desargant, G.J. The dielectric properties of brine in sea ice at microwave frequencies. *IEEE Trans. Antennas Propag.* **1985**, *33*, 523–532. [CrossRef]
27. Denoth, A. Snow dielectric measurements. *Adv. Space Res.* **1989**, *9*, 233–243. [CrossRef]
28. Ulaby, F.T.; Stiles, H.W.; Abdelrazik, M. Snowcover influence on backscattering from terrain. *IEEE Trans. Geosci. Remote Sens.* **1984**, *GE-22*, 126–133. [CrossRef]
29. Winebrenner, D.P.; Bredow, J.; Fung, A.K.; Drinkwater, M.R.; Nghiem, S.; Gow, A.J.; Perovich, D.K.; Grenfell, T.C.; Han, H.C.; Kong, J.A.; et al. Microwave sea ice signature modeling. In *Microwave Remote Sensing of Sea Ice*; Carsey, F., Ed.; American Geophysical Union: Washington, DC, USA, 1992; pp. 137–175.
30. Nandan, V.; Geldsetzer, T.; Yackel, J.J.; Islam, T.; Gill, J.P.; Mahmud, M. Multifrequency Microwave Backscatter from a Highly Saline Snow Cover on Smooth First-Year Sea Ice: First-Order Theoretical Modeling. *IEEE Trans. Geosci. Remote Sens.* **2017**, *55*, 2177–2190. [CrossRef]
31. Lorrain, P.; Corson, D.R.; Lorrain, F. *Electromagnetic Fields and Waves*, 3rd ed.; W.H. Freeman & Company: New York, NY, USA, 1988; p. 754.
32. Geldsetzer, T.; Yackel, J.J. Sea ice type and open water discrimination using dual co-polarized C-band SAR. *Can. J. Remote Sens.* **2009**, *35*, 73–84. [CrossRef]

33. Fung, A.K. *Microwave Scattering and Emission Models and Their Applications*; Artech House: Norwood, MA, USA, 1994.
34. Hallikainen, M.T.; Winebrenner, D.P. The physical basis for sea ice remote sensing. In *Microwave Remote Sensing of Sea Ice*; Carsey, F., Ed.; American Geophysical Union: Washington, DC, USA, 1992; pp. 29–46.

 *remote sensing*

Article

Retrieval of Effective Correlation Length and Snow Water Equivalent from Radar and Passive Microwave Measurements

Juha Lemmetyinen [1,2,*], Chris Derksen [3], Helmut Rott [4,5], Giovanni Macelloni [6], Josh King [3], Martin Schneebeli [7], Andreas Wiesmann [8], Leena Leppänen [1], Anna Kontu [1] and Jouni Pulliainen [1]

[1] Finnish Meteorological Institute, Erik Palménin aukio 1, FI-00560 Helsinki, Finland; leena.leppanen@fmi.fi (L.L.); anna.kontu@fmi.fi (A.K.); jouni.pulliainen@fmi.fi (J.P.)

[2] Institute of Remote Sensing and Digital Earth, Chinese Academy of Sciences, No.9 Dengzhuang South Road, Haidian District, Beijing 100094, China

[3] Environment and Climate Change Canada, 4905 Dufferin Street, Toronto, ON M3H 5T4, Canada; chris.derksen@canada.ca (C.D.); joshua.king@canada.ca (J.K.)

[4] ENVEO IT GmbH, Fürstenweg 176, A-6020 Innsbruck, Austria; helmut.rott@enveo.at

[5] Institute of Atmospheric and Cryospheric Sciences, University of Innsbruck, A-6020 Innsbruck, Austria

[6] Institute of Applied Physics "Nello Carrara", Via Madonna del Piano, 10-50019 Sesto Fiorentino (FI), Italy; g.macelloni@ifac.cnr.it

[7] WSL Institute for Snow and Avalanche Research SLF, Flüelastrasse 11, CH-7260 Davos Dorf, Switzerland; schneebeli@slf.ch

[8] GAMMA Remote Sensing Research and Consulting AG, Worbstr. 225, CH-3073 Gümligen, Switzerland; wiesmann@gamma-rs.ch

* Correspondence: juha.lemmetyinen@fmi.fi; Tel.: +358-407-303-663

Received: 18 January 2018; Accepted: 21 January 2018; Published: 25 January 2018

Abstract: Current methods for retrieving SWE (snow water equivalent) from space rely on passive microwave sensors. Observations are limited by poor spatial resolution, ambiguities related to separation of snow microstructural properties from the total snow mass, and signal saturation when snow is deep (~>80 cm). The use of SAR (Synthetic Aperture Radar) at suitable frequencies has been suggested as a potential observation method to overcome the coarse resolution of passive microwave sensors. Nevertheless, suitable sensors operating from space are, up to now, unavailable. Active microwave retrievals suffer, however, from the same difficulties as the passive case in separating impacts of scattering efficiency from those of snow mass. In this study, we explore the potential of applying active (radar) and passive (radiometer) microwave observations in tandem, by using a dataset of co-incident tower-based active and passive microwave observations and detailed in situ data from a test site in Northern Finland. The dataset spans four winter seasons with daily coverage. In order to quantify the temporal variability of snow microstructure, we derive an effective correlation length for the snowpack (treated as a single layer), which matches the simulated microwave response of a semi-empirical radiative transfer model to observations. This effective parameter is derived from radiometer and radar observations at different frequencies and frequency combinations (10.2, 13.3 and 16.7 GHz for radar; 10.65, 18.7 and 37 GHz for radiometer). Under dry snow conditions, correlations are found between the effective correlation length retrieved from active and passive measurements. Consequently, the derived effective correlation length from passive microwave observations is applied to parameterize the retrieval of SWE using radar, improving retrieval skill compared to a case with no prior knowledge of snow-scattering efficiency. The same concept can be applied to future radar satellite mission concepts focused on retrieving SWE, exploiting existing methods for retrieval of snow microstructural parameters, as employed within the ESA (European Space Agency) GlobSnow SWE product. Using radar alone, a seasonally optimized value of effective correlation length to parameterize retrievals of SWE was sufficient to provide an accuracy of <25 mm (unbiased) Root-Mean Square Error using certain frequency combinations. A temporally dynamic value, derived from e.g., physical snow models, is necessary to further improve retrieval skill, in particular for snow regimes with larger temporal variability in snow microstructure and a more pronounced layered structure.

Keywords: snow water equivalent; passive microwave; radar; snow correlation length

---

1. Introduction

The mass of seasonal snow cover, or snow water equivalent (SWE) remains difficult to estimate on a global scale. Observational needs in terms of spatial resolution and product accuracy cannot be met by present satellite, in situ, or model-based data products at the global or regional-scale [1,2]. Global scale EO (Earth-Observation)-based products [3,4] rely on passive microwave sensors, while watershed-scale SWE has been successfully tracked with airborne LiDAR by relating observed snow height to the snow free DEM (Digital Elevation Model), and inferring SWE from the observations by modeling snow density [5]. The cost of timely airborne LiDAR surveys, however, is prohibitive for large-scale applications beyond individual watersheds or regions, so continuing efforts are made to apply Earth Observing satellite sensors for this purpose. Applying passive microwave observations from space for snow cover detection is appealing due to the availability of a long time series of daily observations with near global coverage, extending back almost 40 years. However, estimation of SWE has proved challenging despite several decades of efforts in developing retrieval approaches [6,7]. The main challenges hampering retrieval accuracy are related to the separation of the effect of increasing snow mass from other varying microstructural properties of the snowpack (density, layering, snow structural properties), and mitigating mixed pixel effects in the coarse scale passive microwave observations over heterogeneous landscapes. Existing active microwave sensors are unable to estimate SWE at the global scale and within user requirements because of the lack of current sensors at frequencies higher than X-band. In order to overcome these limits, a dual-band (X- and Ku) SAR mission called CoReH2O (Cold Regions Hydrology High-resolution Observatory, [8]), was proposed as a candidate for the 7th Earth Explorer mission of the European Agency (ESA), with the objective to provide SWE products at a spatial resolution of 200 m, exceeding that of current passive microwave methods. However, following phase-A CoReH2O was not selected for further development.

A priori characterization of snow structural parameters determining the scattering efficiency of microwaves in snow is of primary importance in the accuracy of SWE retrieval algorithms based on radiometer measurements, as this knowledge is required to resolve the total snow mass from observed signal changes [9–11]. A key parameter defining the scattering of microwaves has conventionally been the snow grain size, an estimate of the average size of snow grains in the snowpack [12]. The snow grain size has been used to empirically define the rate of microwave extinction in snow [13], which in turn has been applied in a radiative transfer model simulating emission from snow covered ground [14]. An effective grain size can be determined directly from passive microwave measurements, using widely available measurements from weather stations to fix the snow depth for the grain size inversion [15]; an effective grain size determined in this fashion can be related to grain sizes observed in the field [16]. Grain size estimates for application in SWE retrievals can also be obtained by means of applying a model to estimate snow grain metamorphism during the snow season [17].

The snow structural parameter itself remains difficult to measure, and the conversion from the 3D structure to effective model grain size is not unique due to the complex nature of snow grain metamorphism [18,19]. Theoretically-based emission and backscattering models based on the Dense Medium Radiative Transfer (DMRT) theory have assumed snow as a collection of spherical particles, introducing a stickiness parameter to emulate the sintering and clustering of snow grains [20–22]. However, it remains difficult to assign properties of snow observed in nature directly to these formulations of DMRT. While measurements of snow specific surface area (SSA) can be applied to estimate the snow optical grain diameter, empirical scaling is required to translate this value to one explaining the observed microwave response [23]. Recent efforts have focused on simulating snow as a bicontinuous medium, simulating the resulting active and passive microwave response with some success [24]. Arguably, statistical parameters such as autocorrelation length of the snow structure in different axial directions are able to describe the snow microstructure with higher fidelity than the conventional measure of grain size [25].

An effective grain size can be used to approximate the scattering behavior of the snowpack using a forward model based on that parameter [15,16]. The term 'effective' refers to the fact that the grain size may compensate for simplifications in the model setup (namely the aggregation of a multiple-layered snowpack to a single layer, a practical implementation for operational SWE retrieval schemes, including the one envisaged for CoReH2O), as well as other deficiencies in model input data or model physics. In this study, we present the retrieval of an effective correlation length, which similarly describes the radiative transfer properties of snow by a single parameter. The effective correlation length is retrieved from active and passive microwave observations of naturally accumulated snow over four winter seasons at a test site in northern Finland. We examine the interchangeability of the retrieved correlation length (derived independently from active and passive measurements at different frequencies) for the purpose of initializing the retrieval of SWE from radar observations. Specifically, the study has the following objectives;

- Retrieve an effective snow correlation length by matching emission and backscattering model predictions to the radiometer and radar measurements, respectively. Examine the seasonal behavior in the observed changes and relate these to physical properties of the snowpack
- Examine the interchangeability of the active and passive microwave effective correlation length, and determine the sensitivity of these estimates to observation frequency and polarization.
- Apply an effective correlation length derived from one sensor type (passive) to initialize the retrieval of SWE using the other (active). Compare the impact on SWE retrievals of applying temporally dynamic effective correlation length versus a seasonally constant value, optimized separately for each winter season.

The third objective is meant as a first feasibility demonstration of synergistic active/passive retrieval methods for potential implementation in proposed satellite mission concepts exploiting co-incident radar and radiometer measurements [26].

The study makes use of data acquired from the NoSREx (Nordic Snow Radar Experiment, [27]). The experiment was initiated in November 2009 in support of CoReH2O geophysical algorithm development and included observations of a boreal forest snowpack using both active and passive microwave instruments over four winter seasons.

Section 2 of the study describes the forward model and the methods applied for retrieval of correlation length, and SWE. Section 3 presents the NoSREx datasets. Section 4 presents the main results of the study. These are discussed and the conclusions of the study are given in Section 5.

2. Forward Model and Retrieval Method

2.1. MEMLS3&a Model

The Microwave Emission Model for Layered Snowpacks (MEMLS, [28,29]) is a semi-empirical model utilizing snow correlation length (i.e., the autocorrelation of snow structural variations in a spatial dimension) to define snow microstructure. This can be related to snow-scattering parameters better than the ambiguous snow grain size [25], and is also a parameter that can be objectively estimated from field and laboratory observations. A recent modification to MEMLS expanded the model to also simulate microwave backscattering from snow (MEMLS3&a, [30]). MEMLS3&a allows, by means of model inversion, determination of an effective correlation length describing the average scattering properties of a snowpack using a unified physical approach for both active and passive microwave observations.

The original MEMLS model considers snow as a vertically stacked system of homogeneous snow layers, with each layer characterized by temperature, thickness, liquid water content, salinity, density and correlation length. Radiative transfer in individual layers is calculated using a two-flux approach (fluxes propagating in forward and backward directions), while scattering and absorption coefficients are functions of a six-flux model (including fluxes in perpendicular directions). The absorption coefficient is obtained from density, frequency, temperature, moisture and salinity; the scattering coefficient is determined based on frequency, density, and correlation length. Several empirical formulations have been derived to relate correlation length with the six-flux scattering coefficient [28]; MEMLS also includes an option to use the Improved Born Approximation (IBA) to estimate correlation length, expanding the model for coarse-grained snow [29]. For multiple layers, a sandwich model is applied to determine the effects of internal scattering and reflections at layer interfaces. In the MEMLS3&a model, a prediction of backscattering from snow is calculated based on the specular and diffuse components of snow reflectivity. The specular component is determined from a mean-square slope of slight surface undulations, assuming a Gaussian distribution. The diffuse component is determined from the diffuse component of the total snow reflectivity, which in turn is related to the total emissivity of the snowpack [30].

2.2. Retrieval of Effective Correlation Length

MEMLS3&a was applied in a one-layer configuration to retrieve effective exponential correlation length $p_{\mathrm{ex,eff}}$ (i.e., a value of correlation length matching model predictions to observations). The approach is analogous to the method introduced by Pulliainen [15] for large-scale passive microwave data using conventional snow grain size. The cost function for iterative inversion can be formulated as

$$CF(p_{\mathrm{ex,eff}}) = \sum_{k=1}^{P} \frac{\left[\Phi_k^s(p_{\mathrm{ex,eff}}, SD, x_1, \ldots, x_n) - \phi_k^s\right]^2}{\sigma_k^s} \tag{1}$$

where SD is the snow depth, $p_{\mathrm{ex,eff}}$ the effective (exponential) correlation length, P the number of observation channels k, Φ_k^s the forward model estimate for sensor types (active or passive), x_n the model parameters excluding SD and $p_{\mathrm{ex,eff}}$, ϕ_k^s the observed microwave response and σ_k^s the variance of the combined error of the sensor and the forward model. The same a priori parameters x were applied in both active and passive forward model simulations. The Improved Born Approximation (IBA) was applied to calculate the snow-scattering coefficient [29]. Ground reflectivity (s0v, s0h) at different frequencies corresponds to bare soil with a surface roughness (h_{rms}) of 1 cm and a ground permittivity ($\varepsilon_{\mathrm{gnd}}$) of 4, calculated using an empirical model [31]. The specular part of reflectivity in MEMLS was considered to be 0.9 across all frequencies. Downwelling sky brightness temperature was estimated using a 55% fractile atmosphere transmissivity model [32]. Snow density (ρ_{snow}) was assigned a constant value of 200 kg m^{-3}, which closely corresponds to the approximate bulk density over the four seasons at the test site [27], and which is also very close to the typical taiga snow density [33]. SD was taken from measured in situ snow depth, and was in baseline retrievals the only temporally variable ancillary input. All other parameters were kept as constants, while $p_{\mathrm{ex,eff}}$ was the sole free parameter in the iterative optimization (minimization) of F. It should be noted that this approach was chosen as the baseline, although measured values for factors such as air temperature (T_{air}), ground temperature (T_{gnd}), ground permittivity and snow density would have been available from the NoSREx experiment. However, our purpose was to emulate potential satellite scale retrievals, where such values are likely not widely available. The retrieved effective correlation length thus accounts also for potential errors arising from the use of these constant ancillary parameters.

The retrievals of effective correlation length using active and passive microwave observations are henceforth labeled $p^{active}_{\mathrm{ex,eff}}$ and $p^{passive}_{\mathrm{ex,eff}}$, respectively. The ancillary parameters x, as well as other setting of MEMLS3&a, are summarized in Table 1.

Table 1. Ancillary input parameters and MEMLS3&a model settings.

Parameter	Description	Value
F	Centre frequency	active: 10.2, 13.3, 16.7 GHz passive: 18.7, 37 GHz
θ_k	Incidence angle	50°
s0v	V-pol reflectivity of snow-ground interface	0.03(@37 GHz) ... 0.06(@10.2 GHz) using ε_{gnd} = 4 and h_{rms} = 1 cm [31]
s0h	H-pol reflectivity of snow-ground interface	0.05(@37 GHz) ... 0.07(@10.2 GHz) using ε_{gnd} = 4 and h_{rms} = 1 cm [31]
ss0v, ss0h	Specular part of reflectivity	0.9
q	Fraction of cross-polarized scattering	(cross polarization not used)
T_{sky}	Downwelling sky brightness temperature	5 (@10.2 GHz) to 35 K (@37 GHz) [32]
T_{gnd}	Ground temperature	−5 °C
T_{snow}	Snow temperature	−5 °C
ρ_{snow}	Snow density	200 kg m^{-3}
	Scattering model	IBA
	Number of layers	1

To analyze the effect of uncertainties of the approach, retrievals of $p^{active}_{ex,eff}$ and $p^{passive}_{ex,eff}$ were repeated using measured values for available ancillary parameters for T_{air}, T_{gnd}, ε_{gnd} and ρ_{snow}. T_{air}, T_{gnd} and ε_{gnd} were taken from automated sensors. The ground reflectivity was recalculated based on the measured ε_{gnd}, considering the measured value valid across all frequencies, which is a reasonable assumption for frozen ground [34]. The measured density was taken from weekly manual observations of bulk snow density (measurements were made twice a week for the first season of NoSREx). To reduce random variability due to variable location of the manual measurements, and to obtain an estimate for each day, a third order fit was performed on the density data, using the fitted values in retrievals.

2.3. Retrieval of SWE

SWE was determined iteratively from active and passive microwave measurements by inverting the MEMLS3&a model. The cost function used in iterative minimization is analogous to Equation (1), so that

$$CF(\text{SWE}) = \sum_{k=1}^{P} \frac{\left[\Phi_k^s(p_{ex,eff}, \text{SWE}, x_1, \ldots, x_n) - \phi_k^s\right]^2}{\sigma_k^s} \tag{2}$$

where SWE is the sole free parameter and $p_{ex,eff}$ is given a priori. The SWE retrieval tests in this study were made using radar backscattering observations; four configurations (labeled Configurations 1 to 4) were used to initialize $p_{ex,eff}$ in the retrieval:

- Configuration 1: An overall average of optimized daily values $<p^{active}_{ex,eff}>$ was calculated from all retrievals of $p^{active}_{ex,eff}$ under dry snow conditions for all four seasons. Averages were calculated separately for each channel and combination. These average values of optimizations were used to initialize the respective retrievals of SWE at time t ($p_{ex,eff}(t) = <p^{active}_{ex,eff}>$).
- Configuration 2: As in Configuration 1, but the average of optimized daily values $<p^{active}_{ex,eff}>$ was calculated and applied in SWE retrieval individually for each of the four winter seasons, thus applying seasonal optimization to the retrieval.

- Configuration 3: For each radar retrieval of SWE at time t, the effective correlation length was acquired from the temporally closest passive microwave retrieval ($p_{ex,eff}(t) = p^{passive}_{ex,eff}(t)$). As a default, $p^{passive}_{ex,eff}$ was obtained from 18.7–37 GHz, V-pol, radiometer retrievals.
- Configuration 4: As in Configuration 3, but the value of $p^{passive}_{ex,eff}$ was scaled so that $p_{ex,eff}(t) = \beta\, p^{passive}_{ex,eff}(t)$. A constant scaling value β was applied in SWE retrieval across all seasons.

The first of the above configurations represents a baseline retrieval where $p_{ex,eff}$ is optimized for the site in question based on a climatological average, but without any seasonal sensitivity to potential changes. The second configuration emulates a retrieval where $p^{active}_{ex,eff}$ has been seasonally optimized, but otherwise kept temporally constant. In practice, this could be achieved e.g., by means of applying e.g., a physical snow model. The contrast with Configuration 1 thus demonstrates the effect of using a seasonally sensitive microstructural indicator, compared to a static climatological average. The third and fourth configurations emulate a scheme where the active microwave retrieval is initialized using effective correlation length derived from passive microwave observations, with the last configuration applying an additional scaling factor, to account for the mean difference in the retrieved $p^{active}_{ex,eff}$ and $p^{passive}_{ex,eff}$. All other input variables of the MEMLS3&a model were kept constant in the retrieval, following Table 1, as in the retrieval of $p_{ex,eff}$.

As with retrievals of $p_{ex,eff}$ (see Section 2.2), SWE retrievals were repeated using measured ancillary data for T_{air}, T_{gnd}, ε_{gnd} and ρ_{snow}. Furthermore, comparative retrievals using passive microwave observations were made adapting Configurations 1 and 2 for SodRad radiometer observations, as the seasonal and overall average of $p^{passive}_{ex,eff}$ to initialize the retrievals.

3. The NoSREx Campaign

3.1. Microwave Observations

The objective of the NoSREx campaign [27] was to provide a continuous time series of active and passive microwave observations of snow cover in a representative location of boreal forest, covering the complete snow accumulation and ablation period. The campaign lasted four full snow seasons, providing near-continuous observations of snow cover microwave signatures. The campaign hosted two tower-based microwave instruments; SnowScat, a stepped-frequency, fully polarized radar operating on frequencies from X- to Ku band, and the SodRad radiometer, a microwave dual-polarization radiometer system operating from X- to W bands. SnowScat was installed on a tower at the height of 9.6 m overlooking a large forest clearing. SodRad was installed on an adjacent platform at the height of 4 m, overlooking the same general test area. However, the footprints of the instruments were not entirely co-incident (see e.g., [27,30], for a detailed description).

The standard measurement of SnowScat during the NoSREx experiment took place every three to four hours (the measurement sequence was changed to four hours after the first season to allow for scanning of a secondary observation area); co- and cross-polarized backscattering was measured at four incidence angles (30°, 40°, 50°, and 60°). Each elevation scan in the main observation area contained 17 azimuth looks at 6° intervals. Observations at 50° angle of incidence were chosen for this study. The data for each incidence angle were averaged over all measured azimuth directions in order to reduce effects of random speckle and to increase the number of independent looks. The stepped-frequency scan was integrated over three bands of 2 GHz, with center frequencies 10.2, 13.3 and 16.7 GHz. Only the co-polarized vertical measurements (VV-polarization) were used in this study. The inherent reason was that the MEMLS3&a model essentially calculates cross-polarized backscattering from co-polarization using a fixed ratio; thus applying cross-polarized observations was not seen to provide additional benefit.

Brightness temperatures from SodRad at 10.65, 18.7, 21, 37, 90 and 150 GHz (H and V pol) were available at incidence angles ranging from 40 to 60°; however, only the vertically polarized 10.65, 18.7 and 37 GHz channels at 50° were used for this study, as initial retrieval tests with H polarization resulted in larger temporal variability and decreased correlation with $p^{active}_{ex,eff}$ and so were omitted from further study. SodRad measurements took place in-between SnowScat scans to avoid RFI (radio frequency interference) from the radar signal.

3.2. In Situ Data

The microwave observations were supported by regular manual snow profile measurements as well as extensive array of automated measurements on snow, ground and meteorological parameters [27]. SWE measured from snow profile measurements served as the main reference for SWE retrievals. For the third and fourth season, collected data included vertical profiles of snow specific surface area (SSA), using an infrared laser and integrating sphere [35]. The commercial IceCube instrument was used; the SSA measurements are described by Leppänen et al. [36]. In addition, the SnowMicroPen (SMP; [37]) was used on several occasions—these measurements allowed the quantitative extraction of snow correlation length from the measured vertical profile of penetration resistance [38]. Field data included also conventional measurements of snow grain size [39]; however, these data were not applied here.

3.3. Campaign Summary

Figure 1 depicts the data measured during the four seasons of NoSREx. The data are presented as channel differences to reveal trends associated with snow cover changes. For SnowScat, the channel differences of VV-pol backscattering for 13.3–10.2 GHz, 16.7–10.2 GHz, and 16.7–13.3 GHz are depicted (the data are shown in linear units). For SodRad brightness temperatures, channel differences of 10.65–18.7 GHz, 10.65–37 GHz, and 18.7–37 GHz are shown for the vertical polarization (V-pol). The main reason for applying a channel difference in passive microwave retrievals is to reduce sensitivity to variations in physical temperature affecting microwave emission; in the case of microwave backscattering, the potential benefits of examining channel differences in backscattering are less obvious, but are presented here to determine any potential improvement in retrieval results compared to individual channels.

Also depicted in Figure 1 are in situ measured air temperature (T_{air}), as well as the 2 cm depth ground temperature (T_{gnd}) and ground permittivity (ε_{gnd}) measured at the NoSREx test site. Manual snow profile measurements were conducted weekly (twice a week during the first season of NoSREx) at the test site. Bulk snow density (ρ_{snow}), measured using a manual snow scale, is shown in the lowest panel, together with the measured snow depth (*SD*) from an automated sensor. A third order fit to measured ρ_{snow} is shown, which can be used to reduce the effect of random errors in the manual measurements, which in the case of density arises from local-scale variability in density conditions (each density sample was taken from a different location due to the destructive nature of the measurement). Bulk SWE values calculated from the density measurements were used as reference to retrieved SWE in this study.

Clear differences in backscattering signatures between the four NoSREx seasons are apparent; the second season (Figure 1b) shows a large dynamic range in the channel differences involving the 16.7 GHz channel, while the third season (Figure 1c) shows comparatively less temporal dynamics; the first and fourth seasons exhibit a dynamic range between these two extremes. The channel difference of the two lower frequencies (13.3–10.2 GHz), on the other hand, shows stable behavior over all four seasons with very low dynamics relative to changing SWE. It should be noted that for the third season, radar measurements were available only after January 23. The microwave emission shows similarly large dynamics for the second season applying the frequency combinations involving the 37 GHz channel, while the overall response during the third season in particular is muted. The channel difference of the two lower frequencies (10.65 and 18.7 GHz) shows notably low dynamic behavior during the observed winter periods.

It is difficult to directly associate observed increases in microwave backscattering or decreases in microwave emission to snowfall events; rather, significant increases in backscattering at Ku-bands (shown as an increase of the difference of the 16.7 and 10.2 GHz channels, as well as the difference between 16.7 and 13.3 GHz channels) occur often in periods with no significant precipitation, indicating that such increases are due to other changes in snow or soil backscattering characteristics (e.g., in March 2011, Figure 1b). Similarly, significant precipitation events, such as those occurring in February 2013, induce little or no increases in backscattering or dampening of microwave emission, indicating low sensitivity of the microwave response to newly fallen snow. The new snow may in fact work as a dampening layer, reducing backscattering from lower layers with coarser snow microstructure (see e.g., [40]). This may also partly account for the early season decrease in the backscattering signal.

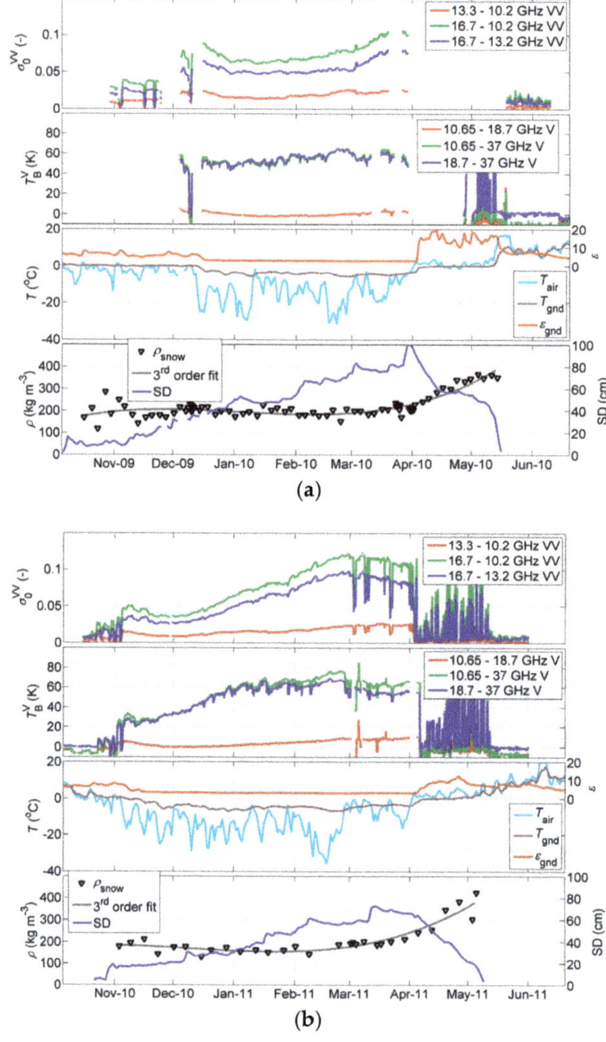

Figure 1. *Cont.*

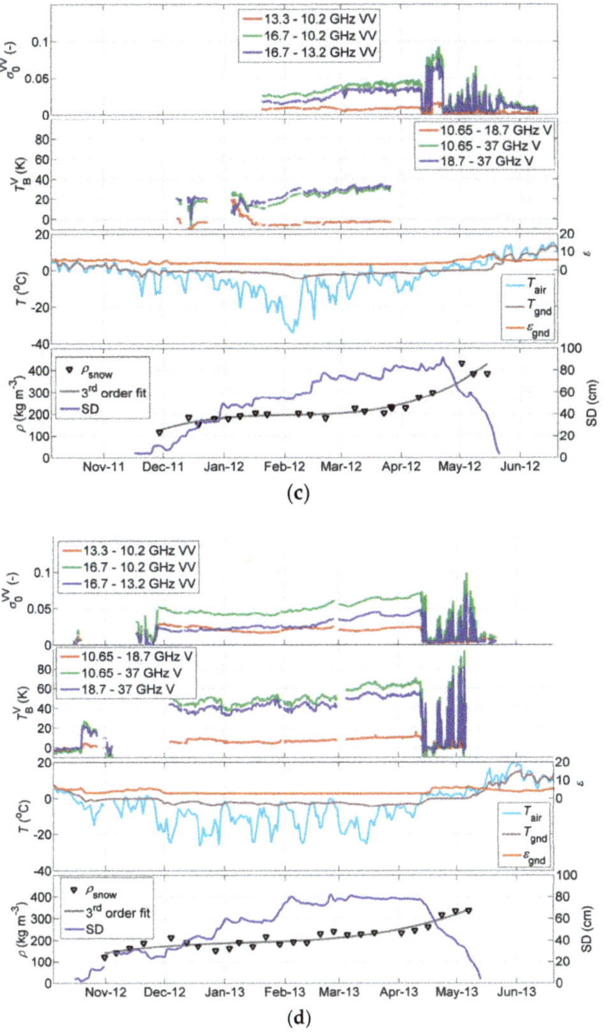

Figure 1. Data from four NoSREx campaigns in 2009–2010 (**a**); 2010–2011 (**b**); 2011–2012 (**c**); 2012–2013 (**d**). Panels from top to bottom: SnowScat channel differences for backscattering measured between 10.2, 13.3 and 16.7 GHz, VV-pol, 50° incidence angle; SodRad brightness temperature channel differences between 10.65, 18.7 and 37 GHz channels, V-pol, 50° incidence angle; air temperature (T_{air}), ground temperature (T_{gnd}) and ground permittivity (ε_{gnd}) at 2 cm depth; snow depth and snow density with 3rd order fit to manual density measurements shown adapted from [27].

4. Results

4.1. Retrieved Active and Passive Correlation Length

The SnowScat and SodRad measurements were applied in Equation (1) to retrieve an effective correlation length $p_{ex,eff}$ for all observation cases in dry snow conditions. These were determined from air temperatures, discarding all cases where air temperature was above -1 °C to allow for margin for unclear cases close to melting point. The measured in situ snow depth was used to initialize the retrieval; as described in Section 2.2, other forward model parameters were kept constant following Table 1. Individual channels, and combinations of two channels were used in the retrieval, mimicking possible satellite sensor configurations. In the case of SnowScat, $p_{ex,eff}$ was retrieved using 10.2, 13.3 and 16.7 GHz observations as individual channels, using the combinations 10.2 & 13.3, 10.2 & 16.7, and 13.3 & 16.7 both independently and as a channel difference in the cost function (total of nine different retrieval combinations). VV polarized backscattering observations for each frequency channel or channel combination were used in all cases. Similarly, in the case of SodRad V-pol radiometer measurements, both individual channels, combinations of channels and channel differences were applied (total nine combinations).

Figure 2 depicts the time series of retrieved $p^{active}_{ex,eff}$ and $p^{passive}_{ex,eff}$ during the four seasons of NoSREx. Retrievals of $p^{active}_{ex,eff}$ using the three combinations of channel differences between the 10.2, 13.3 and 16.7 GHz are shown, as well as retrieved $p^{passive}_{ex,eff}$ using the channel difference combinations between the 10.65, 18.7 and 37 GHz channels. The three time series of $p^{active}_{ex,eff}$ correspond largely to one another in terms of temporal behavior, in particular in the beginning and end of the snow season, but exhibit clear differences in magnitude; also the channel difference of 13.3–10.2 GHz produces a less dynamic time series of $p^{active}_{ex,eff}$. With the exception of 2012–2013, as well as February 2012, the channel difference of 16.7–13.3 GHz gives the largest value of $p^{active}_{ex,eff}$, while the channel difference of 13.3–10.2 GHz results in the smallest values. Interestingly, for the last season these relations are reversed. The passive-microwave derived $p^{passive}_{ex,eff}$ follows some of the temporal behavior of the three $p^{active}_{ex,eff}$ time series, with notable exceptions in the beginnings of the first and second seasons. This may be related to early-season snow melt-refreeze and consequent metamorphism, which caused early-season scattering behavior that was not captured by the MEMLS3&a model when driven directly using in situ snow measurements [27]. It should be noted that radar observations were missing from the beginning of the third snow season of NoSREx. For the last season, both $p^{active}_{ex,eff}$ and $p^{passive}_{ex,eff}$ follow a similar temporal trend; the overall spread in the magnitude of retrievals is also the smallest for this season.

Figure 2 also shows in situ measured values of p_{ex} using the IceCube instrument and the SMP. SMP data were available from only a few days throughout the season, while IceCube measurements were conducted weekly during the third and fourth seasons. The measurements of IceCube SSA, made at approximately 5 cm vertical intervals, were averaged throughout the entire snowpack. Depth-weighted averaging was applied, estimating the layer depth represented by the sample from the recorded snow heights. The average values were converted to optical equivalent grain size D_0, so that [25]

$$D_0 = \frac{6}{\rho_i \cdot \text{SSA}} \tag{3}$$

The following empirical linear relations between D_0 and p_{ex} have been derived [29]

$$p_{ex} = 0.4 \cdot D_0 \text{ for dendritic grains} \tag{4}$$

$$p_{ex} = 0.3 \cdot D_0 \text{ for non} - \text{dendritic grains} \tag{5}$$

Similar scaling values, ranging from 0.25 to 0.4, have been presented elsewhere in literature [41]. Here, the scaling value of 0.4 for dendritic grains (Equation (4)) was applied in all cases. The resulting comparison of p_{ex} measured using IceCube for the third seasons shows relatively good agreement with $p^{passive}_{ex,eff}$ (significant R^2 values for the channel differences depicted in Figure 2c ranged from 0.48 to 0.61), while $p^{active}_{ex,eff}$ generally overestimates measured values and showed a lower coefficient of determination (significant R^2 values ranged from 0.37 to 0.46). Nevertheless, both $p^{passive}_{ex,eff}$ and $p^{active}_{ex,eff}$ as well as in situ measurements show decreasing values of p_{ex} in mid-February; this can be explained by snow accumulation in that period (Figure 1c), which decreased the measured average p_{ex}. For the fourth season, both $p^{active}_{ex,eff}$ and $p^{passive}_{ex,eff}$ overestimate measured values until mid-March 2013 by 20–30% using the channel difference depicted in Figure 2d. However, again the general trend of measured p_{ex} is closely replicated by the retrievals, with a generally decreasing trend in p_{ex}, $p^{active}_{ex,eff}$ and $p^{passive}_{ex,eff}$ until

February 2013. Significant R^2 values against in situ p_{ex} for the fourth season ranged from 0.30 to 0.61 and 0.27 to 0.34, for $p^{active}_{ex,eff}$ and $p^{passive}_{ex,eff}$ respectively.

Measurements of p_{ex} using IceCube generally agree well with SMP measurements of p_{ex} for the third and fourth snow seasons, although the IceCube values show more variability, as depicted by error bars in Figure 2. The results provide some confirmation of the physical representativeness of the retrieved $p_{ex,\,eff}$; it should, however, be noted the method for active/passive SWE retrieval presented in this study (Section 2.3) does not rely on matching of $p_{ex,\,eff}$ with in situ measurements.

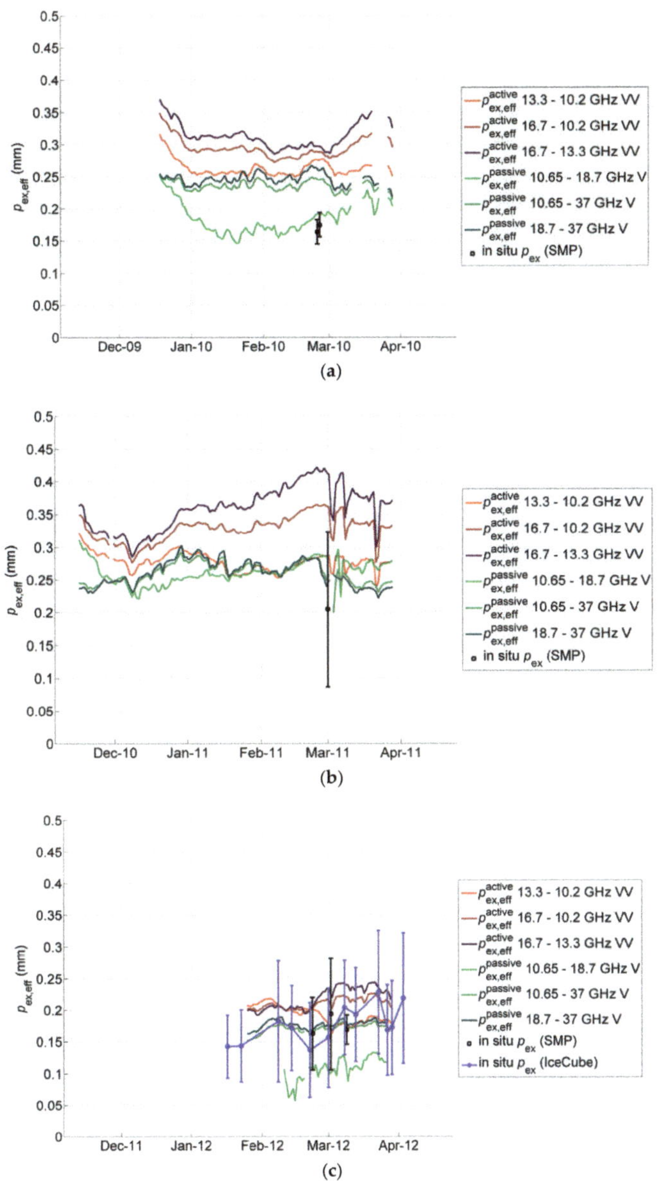

Figure 2. *Cont.*

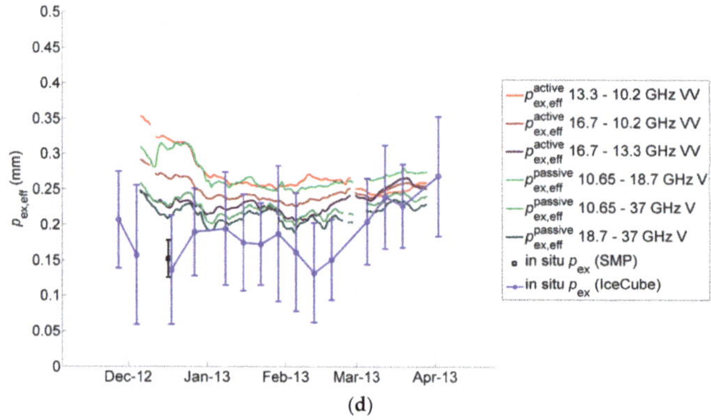

(**d**)

Figure 2. Time series of effective correlation length $p_{ex,eff}$ retrieved for NoSREx campaigns in 2009–2010 (**a**); 2010–2011 (**b**); 2011–2012 (**c**); 2012–2013 (**d**). Retrieved active microwave $p^{active}_{ex,eff}$ for channel differences 13.3–10.2, 16.7–10.2, and 16.7–13.3 GHz (VV-polarization). Retrieved passive microwave $p^{passive}_{ex,eff}$ using the channel differences of 10.65–18.7 GHz, 10.65–37 GHz, and 18.7–37 GHz (V-polarization). In situ measured p_{ex}, estimated from either SMP (black) or IceCube SSA measurements (blue), averaged over whole snowpack (see [36]). Error bars reflect standard deviation of all measured samples.

One objective of this study was to examine the interchangeability of the retrieved correlation length derived independently from active and passive observations. Consequently, the relationships between daily averages of $p^{active}_{ex,eff}$ to $p^{passive}_{ex,eff}$ were analyzed by calculating basic statistics between parameters retrieved using different channel configurations. Table 2 summarizes the coefficient of determination of $p^{active}_{ex,eff}$ relative to $p^{passive}_{ex,eff}$. Coefficients of determination which are not statistically significant ($p > 0.05$) are indicated with asterisks. Results are shown as a matrix of the different active and passive channels and channel combinations. When calculating the daily average, obvious erroneous retrievals were ruled out from the analysis ($p_{ex,eff} < 0.05$ mm or $p_{ex,eff} > 0.5$ mm, which correspond to typical minimum and maximum values measured for natural snow [37]). Passive microwave V-pol and active microwave VV pol combinations are shown; retrievals using in situ measured values for model ancillary parameters are shown in parenthesis. The retrievals were also performed using passive microwave H-pol observations as well as a combination of H- and V-pol observations, with largely similar results, the main difference being that H-pol retrievals of $p^{passive}_{ex,eff}$ exhibited increased temporal variability compared to V-pol retrievals, arising directly from the larger variability of the observed brightness temperature at H-pol. This resulted in most cases in decreased correlation and increased bias and uRMSE against $p^{active}_{ex,eff}$. Therefore, only the results for V-pol are presented and discussed further.

Overall, the results show that retrievals derived from single channels show low correlation between frequencies, with the exception of the highest frequencies of 16.7 GHz for active and 37 GHz for passive. Moderate association is apparent also between retrievals at 13.3 (active) and 37 (passive) GHz. Increasingly better results in terms of the coefficient of determination are obtained using multiple channels, and in particular channel differences. The highest overall R^2 for most active microwave channels were obtained against the channel differences of 18.7–37 GHz; this combination also typically presented the smallest bias or uRMSE when compared to other passive microwave retrievals. Analysis for individual seasons (not shown) provided higher coefficients of determination between $p^{active}_{ex,eff}$ and $p^{passive}_{ex,eff}$ in particular for the third and fourth season of NoSREx, with R^2 values exceeding 0.8. The use of measured in situ data in place of constants increased correlations moderately for most frequencies and combinations (values given in parenthesis in Table 2); a notable exception are correlations of $p^{active}_{ex,eff}$ with $p^{active}_{ex,eff}$ derived from the 10.6–18.7 GHz channel difference, where almost all correlations were deteriorated by use of in situ data.

Table 2. Coefficient of determination between daily averages of $p^{active}_{ex,eff}$ and $p^{passive}_{ex,eff}$ retrieved using different channels and channel combinations. All four seasons of retrievals from NoSREx summarized. Statistically insignificant R^2 values indicated with asterisks (*). Results from retrievals with measured ancillary data given in parenthesis.

R^2	10.2 GHz	13.3 GHz	16.7 GHz	10.2 & 13.3 GHz	10.2 & 16.7 GHz	13.3 & 16.7 GHz	13.3–10.2 GHz	16.7–10.2 GHz	16.7–13.3 GHz
10.65 GHz	0.07 (0.11)	0.13 (0.07)	0.00 (0.20)	0.12 (0.08)	0.00 (0.21)	0.00 (0.20)	0.15 (0.01)	0.01 (0.15)	0.03 * (0.17)
18.7 GHz	0.00 (0.08)	0.10 (0.39)	0.04 (0.02)	0.07 (0.35)	0.04 (0.03)	0.02 (0.06) *	0.27 (0.53)	0.04 (0.01)	0.10 (0.01)
37 GHz	0.00 (0.09)	0.40 (0.47)	0.74 (0.79)	0.33 (0.41)	0.73 (0.79)	0.75 (0.80)	0.59 (0.65)	0.74 (0.79)	0.59 (0.63)
10.65 & 18.7 GHz	0.01 (0.09)	0.13 * (0.41)	0.03 (0.03)	0.10 (0.37)	0.03 * (0.03)	0.01 * (0.06) *	0.29 (0.53)	0.04 (0.02)	0.10 (0.01)
10.65 & 37 GHz	0.00 (0.04)	0.40 (0.47)	0.73 (0.79)	0.33 (0.41)	0.73 (0.78)	0.75 (0.80)	0.59 (0.65)	0.74 (0.79)	0.58 (0.63)
18.7 & 37 GHz	0.00 (0.04)	0.42 (0.49)	0.72 (0.77)	0.34 (0.42)	0.72 (0.77)	0.74 (0.79)	0.62 (0.68)	0.73 (0.78)	0.56 (0.61)
10.65–18.7 GHz	0.03 (0.05)	0.27 (0.10)	0.19 (0.07)	0.19 (0.05)	0.19 (0.06)	0.23 (0.08)	0.62 (0.40)	0.23 (0.09)	0.09 (0.02)
10.65–37 GHz	0.01 (0.01)	0.34 (0.38)	0.74 (0.79)	0.27 (0.32)	0.74 (0.78)	0.75 (0.78)	0.59 (0.60)	0.78 (0.82)	0.62 (0.68)
18.7–37 GHz	0.00 (0.02)	0.29 (0.34)	0.78 (0.82)	0.24 (0.29)	0.78 (0.82)	0.77 (0.80)	0.44 (0.47)	0.79 (0.84)	0.69 (0.76)

Figure 3 shows the retrieved $p^{passive}_{ex,eff}$ against $p^{active}_{ex,eff}$ separately for the four seasons of NoSREx, using the channel differences of 16.7–10.2 GHz (active) and 18.7–37 GHz (passive). This comparison gave the best overall coefficient of determination over the four seasons (0.79). However, a consistent underestimation of 0.03–0.06 mm for $p^{passive}_{ex,eff}$ relative to $p^{active}_{ex,eff}$ was apparent. Against retrievals with 18.7–37 GHz, $p^{active}_{ex,eff}$ was on average 15% larger. This value was applied in SWE retrieval Configuration 4 (see Section 2.3) to scale between $p^{passive}_{ex,eff}$ and $p^{active}_{ex,eff}$.

It is notable in Figure 3 that retrievals from both sensor types similarly resolve the differing magnitudes of $p_{ex,eff}$ between the seasons; the second season gave on average the largest values, while the third season indicated the lowest values of $p_{ex,eff}$. This confirms that snow microstructure strongly depends on meteorologically driven seasonal variability in snowpack temporal evolution, so a temporal (or at least seasonal) dynamic retrieval of correlation length is desirable. Incidentally, these results concur with the inter-seasonal relations of the average size of snow grains derived from conventional grain size observations (Table 4 in [27]). The second season, with the largest average $p_{ex,eff}$, also exhibits the largest scatter and lowest correlation coefficient between active and passive retrievals.

The average of retrieved $p_{ex,eff}$ for the four NoSREx seasons are summarized in Tables 3 and 4 for active and passive retrievals, respectively, using the applied frequencies and combinations. For most cases, the second season exhibits the largest, and the third season the smallest $p_{ex,eff}$ for both active and passive cases; notable exceptions are retrievals involving the lower X-band frequencies (10.2 and 10.65 GHz). The values represent retrievals with constant ancillary model parameters; the use of measured in situ data increased the value of $p^{active}_{ex,eff}$ on average by 15%, 7% and 5% for 10.2, 13.3 and 16.7 GHz, respectively. For $p^{passive}_{ex,eff}$, the induced differences were on average 21%, 10% and −3.0% for 10.65, 18.7 and 37 GHz, respectively. For channel combinations, the effect was typically smaller (2 to 8% increase for $p^{active}_{ex,eff}$, and a change of −3 to 16% for $p^{passive}_{ex,eff}$). In most cases, the use of ancillary data increased the absolute difference between $p^{active}_{ex,eff}$ and $p^{passive}_{ex,eff}$; the average difference between all retrievals increased from 22 to 40%. The difference between $p^{active}_{ex,eff}$ and $p^{passive}_{ex,eff}$ retrieved with the 18.7–37 GHz channel difference increased from 15 to 30%, which was used to scale $p^{passive}_{ex,eff}$ in Configuration 4, when performing comparative SWE retrievals with measured ancillary in situ information.

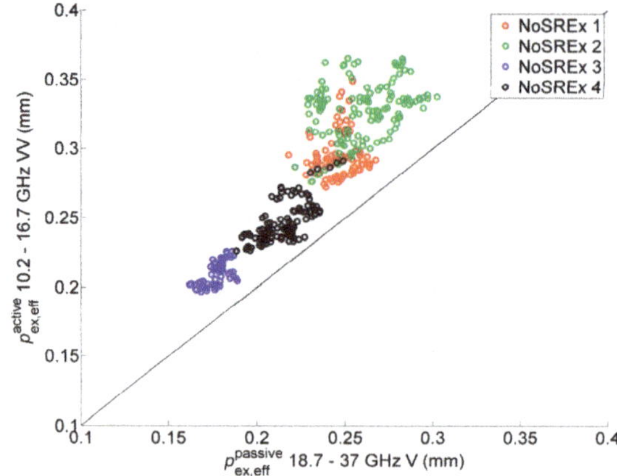

Figure 3. Daily $p^{passive}_{ex,eff}$ retrieved using channel difference of 18.7–37 GHz, V-pol, compared to $p^{active}_{ex,eff}$ retrieved using channel difference of 16.7–10.2 GHz, VV pol. Constant values (Table 1) used for ancillary data. Color codes represent the four different seasons of NoSREx (1 to 4).

Table 3. The average of $p^{active}_{ex,eff}$ (in mm) across seasons for different channels and channel combinations.

Season	10.2 GHz	13.3 GHz	16.7 GHz	10.2 & 13.3 GHz	10.2 & 16.7 GHz	13.3 & 16.7 GHz	13.3–10.2 GHz	16.7–10.2 GHz	16.7–13.3 GHz
NoSREx 1	0.28	0.27	0.29	0.27	0.29	0.29	0.26	0.29	0.31
NoSREx 2	0.21	0.26	0.31	0.25	0.31	0.30	0.28	0.33	0.36
NoSREx 3	0.23	0.21	0.21	0.21	0.21	0.21	0.19	0.21	0.22
NoSREx 4	0.23	0.25	0.24	0.25	0.24	0.25	0.27	0.25	0.23

Table 4. The average of $p^{passive}_{ex,eff}$ (in mm) across seasons for different channels and channel combinations. Retrievals at 10.65 and 18.7 GHz or with the channel difference of these frequencies for NoSREx 3 did not produce realistic values of $p^{passive}_{ex,eff}$ and were discarded.

Season	10.65 GHz	18.7 GHz	37 GHz	10.65 & 18.7 GHz	10.65 & 37 GHz	18.7 & 37 GHz	10.65–18.7 GHz	10.65–37 GHz	18.7–37 GHz
NoSREx 1	0.25	0.16	0.23	0.17	0.23	0.23	0.18	0.24	0.25
NoSREx 2	0.31	0.18	0.24	0.18	0.24	0.24	0.26	0.26	0.26
NoSREx 3	-	-	0.17	-	0.17	0.16	0.11	0.17	0.18
NoSREx 4	0.34	0.22	0.22	0.22	0.22	0.22	0.27	0.23	0.22

4.2. SWE Retrieval Using Radar Observations

SWE was retrieved from SnowScat observations by applying the four retrieval configurations described in Section 2. Retrievals for each season were compared to manual in situ measurements of bulk SWE made weekly at the test site. The retrievals were again constrained to the dry snow season. As described in Section 2, in the first and second configurations, $p_{ex,eff}$ in Equation (2) was set to be the overall average, or the seasonal average, of retrieved $<p^{active}_{ex,eff}>$, respectively. In the third configuration, $p_{ex,eff}$ was taken from the temporally closest retrieval of $p^{passive}_{ex,eff}$, thus testing the suitability of passive- microwave derived $p_{ex,eff}$ in informing the SWE retrieval from radar observations. The fourth configuration applied this same method, but using also an additional linear scaling factor (20%, as described in Section 4.1), so that $p_{ex,eff} = \beta * p^{passive}_{ex,eff}$. A constant value of $\beta = 1.2$ was applied over all seasons, based on the average difference between $p^{passive}_{ex,eff}$ and $p^{active}_{ex,eff}$.

Similarly to the retrieval of $p_{ex,eff}$, SWE retrievals were performed for several channels and channel combinations. In Configurations 1 and 2, $<p^{active}_{ex,eff}>$ was always chosen from the corresponding retrieval at the same channel or combination of radar channels, while in Configurations 3 and 4, $p^{passive}_{ex,eff}$ was always obtained from the retrieval using the 18.7–37 GHz, V-pol channel difference (retrievals from H-pol were not applied). Retrievals using this passive microwave channel combination showed the best overall agreement with $p^{active}_{ex,eff}$ retrievals (Table 2).

Figure 4 shows the time series of retrieved SWE for the four seasons of the experiment using retrieval Configurations 1 and 3 (left panels) and 2 and 4 (right panels), using the 13.3 and 16.7 GHz channel combinations. Red lines represent retrievals with Configurations 1 and 2, applying either the overall, or the seasonal average of $<p^{active}_{ex,eff}>$; green lines represent retrievals using $p^{passive}_{ex,eff}$ (Configurations 3 and 4). In comparison to in situ data, retrievals using Configurations 1 and 3 show large biases for all seasons, in particular for the third season (Figure 4e). Applying the seasonally optimized $<p^{active}_{ex,eff}>$ (Configuration 2, right panels) produces SWE estimates which closely match the in situ measured values, with the most notable discrepancies apparent for the fourth season. Even the weak signal dynamics observed during the third season (see Figure 1c) is sufficient to result in an increase of retrieved SWE which closely matches that of in situ observations (Figure 4f). Similarly, without scaling, the temporally dynamic $p^{passive}_{ex,eff}$ produces estimates which overestimate in situ measured SWE for all seasons (green lines in left panels. Applying the constant scaling factor β corrects for the overestimation, but actually results in underestimation of SWE in particular for the first and second seasons. This can be explained by the scaling factor β being generally applicable to scale $p^{passive}_{ex,eff}$ for all frequency combinations of $p^{active}_{ex,eff}$; a more optimized fit could be achieved by defining β individually for each channel or combination of channels, as well as for each season.

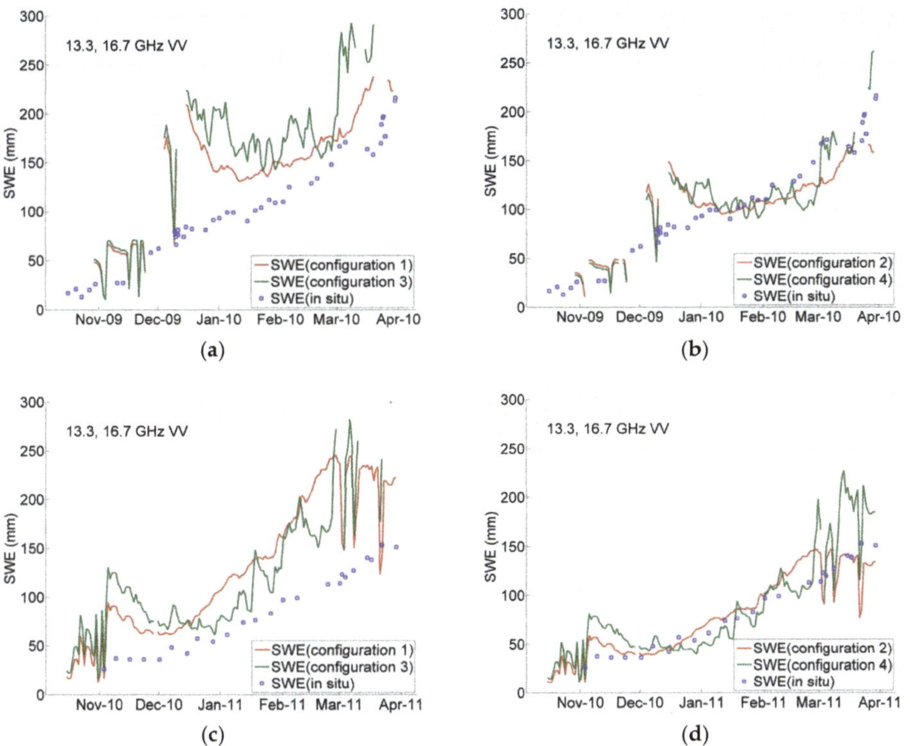

Figure 4. *Cont.*

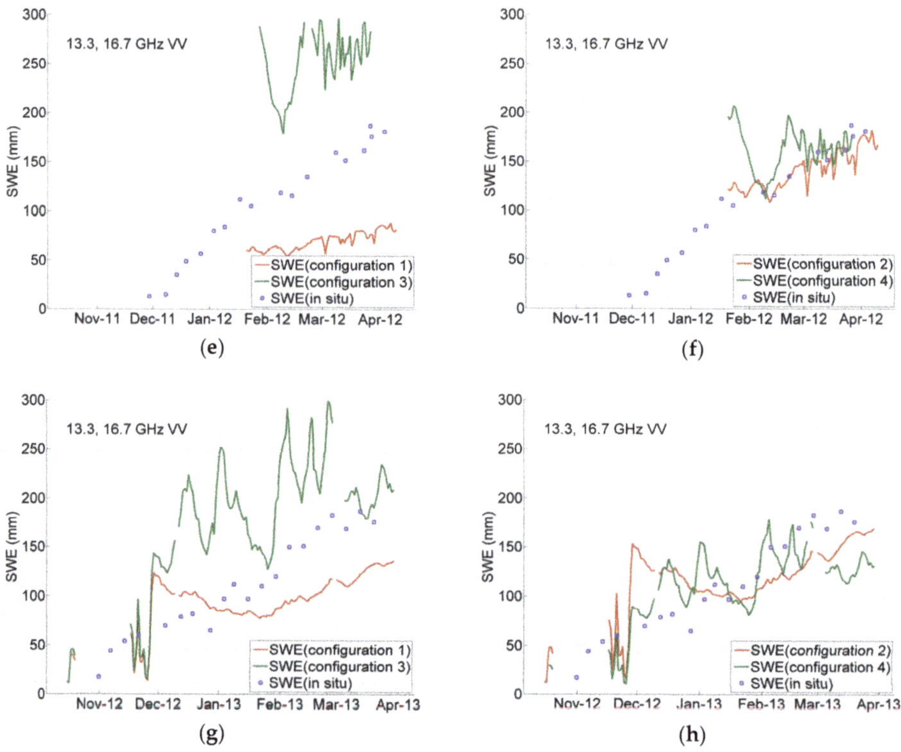

Figure 4. Retrieved SWE from 13.3 and 16.7 GHz, VV pol channels of SnowScat for NoSREx campaigns in 2009–2010 (**a**,**b**); 2010–2011 (**c**,**d**); 2011–2012 (**e**,**f**); 2012–2013 (**g**,**h**). Left panels: retrievals with Configurations 1 and 3; right panels: retrievals with Configurations 2 and 4.

A comparison of SWE retrievals against in situ measured SWE using the three individual SnowScat frequency bands of 10.2, 13.3 and 16.7 GHz is shown in Figure 5. The data represent all four seasons of NoSREx; the red markers represent SWE retrievals with using the seasonally optimized $<p^{active}_{ex,eff}>$ (Configuration 2), while green markers denote retrievals with a temporally dynamic but scaled $p^{passive}_{ex,eff}$ (Configuration 4). Error metrics in terms of the coefficient of determination, bias and uRMS errors are also displayed. Retrievals using 10.2 GHz (Figure 5a) show little or no correlation with measured in situ SWE, denoted by a low coefficient of determination for both configurations. With the higher frequency bands, correlation as well as bias and uRMSE are improved for both configurations. Overall, Configuration 2, using a seasonally optimized, but otherwise temporally static $p_{ex,eff} = <p^{active}_{ex,eff}>$ yields better error metrics compared to Configuration 4. The good performance of Configuration 2 can be expected due to the seasonal optimization of $p_{ex,eff}$; this is reflected in particular by the low retrieval bias. However, the correlation and uRMSE values also indicate good agreement, suggesting that such a static value, if optimized for a given season, is sufficient to account for most of the seasonal dynamics in snow microstructure at the test site.

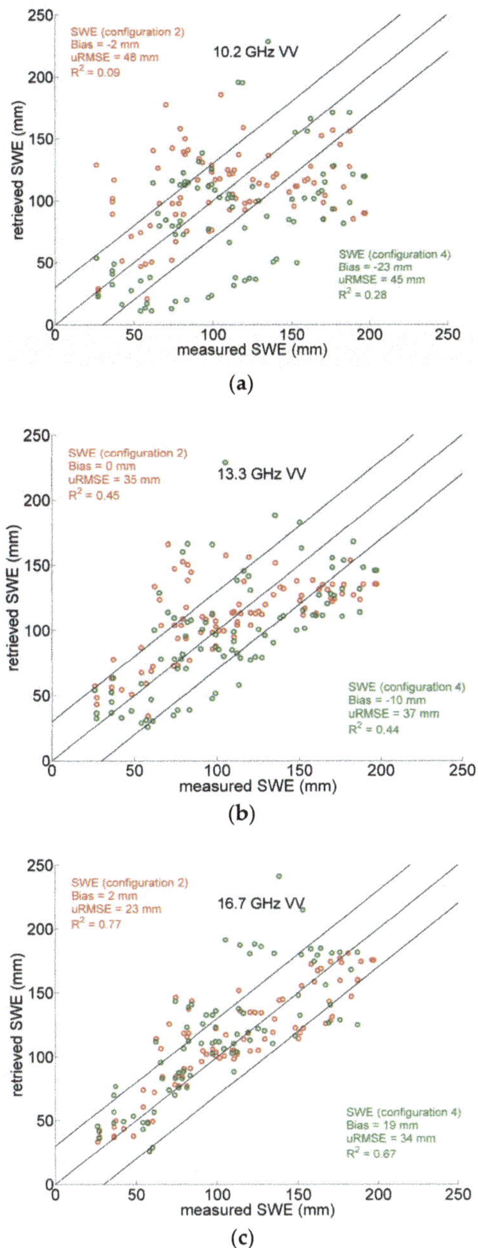

Figure 5. Comparison of retrieved SWE against manual measurements for all four NoSREx seasons. SnowScat channels 10.2 (**a**) 13.3 GHz (**b**) and 16.7 GHz (**c**). Red: retrieval using Configuration 2 with $p_{ex,eff} = <p^{active}_{ex,eff}>$ (seasonal optimization). Green: retrieval using Configuration 4 with $p_{ex,eff} = \beta^* p^{passive}_{ex,eff}$. The diagonal lines represent a 1:1 fit $+/-$ 30 mm SWE; an RMSE of 30 mm was defined as a threshold for retrieval accuracy.

Figure 6 shows a comparison of retrieved vs. measured SWE retrievals using retrieval Configurations 2 and 4 for different channel combinations, again for all seasons of NoSREx. The combination of the two lower bands (Figure 6a) yields the largest discrepancies against in situ measured SWE; the use of the 10.2 GHz channel even slightly degrades the error metrics when compared to using 13.3 GHz alone (Figure 5b). Of the combinations, the use of 10.2 and 16.7 GHz (Figure 6b) gives for both configurations the best overall error metrics. However, the results against using the 16.7 GHz channel alone are practically the same, thus any benefit of applying the 10.2 GHz channel cannot be shown.

Error metrics for SWE retrieval tests using Configurations 1 and 2 are summarized in Table 5. Using a single value for $p_{ex,eff}$ (Configuration 1) results in bias errors for several frequency combinations, particularly notable for the third season. R^2 values against the whole dataset of four seasons were below 0.2 for any frequency combination. For individual seasons using the 16.7 GHz channel, however, reasonable correlations were found (namely for the second and third seasons). Using Configuration 2, obtained bias errors are on average below 10 mm for all frequencies and combinations, demonstrating the sensitivity of the retrieval to $p_{ex,eff}$ and the necessity for seasonal optimization. All frequency combinations show both reduced uRMSE and an increased R^2 value when using Configuration 2, compared to Configuration 1. Furthermore, all retrievals with Configuration 2 using the 16.7 GHz channel give an R^2 value exceeding 0.7 on average for the whole dataset, with values exceeding 0.9 for the second season. The fourth season provides the worst results in terms of R^2. It can be also observed that, in this case, the addition of a secondary channel to 16.7 provides an improvement in R^2 and uRMSE.

Table 6 summarizes similarly the retrievals using Configurations 3 and 4, i.e., by applying $p_{ex,eff}$ from passive microwave observations in the retrieval. Relatively large bias errors are obtained when using $p^{passive}_{ex,eff}$ directly without scaling (Configuration 3); however, R^2 values in the order of 0.6 are still obtained for several frequency combinations on average, with R^2 values exceeding 0.8 for several frequency combinations for the second season, and the 16.7–13.3 GHz combination for the fourth season. This provides an indication that the passive microwave derived $p_{ex,eff}$ is able to improve radar retrievals, at least when compared to applying a single value for the whole dataset (i.e., Configuration 1, Table 5). Applying a scaling factor to $p^{passive}_{ex,eff}$ (Configuration 4) further improves the error metrics against in situ SWE, in particular in terms of reduced bias and uRMSE. R^2 values for some frequency combinations also show improvement (e.g., combinations of 10.2 & 16.7 GHz and 13.3 & 16.7 GHz), while e.g., retrievals using 16.7 GHz and the channel difference of 16.7–10.2 GHz shows slight deterioration.

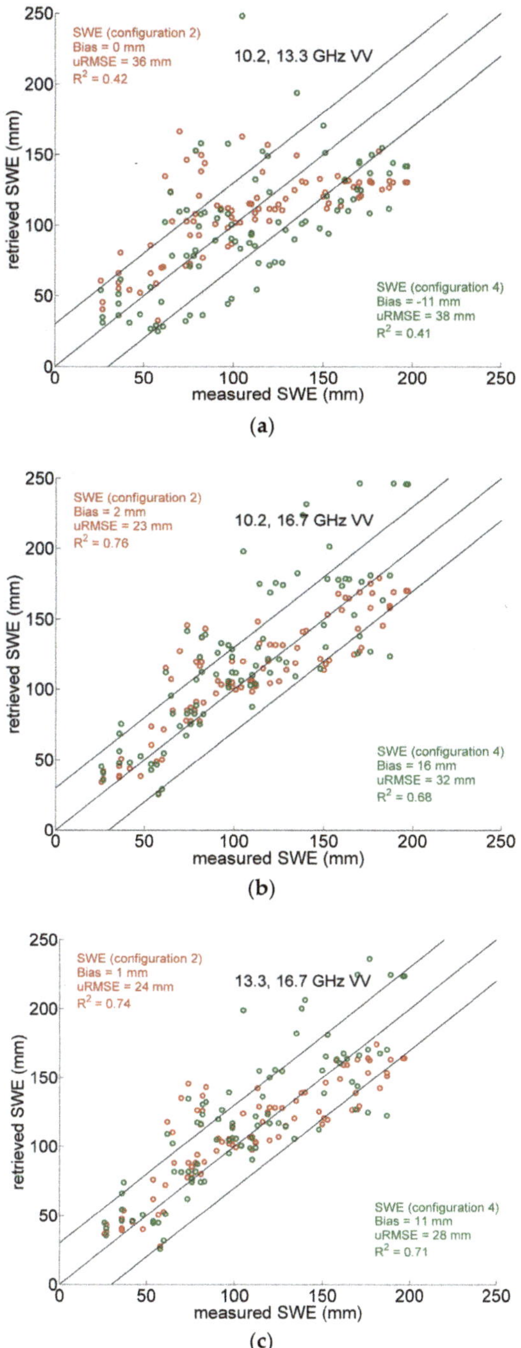

Figure 6. Same as Figure 5 but for channel combinations of 10.2 and 13.3 GHz (**a**); 10.2 and 16.7 GHz (**b**); 13.3 and 16.7 GHz (**c**).

Table 5. Bias, uRMSE and R^2 of retrieved SWE against in situ observations using retrieval Configurations 1 and 2. Error metrics shown for NoSREx individual seasons as well as for the whole dataset.

Configuration 1	NoSREx 1			NoSREx 2			NoSREx 3			NoSREx 4			All		
Frequency	Bias (mm)	uRMSE (mm)	R^2	Bias (mm)	uRMSE (mm)	R^2	Bias (mm)	uRMSE (mm)	R^2	Bias (mm)	uRMSE (mm)	R^2	Bias (mm)	uRMSE (mm)	R^2
10.2 GHz	68	63	0.01	−22	37	0.13	−26	42	0.40	−12	49	0.02	18	68	0.05
13.3 GHz	40	35	0.44	14	14	0.89	−72	32	0.19	14	48	0.03	16	48	0.18
16.7 GHz	53	32	0.70	78	45	0.91	−80	19	0.86	−31	32	0.59	27	64	0.19
10.2 & 13.3 GHz	43	37	0.38	9	17	0.86	−67	33	0.24	10	48	0.03	15	48	0.16
10.2 & 16.7 GHz	51	31	0.69	71	38	0.91	−78	19	0.86	−31	32	0.57	25	61	0.19
13.3 & 16.7 GHz	47	29	0.67	60	29	0.92	−78	20	0.84	−24	34	0.48	22	55	0.20
13.3–10.2 GHz	31	29	0.61	41	15	0.92	−92	29	0.00	36	51	0.02	21	51	0.16
16.7–10.2 GHz	56	40	0.74	72	44	0.70	−87	16	0.87	−34	29	0.70	24	67	0.19
16.7–13.3 GHz	53	42	0.36	93	64	0.71	−85	13	0.81	−59	27	0.74	18	80	0.02

Configuration 2	NoSREx 1			NoSREx 2			NoSREx 3			NoSREx 4			All		
Frequency	Bias (mm)	uRMSE (mm)	R^2	Bias (mm)	uRMSE (mm)	R^2	Bias (mm)	uRMSE (mm)	R^2	Bias (mm)	uRMSE (mm)	R^2	Bias (mm)	uRMSE (mm)	R^2
10.2 GHz	−6	51	0.01	6	39	0.13	−12	44	0.40	0	50	0.02	−2	48	0.09
13.3 GHz	2	34	0.44	3	17	0.88	−11	37	0.19	0	47	0.03	0	35	0.45
16.7 GHz	3	25	0.70	7	13	0.91	−3	11	0.86	−2	29	0.59	2	23	0.77
10.2 & 13.3 GHz	0	36	0.38	3	18	0.86	−11	38	0.24	0	47	0.03	0	36	0.42
10.2 & 16.7 GHz	3	26	0.68	6	12	0.91	−4	11	0.85	−2	30	0.57	2	23	0.76
13.3 & 16.7 GHz	2	26	0.67	5	11	0.92	−6	14	0.84	−2	33	0.48	1	24	0.74
13.3–10.2 GHz	9	29	0.61	4	12	0.92	−11	33	0.00	1	48	0.02	4	32	0.54
16.7–10.2 GHz	8	25	0.74	10	16	0.91	1	13	0.87	−2	25	0.70	6	22	0.78
16.7–13.3 GHz	16	38	0.73	17	28	0.87	13	32	0.80	2	31	0.70	13	34	0.75

Table 6. Same as Table 5 but for Configurations 3 and 4.

Configuration 3	NoSREx 1			NoSREx 2			NoSREx 3			NoSREx 4			All		
Frequency	Bias (mm)	uRMSE (mm)	R^2	Bias (mm)	uRMSE (mm)	R^2	Bias (mm)	uRMSE (mm)	R^2	Bias (mm)	uRMSE (mm)	R^2	Bias (mm)	uRMSE (mm)	R^2
10.2 GHz	41	43	0.29	−34	40	0.07	111	38	0.34	14	46	0.12	21	59	0.29
13.3 GHz	41	26	0.71	6	24	0.65	61	49	0.24	86	57	0.15	43	46	0.49
16.7 GHz	81	41	0.57	59	28	0.81	108	21	0.61	59	35	0.62	73	38	0.69
10.2 & 13.3 GHz	41	27	0.67	0	25	0.61	70	48	0.27	75	54	0.16	40	46	0.48
10.2 & 16.7 GHz	79	40	0.56	68	38	0.84	109	20	0.58	57	35	0.61	74	39	0.68
13.3 & 16.7 GHz	70	36	0.58	58	35	0.81	104	19	0.56	65	39	0.52	69	37	0.70
13.3–10.2 GHz	41	29	0.78	36	33	0.75	32	56	0.26	112	65	0.05	51	51	0.44
16.7–10.2 GHz	82	43	0.34	91	45	0.82	110	30	0.56	67	36	0.63	85	43	0.61
16.7–13.3 GHz	113	52	0.53	87	36	0.67	100	42	0.01	35	33	0.87	84	55	0.43

Configuration 4	NoSREx 1			NoSREx 2			NoSREx 3			NoSREx 4			All		
Frequency	Bias (mm)	uRMSE (mm)	R^2	Bias (mm)	uRMSE (mm)	R^2	Bias (mm)	uRMSE (mm)	R^2	Bias (mm)	uRMSE (mm)	R^2	Bias (mm)	uRMSE (mm)	R^2
10.2 GHz	−12	38	0.31	−51	38	0.08	29	43	0.38	−32	42	0.12	−23	45	0.28
13.3 GHz	−14	25	0.73	−25	25	0.62	1	57	0.43	12	45	0.16	−10	37	0.44
16.7 GHz	23	32	0.77	32	34	0.88	23	27	0.20	−8	28	0.62	19	34	0.67
10.2 & 13.3 GHz	−14	27	0.69	−28	26	0.58	8	61	0.45	5	44	0.16	−11	38	0.41
10.2 & 16.7 GHz	21	31	0.77	26	28	0.88	24	28	0.13	−9	28	0.61	16	32	0.68
13.3 & 16.7 GHz	13	26	0.77	16	22	0.87	18	32	0.02	−4	31	0.52	11	28	0.71
13.3–10.2 GHz	−15	21	0.78	−7	19	0.76	−32	43	0.25	46	55	0.15	−2	42	0.41
16.7–10.2 GHz	19	27	0.64	47	41	0.84	16	19	0.60	−1	26	0.67	21	35	0.57
16.7–13.3 GHz	41	38	0.47	59	38	0.89	55	39	0.73	−25	16	0.88	31	47	0.45

Retrievals are generally overestimated during the early winter, in particular for the first and fourth seasons (see Figure 4). This is a direct result of increased backscattering observed in early winter, possibly associated with formation of basal crusts due to snow melt-refreeze events (Figure 1; see also [27,42]). Applying the channel difference mitigates this feature for some seasons and channel combinations; as an example, Figure 7 demonstrates the retrieval for the fourth season of NoSREx using the channel difference of 16.7–13.3 GHz. The overestimation in early winter retrievals for Configuration 2 (red line) is reduced compared to applying the channels individually in the cost function (compare to Figure 4h). While helpful in this case, applying the channel difference in place of individual channels in the retrieval was not found to universally improve retrieval error statistics (see Table 5).

SWE retrievals were repeated using available measured in situ data (T_{air}, T_{gnd}, ε_{gnd}, and ρ_{snow}). Values of $p_{ex,eff}$ applied in the various SWE retrieval configurations were also derived from retrievals involving measured in situ data. For Configurations 1, 2 and 4 retrievals involving the 16.7 GHz channel (including channel combinations) yielded on average a 27 to 75% reduction in bias errors, a reduction in uRMSE from 5 to 14%, and a 7 to 15% improvement in the coefficient of determination. For Configuration 3, however, the use of in situ data resulted in an increase of retrieval bias and uRMSE by 33 and 16%, respectively, while the coefficient of determination was marginally reduced, compared to retrievals with static ancillary parameters.

Retrievals of SWE using the SodRad radiometer observations generally showed less skill compared to radar retrievals. Figure 8 shows retrieved against measured SWE from the 10.65–36.5 GHz channel difference, which yielded the best overall error metrics. Results from Configurations 1 and 2, adapted for radiometer observations, are shown. Use of the seasonal average of $p^{passive}_{ex}$ (Configuration 2) can be seen to reduce both the bias and uRMSE, and increase the coefficient of determination compared to Configuration 1, where the overall average of $p^{passive}_{ex}$ was used. However, the error metrics are inferior compared to any radar channel combination involving the 16.7 GHz channel and using the seasonal average of p^{active}_{ex} (Configuration 2 in Table 5), with the exception of the larger bias error of the 13.3–16.7 GHz channel difference retrievals.

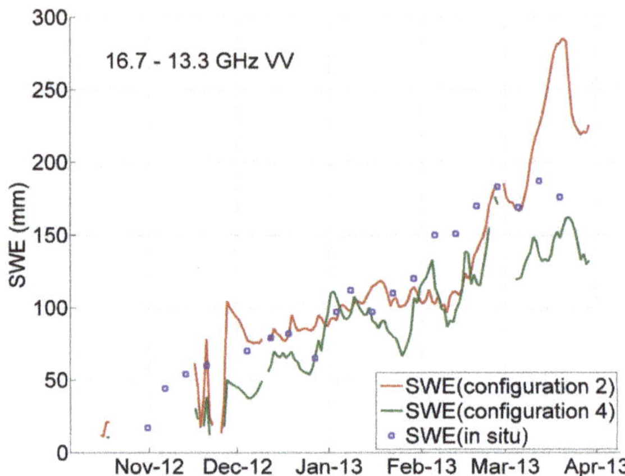

Figure 7. SWE retrieved with channel difference of 16.7–13.3 GHz for NoSREx season 4.

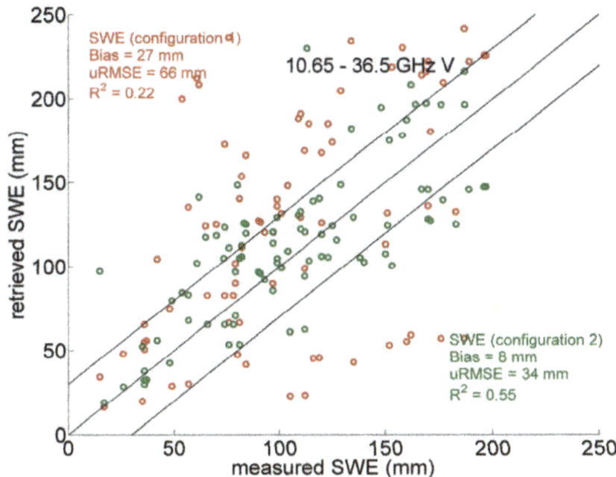

Figure 8. SWE retrieved with channel difference of 10.65–37 GHz using Configurations 1 and 2 adapted for SodRad passive microwave observations from all four NoSREx seasons.

5. Discussion

Results in Section 4.1 demonstrated that at certain frequencies and combinations of frequencies, active and passive -derived $p_{ex,eff}$ were shown to be temporally correlated (Table 2). This can be expected, as similar physics drive the backscattering and emission signal behavior, linked to varying scattering and absorption properties induced by precipitation events, and snowpack metamorphism over time. On average, obtained correlations were higher between higher frequencies which are generally more sensitive to the volume properties of snow cover, while lower X-band active and passive frequencies were found to be mostly uncorrelated, affirming that the underlying soil and not dry snow-scattering properties govern the signal at these frequencies and snow depths. However, even at higher frequencies a seasonally consistent bias between active and passive -derived $p_{ex,eff}$ was found (Figure 3), indicating deficiencies in either the model physics or the applied ancillary data. Possible reasons for the bias include the omission of snow layering in the applied one-layer model configuration which affects in particular the highest 37 GHz passive microwave channel simulations, as well as deficiencies in the applied ground reflectivity model [31], which is not verified for active microwave observations beyond this paper and tests conducted by Proksch et al. [30].

Considering practical satellite scale retrievals, obtaining seasonal characteristics for snow microstructure and thus scattering properties could be realized by assigning e.g., seasonally or regionally dynamic values for correlation length, derived from snow climatology or physical snow models. However, as demonstrated in this study, passive microwave observations may also be of use in providing a priori information on snow microstructure for radar retrievals. This implies a method analogous to the one introduced in [15], characterizing snow-scattering efficiency over regions where in situ data is available and extending this over larger areas by means of spatio-temporal interpolation. In such a scheme, an effective correlation length derived at large scales with daily coverage from passive microwave observations could initialize active microwave retrievals of SWE from backscattering observations at suitable frequencies. This naturally implies assuming that differences in the scale and geometry of observations do not result in detrimental differences affecting the interchangeability of $p^{passive}_{ex,eff}$ and $p^{active}_{ex,eff}$; this matter was not accounted for in this study and should be investigated before the method proposed here could be applied at the satellite scale. Forest cover in particular poses a challenge in relating coarse scale passive microwave observations to high resolution SAR. However, new methods for mitigating forest canopy effects in snow parameter retrieval from both radar and passive microwave measurements have been developed using recent experimental data [43].

Taiga snow typically has a poorly defined internal layer structure compared to other classes of seasonal snow: snow grains near the base are larger than snow grains near the surface (with depth hoar by the end of the season) but the internal boundaries between layers are weak. This means a single snow layer was sufficient for the MEMLS3&a simulations in this study, in particular for late winter. This may not be the case for strongly layered snowpacks that are found in open tundra and prairie environments where wind processes strongly influence snow distribution. Tundra snow is typically composed of fine-grained wind slab layers overlying depth hoar composed of large faceted grains. Under these conditions, it is likely that two layer snow simulations would be necessary [44]. However, also in the case of NoSREx data, early season radar observations, and thus retrievals of both $p^{active}_{ex,eff}$ and SWE, were assumed to be influenced by melt-refreeze crusts in the base of the snowpack; causing the high backscattering values seen in the early season (Figure 1). Accounting for these would require use of a multi-layer model to properly represent the snow-scattering properties.

Reasonable estimates of SWE (Section 4.2) were achieved by inverting active microwave backscattering observations, provided that the mean level of snowpack scattering efficiency for each season was known (i.e., retrieval Configuration 2, see Table 5). This is partly incidental, as increases in snow mass and evolution of snow microstructure both occur over the season, while the direct impact of newly precipitated snow on the microwave signature was in many cases very small (see Figure 1); the new snow may also dampen the backscattering signature from lower snow layers thus compensating for any potential increase in total volume scattering from the new snow layer [40]. Snow regimes with more short-term temporal changes in microstructure may necessitate temporally dynamic estimates of the microstructure to achieve the same retrieval skill. In an operational context, assigning seasonally constant estimates of scattering efficiency is unrealistic if short latency is required (a seasonal average can be assigned only after the snow season is over). Tests with Configuration 2 nevertheless demonstrate that temporal dynamics may not in all cases be critical for SWE retrieval performance, if the average magnitude of scattering efficiency is correctly estimated. Retrievals of SWE using passive microwave observations produced generally inferior results (Figure 8), indicating that radar retrievals may have the potential to improve the absolute accuracy of current SWE products from space.

The use of temporally variant, measured ancillary information for snow and soil properties was shown to generally improve the coefficient of determination between correlation length derived active and passive measurements (Table 2), as well as error metrics of SWE retrievals, compared to retrievals using constant values. The exception was retrieval Configuration 3, where $p^{passive}_{ex}$ was used without scaling to initialize SWE retrievals from radar. However, the average improvement in e.g., the uRMSE were less than 14% in any configuration when applying the 16.7 GHz channel. The results show that while the use of accurate ancillary information is preferable, retrieval skill may still be acceptable when using constant values derived from e.g., climatology, when measured or modeled data is unavailable.

Similar retrieval results for SWE were obtained by using an absorption-loss-based methodology using the same NoSREx dataset [45]; that study also showed that a priori definition of snow-scattering characteristics (in the form of single scattering albedo) was essential to achieve reasonable retrievals. Nevertheless, further improvement may be achieved by applying more sophisticated retrieval schemes, including balancing or constraining the parameterization of the retrieval (following e.g., [45]), as well as by estimating the microstructural evolution of snow based on a physical model. Direct coupling with physical snow models has been previously demonstrated using the same dataset [46], including retrieval of SWE using a coupled model. A demonstration of different microstructural parameterizations and their use in emission models using a suite of different emission and physical models is given in [47]. Similar schemes have been demonstrated previously for SWE estimation [48] and for forward model simulations [41,49] using other datasets.

Information on SWE can be coupled with land surface models by assimilation of snow mass estimates [50] or by direct radiance assimilation [51–53]. With respect to direct assimilation of radiances, this study indicates that the MEMLS3&a model may be applied to simulate both active and passive microwave data streams, provided a suitable proxy of snow microstructure (i.e., correlation length) is produced by the land surface model (e.g., [54]).

6. Conclusions

This study investigated the seasonal behavior of snow bulk scattering properties by means of retrieving a proxy variable, the effective snow correlation length, from active and passive microwave observations. Tower-based measurements from the NoSREx campaign, representing four years of differing snow conditions in a taiga environment, were used. A numerical inversion of the MEMLS3&a backscattering and emission model was used to retrieve an effective correlation length matching model estimates to observations. Snow depth was used as the only temporally variable ancillary input, mimicking a similar scheme applied for satellite passive microwave SWE retrievals, in which conventional snow grain size is used as a proxy [4]. The effective snow-scattering efficiency derived in this fashion accounts also for deficiencies in e.g., ancillary data supplied to the forward model used in the inversion, as well as physics of the forward model itself. If these deficiencies are severe, the retrieved effective snow parameters (either effective grain size or correlation length) may lose any physical relation to actual snow microstructural conditions. Nevertheless, for the last two seasons when weekly observations of SSA profiles in snow were available, the retrieved values showed similar temporal trends with MEMLS3&a retrievals of $p_{ex,eff}$, (Figure 2c,d). A bias was apparent for the fourth NoSREx season (Figure 2d), but these discrepancies can be due to uncertainties in the SSA measurement, the empirical relation applied to convert SSA to p_{ex}, and the use of a one-layer model for the snowpack to retrieve $p_{ex,eff}$, which complicates the relation to physical snow microstructure in the presence of a vertically inhomogeneous snowpack. Similar results for the second season of NoSREx were obtained using the Helsinki University of Technology (HUT) snow emission model [14] and conventional snow grain size measured in situ (see Figure 9 in [16]). This indicates that effective snow microstructural parameters may under certain conditions bear relation to measured physical snow characteristics.

This study also demonstrated SWE retrievals from radar observations, parameterizing the retrieval by a temporally static effective value for snow-scattering efficiency (Configurations 1 and 2), and by a temporally dynamic estimate, derived from passive microwave observations (Configurations 3 and 4). Applying seasonal scaling to the average scattering efficiency (Configuration 2), retrievals yielded on average highly correlated estimates against in situ observations, when higher Ku band radar (16.7 GHz) observations were applied either in a single frequency retrieval (R^2 values between 0.59 and 0.91 for different seasons), or in combination with lower frequencies (R^2 values between 0.48 and 0.92). It is notable that high correlations were achieved despite the value of snow correlation length being static for each season. However, when attempting to apply a single value for the whole dataset (Configuration 1), error metrics for the whole dataset deteriorated with the coefficient of determination falling below 0.2 regardless of the frequencies applied in the retrieval, highlighting the need for at least seasonal characterization of the scattering efficiency. For individual seasons, R^2 values of up to 0.92 were still found, but retrieval skill was hampered by large biases (Table 5). This is consistent with different bulk and microstructural properties of the snowpack influencing the radar backscatter response each season. Applying an effective correlation length derived from passive microwave measurements to initialize the active microwave retrievals (Configurations 3 and 4) was shown to provide improvement compared to using a single constant value (i.e., Configuration 1). This indicates that passive microwave observations may, in some cases, serve to initialize retrievals using radar backscattering, in particular at Ku band frequencies.

In summary, the presented results suggest that synergistic active-passive microwave observations, such as those provided by a new radar mission concept [26], may enable the retrieval of terrestrial SWE via the method of assigning an effective parameter for snow-scattering efficiency based on passive microwave retrievals. Such retrieval products are available already from current passive microwave satellite estimates [4]. Further work includes demonstrating the synergistic retrieval method beyond plot scale observations. Being sensitive to snow volumetric properties, synergistic passive microwave and Ku-band radar measurements also support land surface modeling within operational forecast systems through radiance-based assimilation of backscatter.

Acknowledgments: The staff at FMI-ARC are acknowledged for collection of in situ data and operation of microwave instruments. Staff from WSL Institute for Snow and Avalanche Research SLF participated in field measurements and performed analysis of SMP measurements. The work was supported by the European Space Agency (ESA ESTEC contracts 22671/09/NL/JA/ef and 22829/09/NL/JC). We thank the three anonymous reviewers for providing constructive comments which helped to improve the manuscript.

Author Contributions: Juha Lemmetyinen, Helmut Rott, Andreas Wiesmann, Martin Schneebeli, Leena Leppänen, Anna Kontu and Jouni Pulliainen designed and performed the experiments; Chris Derksen, Giovanni Macelloni and Josh King contributed to data analysis; Juha Lemmetyinen wrote the paper with contributions from all authors.

Conflicts of Interest: The authors declare no conflicts of interest.

References

1. Mudryk, L.R.; Derksen, C.; Kushner, P.J.; Brown, R. Characterization of Northern Hemisphere snow water equivalent datasets, 1981–2010. *J. Clim.* **2015**, *28*, 8037–8051. [CrossRef]

2. Larue, F.; Royer, A.; De Sève, D.; Langlois, A.; Roy, A.; Brucker, L. Validation analysis of the GlobSnow-2 database over an eco-climatic latitudinal gradient in Eastern Canada. *Remote Sens. Environ.* **2016**, *194*, 264–277. [CrossRef]

3. Kelly, R.E.J. The AMSR-E Snow depth algorithm: Description and initial results. *J. Remote Sens. Soc. Jpn.* **2009**, *29*, 307–317.

4. Takala, M.; Luojus, K.; Pulliainen, J.; Derksen, C.; Lemmetyinen, J.; Kärnä, J.-P.; Koskinen, J.; Bojkov, B. Estimating northern hemisphere snow water equivalent for climate research through assimilation of space-borne radiometer data and ground-based measurements. *Remote Sens. Environ.* **2011**, *115*, 3517–3529. [CrossRef]

5. Painter, T.H.; Berisford, D.F.; Boardman, J.W.; Bormann, K.J.; Deems, J.S.; Gehrke, F.; Hedrick, A.; Joyce, M.; Laidlaw, R.; Marks, D.; et al. The Airborne Snow Observatory: Fusion of scanning LiDAR, imaging spectrometer, and physically-based modeling for mapping snow water equivalent and snow albedo. *Remote Sens. Environ.* **2016**, *184*, 139–152. [CrossRef]

6. Foster, J.; Sun, C.; Walker, J.; Kelly, R.; Chang, A.; Dong, J.; Powell, H. Quantifying uncertainty in passive microwave snow water equivalent observations. *Remote Sens. Environ.* **2005**, *94*, 187–203. [CrossRef]

7. Clifford, D. Global estimates of snow water equivalent from passive microwave instruments: History, challenges and future developments. *Int. J. Remote Sens.* **2010**, *31*, 3707–3726. [CrossRef]

8. Rott, H.; Yueh, S.H.; Cline, D.W.; Duguay, C.; Essery, R.; Haas, C.; Hélière, F.; Kern, M.; Macelloni, G.; Malnes, E.; et al. Cold regions hydrology high-resolution observatory for snow and cold land processes. *Proc. IEEE* **2010**, *98*, 752–765. [CrossRef]

9. Dai, L.Y.; Che, T.; Wang, J.; Zhang, P. Snow depth and snow water equivalent estimation from AMSR-E data based on a priori snow characteristics in Xinjiang, China. *Remote Sens. Environ.* **2012**, *127*, 14–29. [CrossRef]

10. Davenport, I.J.; Sandells, M.J.; Gurney, R.J. The effects of variation in snow properties on passive microwave snow mass estimation. *Remote Sens. Environ.* **2012**, *118*, 168–175. [CrossRef]

11. Durand, M.; Liu, D. The need for prior information in characterizing snow water equivalent from microwave brightness temperatures. *Remote Sens. Environ.* **2012**, *126*, 248–257. [CrossRef]

12. Fierz, C.; Armstrong, R.L.; Durand, Y.; Etchevers, P.; Greene, E.; McClung, D.M.; Nishimura, K.; Satyawali, P.K.; Sokratov, S.A. *The International Classification for Seasonal Snow on the Ground*; IACS Contribution no. 1; UNESCOIHP, I HP-VII Technical Documents in Hydrology 83; UNESCO: Paris, France, 2009.

13. Hallikainen, M.T.; Ulaby, F.; Deventer, T. Extinction behavior of dry snow in the 18- to 90-GHz range. *IEEE Trans. Geosci. Remote Sens.* **1987**, *GE-25*, 737–745. [CrossRef]

14. Pulliainen, J.T.; Grandell, J.; Hallikainen, M.T. HUT snow emission model and its applicability to snow water equivalent retrieval. *IEEE Trans. Geosci. Remote Sens.* **1999**, *37*, 1378–1390. [CrossRef]

15. Pulliainen, J. Mapping of snow water equivalent and snow depth in boreal and sub-arctic zones by assimilating space-borne microwave radiometer data and ground-based observations. *Remote Sens. Environ.* **2006**, *101*, 257–269. [CrossRef]

16. Lemmetyinen, J.; Derksen, C.; Toose, P.; Proksch, M.; Pulliainen, J.; Kontu, A.; Rautiainen, K.; Seppänen, J.; Hallikainen, M. Simulating seasonally and spatially varying snow cover brightness temperature using HUT snow emission model and retrieval of a microwave effective grain size. *Remote Sens. Environ.* **2015**, *156*, 71–95. [CrossRef]

17. Kelly, R.; Chang, A.; Tsang, L.; Foster, J. A prototype AMSR-E global snow area and snow depth algorithm. *IEEE Trans. Geosci. Remote Sens.* **2003**, *41*, 230–242. [CrossRef]

18. Pinzer, B.; Schneebeli, M. Snow metamorphism under alternating temperature gradients: Morphology and recrystallization in surface snow. *Geophys. Res. Lett.* **2009**, *36*, L23503. [CrossRef]

19. Pinzer, B.; Schneebeli, M.; Kaempfer, T. Vapor flux and recrystallization during dry snow metamorphism under a steady temperature gradient as observed by time-lapse micro-tomography. *Cryosphere* **2012**, *6*, 1141–1155. [CrossRef]

20. Tsang, L.; Chen, C.-T.; Chang, A.T.C.; Guo, J.; Ding, K.-H. Dense media radiative transfer theory based on quasi-crystalline approximation with applications to microwave remote sensing of snow. *Radio Sci.* **2000**, *35*, 731–749. [CrossRef]

21. Tsang, L.; Pan, J.; Liand, D.; Li, Z.; Cline, D.W.; Tan, Y. Modeling active microwave remote sensing of snow using dense medium radiative transfer (DMRT) theory with multiple-scattering effects. *IEEE Trans. Geosci. Remote Sens.* **2007**, *45*, 990–1004. [CrossRef]

22. Picard, G.; Brucker, L.; Roy, A.; Dupont, F.; Fily, M.; Royer, A. Simulation of the microwave emission of multi-layered snowpacks using the dense media radiative transfer theory: The DMRT-ML model. *Geosci. Model Devel.* **2013**, *6*, 1061–1078. [CrossRef]

23. Löwe, H.; Picard, G. Microwave scattering coefficient of snow in MEMLS and DMRT-ML revisited: The relevance of sticky hard spheres and tomography-based estimates of stickiness. *Cryosphere* **2015**, *9*, 2101–2117. [CrossRef]

24. Tan, S.; Chang, W.; Tsang, L.; Lemmetyinen, J.; Proksch, M. Modeling both active and passive microwave remote sensing of snow using dense media radiative transfer (DMRT) theory with multiple scattering and backscattering enhancement. *IEEE J. Sel. Top. Appl. Earth Obs. Remote Sens.* **2015**, *8*, 4418–4430. [CrossRef]

25. Mätzler, C. Relation between grain-size and correlation length of snow. *J. Glaciol.* **2002**, *162*, 461–466. [CrossRef]

26. Derksen, C.; Lemmetyinen, J.; King, J.; Garnaud, C.; Belair, S.; Lapointe, M.; Crevier, Y.; Girard, R.; Burbidge, G.; Marquez-Martinez, J.; et al. A new dual-frequency Ku-band radar mission concept for cryosphere applications. *Proc. EUSAR* **2018**. submitted.

27. Lemmetyinen, J.; Kontu, A.; Pulliainen, J.; Vehviläinen, J.; Rautiainen, K.; Wiesmann, A.; Mätzler, C.; Werner, C.; Rott, H.; Nagler, T.; et al. Nordic snow radar experiment. *Geosci. Instrum. Methods Data Syst.* **2016**, *5*, 403–415. [CrossRef]

28. Wiesmann, A.; Mätzler, C. Microwave emission model of layered snowpacks. *Remote Sens. Environ.* **1999**, *70*, 307–316. [CrossRef]

29. Mätzler, C.; Wiesmann, A. Extension of the microwave emission model of layered snowpacks to coarse grained snow. *Remote Sens. Environ.* **1999**, *70*, 318–326. [CrossRef]

30. Proksch, M.; Mätzler, C.; Wiesmann, A.; Lemmetyinen, J.; Schwank, M.; Löwe, H.; Schneebeli, M. MEMLS3&a: Microwave emission model of layered snowpacks adapted to include backscattering. *Geosci. Model Dev.* **2015**, *8*, 2611–2626. [CrossRef]

31. Wegmüller, U.; Mätzler, C. Rough bare soil reflectivity model. *IEEE Trans. Geosci. Remote Sens.* **1999**, *37*, 1391–1395. [CrossRef]

32. Pulliainen, J.; Kärnä, J.-P.; Hallikainen, M.T. Development of geophysical retrieval algorithms for the MIMR. *IEEE Trans. Geosci. Remote Sens.* **1993**, *31*, 268–277. [CrossRef]

33. Sturm, M.; Taras, B.; Liston, G.E.; Derksen, C.; Jonas, T.; Lea, J. Estimating snow water equivalent using snow depth data and climate classes. *J. Hydrometeorol.* **2010**, *11*, 1380–1394. [CrossRef]

34. Hallikainen, M.T.; Ulaby, F.T.; Dobson, M.C.; El-Rayes, M.A.; Wu, L.-K. Microwave dielectric behavior of wet Soil—Part I: Empirical models and experimental observations. *IEEE Trans. Geosci. Remote Sens.* **1985**, *23*, 25–33. [CrossRef]

35. Gallet, J.-C.; Domine, F.; Zender, C.; Picard, G. Measurement of the specific surface area of snow using infrared reflectance in an integrating sphere at 1310 and 1550 nm. *Cryosphere* **2009**, *3*, 167–182. [CrossRef]

36. Leppänen, L.; Kontu, A.; Vehviläinen, J.; Lemmetyinen, J.; Pulliainen, J. Comparison of traditional and optical grain size field measurements with SNOWPACK simulations in a taiga environment. *J. Glaciol.* **2015**, *61*, 151–162. [CrossRef]

37. Schneebeli, M.; Pielmeier, C.; Johnson, J.B. Measuring snow microstructure and hardness using a high resolution penetrometer. *Cold Reg. Sci. Technol.* **1999**, *30*, 101–114. [CrossRef]

38. Proksch, M.; Löwe, H.; Schneebeli, M. Density, specific surface area, and correlation length of snow measured by high-resolution penetrometry. *J. Geophys. Res. Earth Surf.* **2015**, *120*, 346–362. [CrossRef]

39. Leppänen, L.; Kontu, A.; Hannula, H.-R.; Sjöblom, H.; Pulliainen, J. Sodankylä manual snow survey program. *Geosci. Inst. Methods Data Syst. Discuss.* **2016**, *5*, 163–179. [CrossRef]

40. Paloscia, S.; Pettinato, S.; Santi, E.; Valt, M. COSMO-SkyMed image investigation of snow features in alpine environment. *Sensors* **2017**, *17*, 84. [CrossRef] [PubMed]

41. Brucker, L.; Royer, A.; Picard, G.; Langlois, A.; Fily, M. Hourly simulations of the microwave brightness temperature of seasonal snow in Quebec, Canada, using a coupled snow evolution-emission model. *Remote Sens. Environ.* **2011**, *115*, 1966–1977. [CrossRef]

42. Lin, C.; Rommen, B.; Floury, N.; Schüttemeyer, D.; Davidson, M.W.; Kern, M.; Kontu, A.; Lemmetyinen, J.; Pulliainen, J.; Wiesmann, A.; et al. Active Microwave Scattering Signature of Snowpack-Continuous Multiyear SnowScat Observation Experiments. *IEEE J. Sel. Top. Appl. Earth Obs. Remote Sens.* **2016**, *9*, 3849–3869. [CrossRef]

43. Cohen, J.; Lemmetyinen, J.; Pulliainen, J.; Heinilä, K.; Montomoli, F.; Seppänen, J.; Hallikainen, M.T. The effect of boreal forest canopy in satellite snow mapping—A multisensor analysis. *IEEE Trans. Geosci. Remote Sens.* **2015**, *52*, 3275–3288. [CrossRef]

44. King, J.; Derksen, C.; Toose, P.; Langlois, A.; Larsen, C.; Lemmetyinen, J.; Marsh, J.; Montpetit, B.; Roy, A.; Rutter, N.; et al. The influence of snow microstructure on dual-frequency radar measurements in a tundra environment. *Remote Sens. Environ.* **2018**. submitted.

45. Cui, Y.; Xiong, C.; Lemmetyinen, J.; Shi, J.; Jiang, L.; Peng, B.; Li, H.; Zhao, T.; Ji, D.; Hu, T. Estimating Snow Water Equivalent with backscattering at X and Ku band based on absorption loss. *Remote Sens.* **2016**, *8*, 505. [CrossRef]

46. Kontu, A.; Lemmetyinen, J.; Vehviläinen, J.; Leppänen, L.; Pulliainen, J. Coupling SNOWPACK-modeled grain size parameters with the HUT snow emission model. *Remote Sens. Environ.* **2017**, *194*, 33–47. [CrossRef]

47. Sandells, M.; Essery, R.; Rutter, N.; Wake, L.; Leppänen, L.; Lemmetyinen, J. Microstructure representation of snow in coupled snowpack and microwave emission models. *Cryosphere* **2017**, *11*, 229–246. [CrossRef]

48. Langlois, A.; Royer, A.; Derksen, C.; Montpetit, B.; Dupont, F.; Goïta, K. Coupling the snow thermodynamic model SNOWPACK with the microwave emission model of layered snowpacks for subarctic and arctic snow water equivalent retrievals. *Water Resour. Res.* **2012**, *48*, 1–14. [CrossRef]

49. Picard, G.; Brucker, L.; Fily, M.; Gallée, H.; Krinner, G. Modelling time series of microwave brightness temperature in Antarctica. *J. Glaciol.* **2009**, *55*, 537–551. [CrossRef]

50. De Lannoy, G.J.M.; Reichle, R.H.; Houser, P.R.; Arsenault, K.R.; Verhoest, N.E.C.; Pauwels, V.N.R. Satellite-scale snow water equivalent assimilation into a high-resolution land surface model. *J. Hydrometeorol.* **2010**, *11*, 352–369. [CrossRef]

51. Durand, M.; Kim, E.J.; Marguilis, S.A. Radiance assimilation shows promise for snowpack characterization. *Geophys. Res. Lett.* **2009**, *36*, L02503. [CrossRef]

52. Kwon, Y.; Yang, Z.-L.; Hoar, T.J.; Toure, A.M. Improving the radiance assimilation performance in estimating snow water storage across snow and land-cover types in North America. *J. Hydrometeorol.* **2017**, *18*, 651–668. [CrossRef]

53. Andreadis, K.M.; Lettenmaier, D.P. Implications of representing snowpack stratigraphy for the assimilation of passive microwave satellite observations. *J. Hydrometeorol.* **2012**, *13*, 1493–1506. [CrossRef]

54. Roy, A.; Royer, A.; Montpetit, B.; Bartlett, P.A.; Langlois, A. Snow specific surface area simulation using the one-layer snow model in the Canadian LAnd Surface Scheme (CLASS). *Cryosphere* **2013**, *7*, 961–975. [CrossRef]

 *remote sensing*

Article

Soil Moisture from Fusion of Scatterometer and SAR: Closing the Scale Gap with Temporal Filtering

Bernhard Bauer-Marschallinger [1,*], Christoph Paulik [1], Simon Hochstöger [1], Thomas Mistelbauer [2], Sara Modanesi [3], Luca Ciabatta [3], Christian Massari [3], Luca Brocca [3] and Wolfgang Wagner [1]

[1] Remote Sensing Research Group, Department of Geodesy and Geoinformation, TU Wien, 1040 Vienna, Austria; cpaulik@vandersat.com (C.P.); simon.hochstoeger@siemens.com (S.H.); wolfgang.wagner@geo.tuwien.ac.at (W.W.)

[2] Earth Observation Data Centre for Water Resources Monitoring (EODC), 1030 Vienna, Austria; thomas.mistelbauer@eodc.eu

[3] Research Institute for Geo-Hydrological Protection (IRPI), National Research Council (NRC), 06128 Perugia, Italy; sara.modanesi@irpi.cnr.it (S.M.); luca.ciabatta@irpi.cnr.it (L.C.); christian.massari@irpi.cnr.it (C.M.); luca.brocca@irpi.cnr.it (L.B.)

* Correspondence: bbm@geo.tuwien.ac.at

Received: 17 May 2018; Accepted: 25 June 2018; Published: 29 June 2018

Abstract: Soil moisture is a key environmental variable, important to e.g., farmers, meteorologists, and disaster management units. We fuse surface soil moisture (SSM) estimates from spatio-temporally complementary radar sensors through temporal filtering of their joint signal and obtain a kilometre-scale, daily soil water content product named SCATSAR-SWI. With 25 km Metop ASCAT SSM and 1 km Sentinel-1 SSM serving as input, the SCATSAR-SWI is globally applicable and achieves daily full coverage over operated areas. We employ a near-real-time-capable SCATSAR-SWI algorithm on a fused 3 year ASCAT-Sentinel-1-SSM data cube over Italy, obtaining a consistent set of model parameters, unperturbed by coverage discontinuities. An evaluation of a therefrom generated SCATSAR-SWI dataset, involving a 1 km Soil Water Balance Model (SWBM) over Umbria, yields comprehensively high agreement with the reference data (median R = 0.61 vs. in situ; 0.71 vs. model; 0.83 vs. ASCAT SSM). While the Sentinel-1 signal is attenuated to some extent, the ASCAT's signal dynamics are fully transferred to the SCATSAR-SWI and benefit from the Sentinel-1 parametrisation. Using the SM2RAIN approach, the SCATSAR-SWI shows excellent capability to reproduce 5 day-accumulated rainfall over Italy, with R = 0.89 against observed rainfall. The SCATSAR-SWI is currently in preparation towards operational product dissemination in the Copernicus Global Land Service (CGLS).

Keywords: soil moisture; SAR; scatterometer; data fusion; scale gap

1. Introduction

Dynamics in soil moisture (SM) are important in the understanding of processes in many environmental and socio-economic fields, e.g., its impact on vegetation vitality, crop yield, droughts or exposure to flood threats. SM is a key driver of water and heat fluxes between the ground and the atmosphere, regulating air temperature and humidity [1]. Vice versa, Surface Soil Moisture (SSM), which is the water content in the soil's top centimetres, is very sensitive to external forcing in the form of precipitation, temperature, solar irradiation, humidity, and wind. SM is thus both an integrator of climatic conditions [2] and a driver of local weather and climate [3], and plays a major role in global water-, energy- and carbon- cycles [4].

Consequently, knowledge of the spatial and temporal variation of SM is crucial to users in meteorology, climatology, hydrology, and agronomy. More specifically, estimation of SM can be a critical skill in numerical weather prediction [5–7], precipitation estimation [8,9], flood risk modelling [10,11], runoff prediction [12], groundwater recharge modelling [13], and irrigation assessment [14]. Furthermore, estimation of SM is crucial for creating understanding of heatwaves [15], droughts [16], ocean–land feedback [17], and long-term trends in hydrology [18]. On these grounds, soil moisture is listed as essential climate variable (ECV) within the framework of the Global Climate Observing System [19].

With increasing expectations towards these applications, the demand for observational data with high spatio-temporal resolution, as well as complete coverage, is steadily growing. While in situ techniques [20] allow for accurate and temporally dense measurements at point scale, they lack spatial representativeness and require equipment, maintenance and ground access, leaving many parts of the world unobserved. For extensive or global undertakings, Earth observation, and more specifically, spaceborne microwave remote sensing has proven its capability to retrieve soil moisture comprehensively. Satellites carrying microwave sensors record surface properties, globally and independently from daylight and cloud cover, and SM products therefrom reached maturity in the last decade [21]. Coarse-scale SM products (12–50 km) from active (radars) or passive sensors (radiometers) like SM from Metop ASCAT [22], Windsat [23], SMOS [24], and SMAP [25] have been thoroughly evaluated and have found widespread use [26,27]. These products have a large swath/footprint and can well capture temporal SM dynamics with their daily or sub-daily revisit times. However, they lack spatial details. They do not support analysis of local hydrological patterns below the 10 km scale, such as effects from convectional rains or topography and thus do not meet the requirements of many users.

On the other hand, Synthetic Aperture Radar (SAR) remote sensing systems like the (elapsed) Envisat ASAR and the current Sentinel-1 (S-1) CSAR radar provide sub-antenna footprint resolution through advanced radar signal processing, involving range and Doppler discrimination [28]. With this imaging technique, the sensors deliver high-resolution radar imagery of the ground's geometric and dielectric properties, potentially resolving objects at the field level. Envisat ASAR was already successfully employed for SSM retrieval [29] during the period 2004–2012, showing capability to monitor soil moisture dynamics at the continental level and potential for model assimilation [30]. Recently, the approach was adapted to its successor SAR mission Sentinel-1 [31], aiming for global monitoring of SSM at the kilometre scale. However, SAR, and in general all high resolution systems, observe individual locations less frequently and thus often fail to capture short-term variations, like during rainfall events.

However, for a single remote sensing system, there always exists a trade-off between spatial and temporal resolution of the observations, leading to missed dynamics either in the spatial or temporal domain. We exemplify the scale gap in remote sensing with the illustration in Figure 1, sketching the characteristics of the SSM products from Metop ASCAT [22] and Sentinel-1 [31]. While the SAR product does not measure often enough to fully capture the SM variations induced by rainfall, the ASCAT product has a much coarser resolution.

With the aim to improve the spatial resolution of the short-repeating coarse-scale SM products, several approaches to estimate sub-pixel information can be found. These methods range from geostatistical analysis [32], downscaling algorithms applying subpixel-patterns [33–35], to multi-band products [36,37]. They have in common the need for apt auxiliary data, either temporally concurring observations from different bands, or a-priori local data as e.g., knowledge on the local land cover or soil properties, which are commonly neither available everywhere nor of consistent quality. A comprehensive survey on soil moisture downscaling methods is given, e.g., by [38].

In the present study, we suggest a data fusion approach in order to generate a high-resolution, high-frequency SSM product, taking up the concept of temporal stability of soil moisture [39]. The concept suggests that a local SM signal is highly correlated with the regional SM signal from its larger surrounding area. Hence, where this assumption holds true, a region's individual zones are subject to persistent regional SM patterns and thus show similar soil moisture dynamics. In this case, a fusion of datasets that are recorded at different scales but describe SM in the same way is valid. In that sense, we identified SSM from scatterometers (coarse-scale) and SAR (fine-scale) as a suitable set of datasets for fusion.

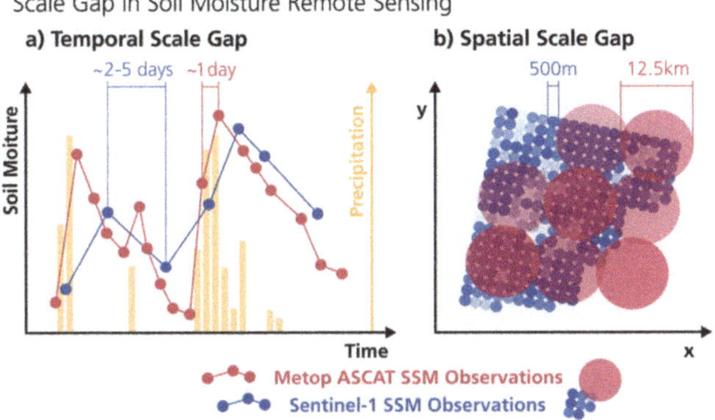

Figure 1. Schematic display of the scale gap by means of radar remote sensing for soil moisture retrieval, exemplifying that sensors with high temporal resolution have commonly low spatial resolution, and vice versa. (**a**) superior temporal resolution of scatterometer (SCAT) sensors; (**b**) superior spatial resolution of Synthetic Aperture Radar (SAR) sensors. Displayed measures (1 day vs. 2–5 days; 12.5 km vs. 500 m) relate to the input data (Metop ASCAT SSM and Sentinel-1 SSM) of the Scatterometer Synthetic Aperture Radar Soil Water Index (SCATSAR-SWI) product presented in this study.

In our method, the actual fusion of the SM signals is realised through temporal filtering at each gridded pixel, following the Soil Water Index (SWI) approach [40] that estimates the water content of soil profile layer $W_{PL}(t)$ from the local history of the surface layer's water content $W_{SL}(\tau)$:

$$W_{PL}(t) = \frac{1}{T} \int_{t}^{-\infty} W_{SL}(\tau) e^{\frac{-(t-\tau)}{T}} d\tau. \tag{1}$$

This approach represents a simple two-layer water balance model, consisting of a top layer (SL), which is accessible to C-band sensors, and the profile layer (PL), which extends downwards. The PL is assumed to serve as a reservoir, connected to the atmosphere only via SL. The SL is in direct contact with the atmosphere and its water content is thus temporally highly dynamic. For the underneath PL, the amount of stored water depends on the infiltration of water added to SL during precipitation events. Hence, the SM in PL can be solely explained by the past dynamics of SL, with more recent events having a stronger impact, which is accounted for by the exponential weighting function in Equation (1). Here, T is the so-called characteristic time length (the T-value) and acts as a local and soil-dependent measure for the infiltration time, hence increasing with depth of PL. Naturally, the temporal SM dynamics are decreasing with depth, as with increasing depth more antecedent SSM observations are effectively integrated. With remotely sensed $SSM(t_i)$ as measurements of W_{SL}, the SWI is formulated by [40] accordingly as

$$\mathrm{SWI}_T(t_n) = \frac{\sum_i^n SSM(t_i) e^{-\frac{t_n - t_i}{T}}}{\sum_i^n e^{-\frac{t_n - t_i}{T}}} \quad \text{for } t_i \leq t_n, \tag{2}$$

where t_n is the observation time of the current SSM measurement and t_i are the observations times of the previous SSM measurements. All SSM observations before t_n are summed up and exponentially weighted and the T-value determines how fast the weights become smaller, thus modelling how strongly past SSM observations influence the SWI value. An examination of how the T-values correspond to soil depth can be found in [41].

The SWI formulation allows to propagate SSM values forward in time, modelling the infiltration into the soil's deeper layers. When we now combine coexisting streams of SSM signals to a fused SSM-history $W_{SL}(\tau)$, we obtain, through the temporal filtering, a SWI value that carries the SM information from all the input streams. When we fuse SSM from SCAT and SAR systems, and when we do the SWI-filtering for each individual location of the SAR's fine-scale grid, we may transfer the high spatial resolution of the SAR SSM, as well as the high temporal resolution of the SCAT SSM, to the SWI product.

Based on these considerations, we develop a near-real-time (NRT) capable data fusion algorithm to overcome the remote sensing scale gap, by ingesting two streams of radar observations from complementary spatio-temporal scale. Assuming that they describe soil moisture processes similarly, we merge coarse-scale SSM signals from Metop ASCAT (scatterometer, 25 km resolution) and fine-scale SSM signals from Sentinel-1 (SAR, 1 km), and yield a high-resolution, high-frequency soil water index product named Scatterometer Synthetic Aperture Radar Soil Water Index (SCATSAR-SWI). Benefiting from the input's either high temporal or spatial resolution, respectively, this kilometre-scale daily product provides observational data for operational use, at full spatial extent each day.

In the following Section 2, we describe the input datasets of the SCATSAR-SWI in detail while Section 3 presents the algorithmic implementation of its retrieval. In Section 4, we explain the evaluation experiments and reference datasets, which is followed by the results and discussions in Section 5. The articles closes with general conclusions and an outlook in Section 6.

2. Input Datasets

The here presented SCATSAR-SWI fuses 25 km Metop ASCAT SSM and 1 km Sentinel-1 SSM data to one SWI dataset. Both input datasets are generated with an SSM retrieval method employing the TU Wien Change Detection Model [40], which derives SSM directly from observed radar backscatter, measured as backscatter coefficient σ^0. In this model, changes in backscatter are interpreted as changes in soil moisture while other surface properties, such as geometry, roughness, and vegetation type, are interpreted as static parameters. The parameters describe maximum dry- and wet- conditions as well as average vegetation and surface geometry signal contributions. The model is self-calibrated at the pixel scale as the model parameters are estimated through statistical analysis of long-term backscatter time series. The obtained SSM values represent the degree of saturation of the topmost soil layer (first $\sim$5 cm) and are given in relative units ranging from 0% (dry) to 100% (wet). The TU Wien model has proven well its capability to produce reliable soil moisture estimates on a global basis, ingesting backscatter measurements from several C-band sensors, including scatterometers as ERS-1/2 [42,43] and Metop ASCAT [44–46], and SARs as Envisat ASAR [30,47] and Sentinel-1 CSAR [31].

2.1. Scatterometer SSM Input: Metop ASCAT

The ASCAT SSM product [48] is an evolution of the first algorithm realisation for ERS-1/2 satellite datasets [40], which was later successfully migrated to the ASCAT sensors onboard the Metop satellites [49]. The European ASCAT mission has been delivering scatterometer data since 2007 (launch of Metop-A), and since 2013 (launch of Metop-B), effectively doubling its coverage frequency (and Metop-C is foreseen to be launched in 2018, replacing Metop-A). The ASCAT sensors have proven to be a valuable instrument monitoring soil moisture changes over land [41,50,51].

For the ASCAT SSM, the TU Wien model is extended with a dynamic vegetation correction to account for σ^0-signal variations induced by vegetation phenology, making use of the multi-angular observation geometry of the ASCAT sensor (recently analysed in [52]). The current retrieval algorithm is implemented within a software package called Soil Water Retrieval Package (WARP). A detailed description and latest improvements of the retrieval algorithm are described in [53].

For this study, input Metop ASCAT SSM data is accessed from the "Satellite Application Facility on Support to Operational Hydrology and Water Management" (H-SAF) service led by Eumetsat [54]. More specifically, we used the Metop-A/B ASCAT NRT SSM image products H101 and H16, delivered as binary BUFR (Binary Universal Form for the Representation of meteorological data)-files in swath orbit geometry. The soil moisture values (in degree of soil saturation) are complemented by the ASCAT normalised radar backscatter $\sigma°(30)$, the estimated noise values of SSM and $\sigma°(30)$ (covering instrument noise, speckle and azimuthal effects), as well as probability values for frozen and snow soil conditions. These by-products allow the generation of the surface state flag (SSF, [55]) expressing the state of the surface with respect to frozen soil conditions. The Metop ASCAT SSM data has been available since January 2007.

2.2. SAR SSM Input: Sentinel-1

The Sentinel-1 mission [56] of the European Earth observation programme Copernicus operates on two identical spacecrafts, Sentinel-1A (S-1A, launched in April 2014) and Sentinel-1B (S-1B, launched in April 2016), each carrying a SAR sensor with unprecedented spatio-temporal coverage as well as radiometric accuracy and stability.

The Sentinel-1 SSM dataset (S-1 SSM) is provided by the Remote Sensing Group of the TU Wien Department of Geodesy and Geoinformation (TU Wien GEO [57]). The employed SAR SSM retrieval algorithm [31,58] ingests S-1 data from Interferometric Wide Swath (IW) mode in VV-polarisation. It is an adaptation of the TU Wien Change Detection method, most closely akin with the method's earlier adaptation to Envisat ASAR Global Monitoring (GM) mode [29].

Analogous to the ASCAT SSM, the method scales the backscatter acquisitions between the dry- and wet-references that correspond respectively to the soil wilting point and saturation level, yielding soil saturation in percent. However, due to the SAR's lack of simultaneous multi-angle σ^0 observations, this product does not model vegetation seasonality in the current version. The SSM retrieval algorithm is implemented within a fully-fledged, parallel-operating, S-1 processing chain called the SAR Geophysical Retrieval Toolbox (SGRT) [59]. The S-1 SSM data has been available since October 2014.

3. The SCATSAR-SWI Fusion Algorithm

The SCATSAR-SWI product is built on a fused data cube comprising SSM data from ASCAT and S-1. In general, the SCATSAR-SWI retrieval algorithm consists of two components: (1) an offline operated data fusion parameter generation, and (2) a NRT-capable SWI product retrieval, taking up the incoming SSM and the beforehand generated data fusion parameters. To give an overview, Figure 2 outlines the general structure of the algorithm.

3.1. SCATSAR Data Cube

Modern remote sensing places high demands on processing- and storage- facilities. It was found, for example, for S-1 that global-scale services in near-real-time (NRT) or reprocessing activities are only technically feasible when employed in High Performance Computing (HPC) environments due to the large daily data volume [60]. As a consequence, a modern remote sensing product retrieval algorithm is then required to support parallelised processing.

In recognition of these findings, the TU Wien GEO team developed an optimised data cube architecture for raster data from satellite sensors [61]. Its basic framework is the Equi7Grid [62], a spatial reference system designed to efficiently handle the archiving, processing, and displaying of high resolution raster data over land while preserving geometric accuracy. It is defined for the entire Earth and consists of seven planar subgrids for each continent, referenced to the ellipsoidal WGS84 datum. The coordinates are defined by individual realisations of the Equidistant Azimuthal projection, given as eastings and northings in metres. The Equi7Grid not only allows to conveniently handle large remote sensing spatio-temporal data, but also to relate and transform data of diverse spatial scales [63]. The Equi7Grid definitions and tools are available at [64].

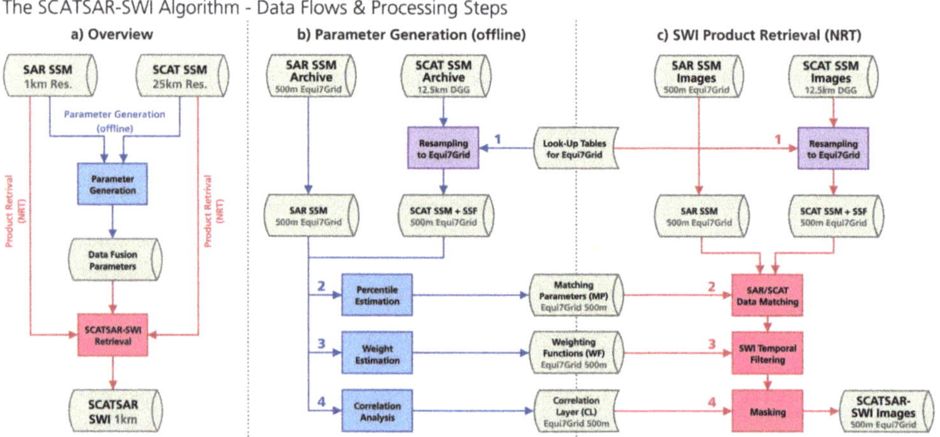

Figure 2. (**a**) overview on the SCATSAR-SWI retrieval algorithm and its data flows, with the offline parameter generation in blue and the near-real-time (NRT) product retrieval in red; (**b**) the parameter generation preparing the data fusion parameters (Matching Parameters, Weighting Functions and Correlation Layer) for the product retrieval, using surface soil moisture (SSM) and surface state flag (SSF) from the input datasets; (**c**) the NRT-capable SWI product retrieval, taking up incoming SSM data as well as prepared data fusion parameters.

Naturally, such a data cube approach is also well suited for the SCATSAR-SWI data fusion, since not only parallel image operations as well as time series analyses are facilitated, but also the combination of datasets from different scales is supported. Consequently, the spatial reference system for SCATSAR-SWI data cube is a Equi7Grid, realised with a sampling distance of 500 m (to allow a nominal spatial resolution of 1 km) and a regular square tiling with a 600 km extent. This choice was found to be optimal for the data handling in terms of file number and size, using LZW-compressed (Lempel–Ziv–Welch lossless data compression) geotiff-files organised per Equi7Grid-tile in image stacks.

The data fusion parameter generation requires a priori SSM datasets from SCAT and SAR, which together cover a multi-year period in order to allow robust estimates of the parameters. The SCATSAR-SWI data cube formed for this study (Section 5.1) holds SSM archives from ASCAT and S-1 for the period October 2014 to October 2017. For the same period, we generated the SWI product that is evaluated in Section 5.

SCAT Resampling

The S-1 SSM product is natively in 500 m-sampled Equi7Grid, hence this data can be directly ingested in the SCATSAR-SWI database, with no resampling action required. Contrary, the ASCAT SSM data is provided per satellite orbit segments as highly-compressed Eumetsat-BUFR files that are referenced to the WGS84 ellipsoid. Hence, a decoding and resampling of the ASCAT data is necessary. For the ingestion of ASCAT SSM data into the data cube, an oversampling of the 12.5 km-sampled data to the 500 m Equi7Grid-tiles is performed, employing the radial basis interpolation module of scipy using the "thin_plate"-function [65]. Despite the large oversampling factor of 25 (from 12.5 km to 500 m), the data volume is not inflated thanks to the efficient file compression. The mapping functions of ellipsoidal- to Equi7Grid- coordinates are constant and are read from precomputed Look-Up-Tables (LUTs) in order to speed up the ASCAT data resampling.

3.2. Data Fusion Parameter Generation

To ensure a valid fusion of the SCAT and SAR inputs, data fusion parameters are required to account for their different observation scale, accuracy, and local dynamics. We realise this with a set of static parameters retrieved from the S-1 and ASCAT archives covering 2014–2017, which are presented in the following subsections.

3.2.1. Matching Parameters

Both SSM products from SCAT and SAR are supposed to describe the same physical parameter, and they are retrieved from radar backscatter data that is recorded by comparable technologies. However, systematic biases are likely to occur due to sensor- and retrieval- specifics and the different complexity at the original scale, evoking land-cover-like patterns. The fusion of SCAT with SAR SSM should account for these systematic differences in order to obtain a consistent SM dataset.

Here, we use data matching per each grid point to obtain SSM signals with compatible levels and dynamics. Data matching techniques are methods to resolve signal biases and different signal variances in a dataset containing disparate sources [66] and thus meet our needs. There is a large set of data matching methods, among which the Cumulative Distribution Function (CDF) matching is widely used for soil moisture applications [43,67]. More specifically, CDF-matching can transform the mean, variance, skewness and kurtosis of a time series to approximately match the ones of the reference time series at relative low computational costs.

We realise the CDF-matching through a piecewise-linear matching of the SSM percentiles at each grid point. This requires a per-pixel signal analysis of the SAR and SCAT SSM archives, estimating the SSM percentiles in intervals of ten percent from 10% to 90%. These values form the Matching Parameters (MP) and are used then later during the SWI retrieval (Section 3.3.1).

3.2.2. Weighting Functions

The quality of input SSM data from SCAT and SAR is not necessarily equal and can vary over time. To regulate the impact of the input quality on the SCATSAR-SWI, we introduce weights for SCAT ($w_{SCAT}(t_p)$) and SAR ($w_{SAR}(t_p)$) that define the weight of the individual SSM values in the SSM history for the SWI estimation.

The quality can be characterised either by empirical estimates of the operator for the data quality, or by quantitative measures, e.g., through their signal error noise values $\epsilon(t)$. The weights can be then realised through the inverse values of $\epsilon(t)$, which form the Weighting Functions (WF) for SCAT and SAR SSM.

Alternatively, as a most simple implementation, only two static weights are defined for all SCAT and SAR values, respectively, identical for every grid point and for every point in time. In this study, we decided to use such static weights, owing to the large imbalance in temporal coverage between SCAT and SAR during the ramp-up-phase of S-1 in the years 2014 and 2015, giving much uncertainty to the error noise estimation. We set globally for all locations

$$w_{SCAT}(t) \rightarrow w_{SCAT} \quad w_{SAR}(t) \rightarrow w_{SAR}, \tag{3}$$

$$w_{SCAT} = w_{SAR} = 1, \tag{4}$$

which virtually gives ASCAT much weight in the obtained SCATSAR-SWI product because ASCAT (A + B) has a 2–10 times higher revisit frequency than S-1 (A (+B)) over European locations and thus ASCAT SSM constitutes the majority of the SSM observations of the fused SSM data cube. With this setup, the SCATSAR-SWI processing takes in a simple way into account the higher average accuracy of the ASCAT SSM data, which is more mature [31].

3.2.3. Correlation Layer

The SCAT and SAR sensors observe in completely different spatial scales. Consequently, the observed radar signals are likely to describe different geophysical processes when spatial heterogeneity on the SAR-scale is given. Fragmented land cover may affect the finer SAR observations very differently than the coarse SCAT observations. However, if both input SSM signals therefrom feature similar temporal dynamics, it can be assumed that they describe the same SM process (invoking the temporal stability concept). To identify grid locations that fulfil this assumption, a Correlation Layer (CL) is built, holding for each grid point the correlation between the SCAT and the SAR SSM time series. The CL can then act as mask during the SWI production, flagging pixels where the correlation is low and thus not eligible for the SWI data fusion.

We calculate the Spearman Rank Correlation Coefficient ρ at each grid point between the ASCAT and S-1 time series—after temporally matching them (yielding two samples of the same size). Additionally, the correlation significance is calculated to account for the sample size of the investigated time series. The CL holds for each grid point then two values, the ρ-value and the corresponding significance (p-) level.

3.3. SCATSAR-SWI Estimation

With a data fusion parameters ready, the SCATSAR-SWI can be calculated. In the following, we describe the methodology for an ongoing and near-real-time production of daily SWI images, which involves SCAT/SAR data matching, the SWI retrieval, and output masking.

3.3.1. SSM Distribution Matching

Following the argumentation discussed already in Section 3.2.1, prior to temporal filtering, the SSM data from SCAT and SAR should be matched. While the SCAT data is superior in capturing the temporal SM dynamics, the SAR product features much finer spatial dynamics, enabling the resolving of SM patterns on the kilometric scale.

Per 500 m Equi7Grid-grid-pixel, we match the previously oversampled ASCAT SSM (see Section 3.1) to the local S-1 SSM distribution. This is done through CDF-matching employed on the SSM percentiles generated as described in Section 3.2.1. At each grid point, the CDFs of ASCAT and S-1 SSM are approximated piece-wise through linearisation between the 10%-percentiles. Each ASCAT SSM value is then mapped from the local ASCAT CDF to corresponding S-1 CDF segment, resulting in an ASCAT SSM image settled on the spatial definition of the SAR record. In this way, we disaggregate the coarse-scaled SM observation and adapt it to the fine-scaled SM patterns.

3.3.2. Recursive Weighted Temporal Filtering

The calculation of the SWI values using the approach introduced in [40] requires the availability of historic SSM time series data (Equation (2)). A computational adaption of the algorithm towards ongoing, day-to-day SWI production has been proposed already in [68]. They found a recursive reformulation of the SWI calculation that is mathematically equivalent to Equation (2). Using their formulation, recasting the terms and shifting the time indexing by 1, we found an alternative expression of their recursive calculation. The SWI for a given characteristic time length T can then be derived using the formula:

$$\mathrm{SWI}_T(t_{i+1}) = \mathrm{SWI}_T(t_i) + \frac{\mathrm{SSM}(t_{i+1}) - \mathrm{SWI}_T(t_i)}{den_T(t_{i+1})}, \tag{5}$$

where t_i refers to the time of the previous, and t_{i+1} of the current SSM measurement, so that $\mathrm{SWI}_T(t_i)$ is the SWI at the time of the previous SSM measurement and $\mathrm{SSM}(t_{i+1})$ is the new SSM measurement. The so-called denominator factor $den_T(t_{i+1})$ expresses the time-dependent filtering and is calculated from the previous denominator factor as

$$den_T(t_{i+1}) = 1 + e^{-\frac{t_{i+1}-t_i}{T}} den_T(t_i). \tag{6}$$

Hence, the difference of the current SSM to the previous SWI (the nominator in Equation (5)) and the filtered time delta (the denominator in Equation (5)) determine the next SWI value.

Evidently, for the recursive calculation, the last SWI value and the last denominator factor need to be stored. Then, the calculation can be continued at any time, practically when new input SSM data are available from SCAT or SAR, or as we implement it, everyday at 12:00 UTC. In our case, all SSM values from the past 24 h are consecutively converted to SWI values with Equation (5), whereas the last SWI value of the 24 h is stored in the output SWI image of that day. The calculation of SWI is initiated at time $t = 0$ with the following values:

$$t_i = t_{i+1} = 0, \quad \text{SWI}_T(0) = \text{SSM}(0), \quad den_T(0) = 1. \tag{7}$$

Our alternative recursive SWI calculation introduced in Equation (5) is convenient when it comes to (quality-) weighted SSM values, which we implement for the SCATSAR-SWI to weight individually the SCAT SSM and SAR SSM input values. To begin with, we add $w(t_i)$ as weight for $\text{SSM}(t_i)$ and n for the cumulative total number observations, and expand the Equation (2) to the Weighted SWI formulation:

$$\text{SWI}_T^w(t_n) = \frac{n \sum_{i=1}^{n} w(t_i) \, \text{SSM}(t_i) \, e^{-\frac{t_n-t_i}{T}}}{\sum_{i=1}^{n} w(t_i) \, \sum_{i=1}^{n} e^{-\frac{t_n-t_i}{T}}} \quad \text{for } t_i \leq t_n. \tag{8}$$

The above equation uses the complete SSM history. For SWI retrieval in ongoing, day-to-day processing, we need, however, the recursive calculation of the Weighted SWI. The SWI_T^w can be recursively calculated as:

$$\text{SWI}_T^W(t_{i+1}) = \frac{\text{SWI}_T^w(t_i) \, \frac{n+1}{n} \, \sum w_i \, (den_T(t_{i+1}) - 1) \; + \; \text{SSM}(t_{i+1}) \, (n+1) \, w(t_{i+1})}{\sum w_{i+1} \, den_T(t_{i+1})}, \tag{9}$$

where $\sum w_i$ and $\sum w_{i+1}$ represent the previous and current cumulative sum of weights, respectively (with $\sum w_{i+1} = \sum w_i + w(t_{i+1})$). Consequently, $\sum w_{i+1}$ needs to be stored together with the denominator factor $den_T(t_{n+1})$ and the number of total observations $(n + 1)$ for the next calculation at time t_{i+2}. It is worth noting that the weights are not required to be normalised.

3.4. Output Masking

There are two maskings in the SCATSAR-SWI. First, the surface state flag, coming from the ASCAT SSM product (SSF, [55]), identifies frozen conditions. The SCATSAR-SWI algorithm discards SSM estimates when the SSF is not "unfrozen" and reduces an internal, T-value-dependent quality flag. An ongoing period of no valid SSM values leads to gaps in the dataset also after frozen conditions, as the quality flag needs to recover, assuring that no values measured under frozen conditions are part of the used SSM history $W_{SL}(\tau)$. All this is done on the ASCAT grid, hence the data gaps are block-shaped and ca. 12.5 km large. SSF-masked gaps mostly occur in alpine or cold climates during the cold season.

Second, we apply the Correlation Mask (CM) that is retrieved from the Correlation Layer (CL), as the SCATSAR-SWI should only be used where the SCAT and SAR reasonably agree. Following Section 3.2.3, the CM is defined through a threshold of minimum correlation and significance, yielding a logical mask on the 500 m-pixel-grid.

4. Evaluation Datasets and Methods

4.1. SCATSAR-SWI Production

For evaluating the SCATSAR-SWI retrieval method, we generated a 3-year dataset over Italy, using all available Metop-A/B-ASCAT and S-1A/B observations from October 2014 to October 2017. From the fused SSM data cube, we generated the parameters as described in Section 3.2, yielding a set of matching parameters, correlation- layer, and -mask per Equi7Grid-tile. With that, we retrieved the SCATSAR-SWI images for each day in the period October 2014–October 2017. We masked the output data for water bodies using the mask included in the S-1 SSM product, and for low SCAT-SAR correlation using the Correlation Layer with $\rho = 0.3$ as lower limit.

4.2. Layout of Experiments

A comprehensive evaluation of remotely sensed SM data requires reference datasets of comparable spatio-temporally density and coverage [50,69]. Having said this, comparisons of remotely sensed observations with in situ data is often troubled by dataset inconsistencies and a scale-mismatch [70]. In order to complement the evaluation against in situ data, we used in this study the Soil Water Balance Model (SWBM, [71]) and carried out an in-depth analysis of the signal quality over the Umbria region in central Italy, validating the SCATSAR-SWI data against the model data, as well as against reference from ASCAT and available in situ observations. Furthermore, we used the SM2RAIN approach [8] to evaluate the skill of the SCATSAR-SWI to record the impact of rainfall on SM. For this, we compared daily mean rainfall observations over Italy, and in detail over Umbria, against estimates from the SM2RAIN algorithm ingesting our SCATSAR-SWI data.

4.3. Study Area: Umbria Region

The Umbria region is located in central Italy and it is characterised by a complex landscape topography. Land uses are mainly forests (42.6%), and crops (49.2%), with urban areas covering only 3.5% of the territory. See in Figure 3a the region's topography, and in Figure 3b its land cover as from CORINE 2012 [72]. The climate is Mediterranean, with mean annual rainfall of 950 mm and mean annual air temperature ranging between 3.5 °C and 14.0 °C.

A dense real-time hydro-meteorological network (90 rain gauges, 77 thermometers, one station every 150 km^2) has been operating for more than 20 years and 12 soil moisture stations distributed throughout the territory [73]. The monitoring network provides semi-hourly data for which a quality-check step is performed in order to remove anomalous values and to fill any temporal gaps.

4.4. Model SM Data: SWBM-SA Umbria

We used modelled 1 km soil moisture (Model SM) estimated with semi-analytical SWBM (SWBM-SA [74]) to test the ability of the SCATSAR-SWI data to reproduce the temporal and the spatial variation of SM at the 1 km-scale. The SWBM was specifically developed to reproduce in situ SM observations in the Umbria region [71], and has been successively applied also in different test-sites across Europe yielding satisfactory results, with correlations higher than 0.8–0.9 and Root Mean Square Differences (RMSDs) lower than 0.025 m^3/m^3 [45,74–76].

For this study, interpolated 1 km rainfall and temperature data drove the SWBM-SA model, yielding hourly-based SM estimates on a 1 km-based regular grid for an area of 8991 km^2 (Figure 3a). The mountainous eastern part of the region was intentionally excluded as the density of rain gauges here is much lower and the satellite observations are expected to be largely impacted by the topography. Input precipitation and temperature data cover the period 2014–2017. Model parameters are obtained from soil texture information available over the study region (outlined in [73]). SM estimates were modelled for the topmost 10 cm of the soil to match the depth of the SCATSAR-SWI at T-value of 1 and 5, following the study of [41].

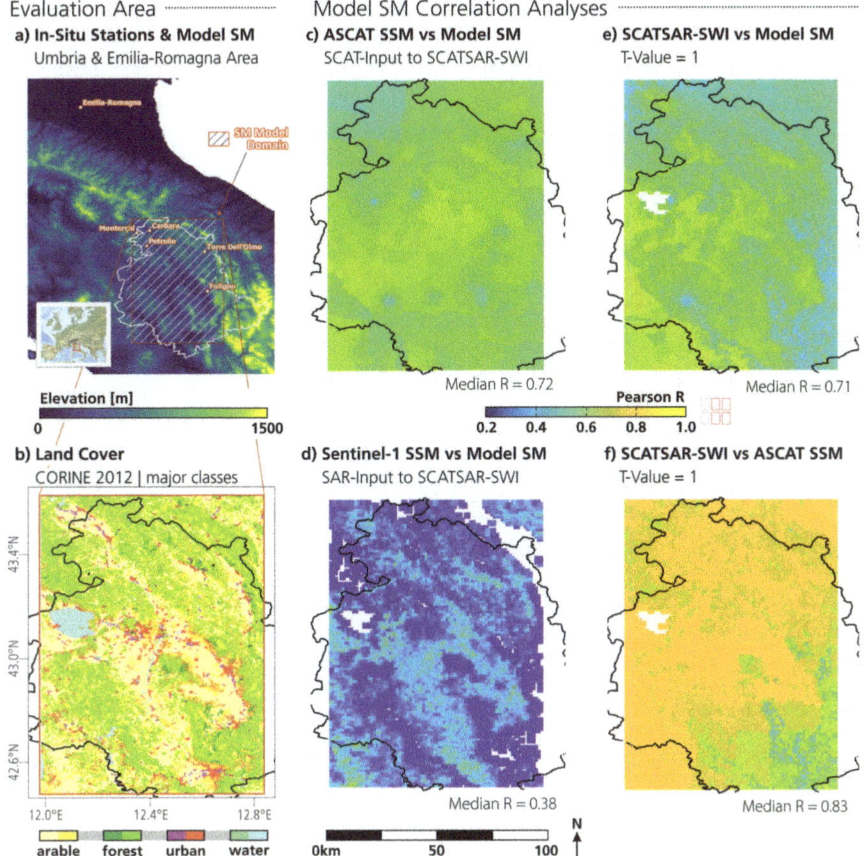

Figure 3. (**a**) evaluation area over Umbria, with Model SM domain, in situ stations, and COSMOS-station in Emilia Romagna. Elevation map as background; (**b**) CORINE Land Cover, grouped as main classes occurring in the Umbria area. Results as discussed in Section 5.2: (**c**,**d**) comparison of the input SSM datasets with Model SM, showing the Pearson R over the model domain; (**e**) comparison of the SCATSAR-SWI with Model SM; and with ASCAT SSM (**f**).

4.5. In Situ SM Data: ISMN

In situ data from the International Soil Moisture Network (ISMN, [77]) were used for evaluation (In Situ SM). The ISMN provides a harmonised repository of in situ SM observations all over the world and is well-evaluated and established in the community [41,78,79]. Here, we use five UMBRIA-stations from the Umbria region and one COSMIC-station, close-by in the Emilia-Romagna region (see Figure 3a).

During the analysis period, only five out of twelve UMBRIA-stations measured soil moisture, some of them discontinuously, or not during the full period. The five instruments consist of four Frequency Domain Reflectometry (FDR) stations and one Time Domain Reflectometry (TDR) station with sensors located at different depths (5–40 cm). Three stations (Cerbara and Torre dell'Olmo and Petrelle) are operating in real-time and provide measurements every 30 min. For the validation of SCATSAR-SWI data, we used only the at all stations available topmost measurements at a depth of

10 cm. The two stations in the east of Umbria are located in a section of good coverage by S-1 (Foligno, Torre Dell'Olmo with 401 measurements), the other three in the northwest were measured only half as often (Cerbara, Monterchi, Petrelle with 198 measurements).

The COsmic-ray Soil Moisture Observing System (COSMOS) is a network of stations that measure soil moisture through a newly-developed cosmic-ray method [80]. The stationary cosmic-ray soil moisture probe measures the neutrons that are generated by cosmic rays within air and soil and other materials, moderated by mainly hydrogen atoms located primarily in soil water, and emitted to the atmosphere where they mix instantaneously at a scale of hundreds of metres and whose density is inversely correlated with soil moisture. This measurement scale makes it a valuable reference for the remotely sensed and 500 m-sampled SCATSAR-SWI.

We used the station Water4Crops Budrio IT (098), located in Emilia-Romagna in the Po Valley in an extremely flat area with intensive, irrigated agriculture. It is a test site for different irrigation techniques and different water sources (reuse of treated wastewater) and has a mean annual precipitation of about 750 mm with a mean annual temperature of 16.0 °C. The station provides measurements of soil moisture in the range of 0–21 cm and data are available from about 2014 to 2016. This station is located in section of medium S-1 coverage (286 measurements).

4.6. Rainfall Observations

The assessment of the SCATSAR-SWI through SM2RAIN has been carried out at national scale (Italian territory) and local scale (Umbria region). The observational rainfall dataset for the Umbria region is already described in Sections 4.3 and 4.4, and the Italy-dataset is obtained from an interpolation of 3000 rain gauges provided by the monitoring network of the National Civil Protection Department [81]. The hourly observations are spatially interpolated over the analysis grid at 3 km resolution by using the Random Generator of Space Interpolations from Uncertain Observations algorithm [82] and aggregated at a daily time step. The dataset is available from April 2015 to April 2016.

4.7. SM2RAIN from SCATSAR-SWI

The SM2RAIN algorithm was developed in [8] to obtain rainfall estimates from in situ and satellite SM observations. The algorithm has been applied extensively to several satellite SM products obtaining good performance in the reproduction of ground-based rainfall observations (see for most recent applications [83,84]). In this study, we applied SM2RAIN to SCATSAR-SWI with T-value = 1 by following the same approach as in [84] (the reader is referred to this paper for details). The SM2RAIN parameter values are calibrated against rain gauge observations described in Section 4.6 to obtain two rainfall products at daily time scale: (1) Italian scale product at 3 km spatial resolution for the period from April 2015 to April 2016, and (2) Umbria scale product at 1 km spatial resolution for the period from October 2015 to September 2016. For comparisons, a second SM2RAIN dataset was created with ASCAT SSM as input, using the same methods but a rainfall product calibrated at 10 km spatial resolution.

4.8. Data Preparations for Evaluation

The different SM products that are examined in the evaluation experiments are characterised by different measurement units and have different grids. Thus, we reprocessed them for a meaningful comparison. More specifically, the comparisons of SCATSAR-SWI and ASCAT SSM data with the reference data (in situ and model) required the following steps: (1) Spatial matching: The grid pixel of the SCATSAR-SWI and ASCAT SSM whose centroid is nearest to the reference location has been selected, and the corresponding relative SM time series has been extracted; (2) Temporal matching: The site-specific soil moisture data closest to the corresponding acquisition time of SCATSAR-SWI and ASCAT SSM were extracted from their time series. Flagged data due to water, frozen conditions (through ASCAT's SSF), or out-of-bound values were masked out; (3) For the

ISMN-comparison, a conversion to volumetric SM units is necessary: per pixel, the mean and the variance of SCATSAR-SWI and ASCAT SSM (both in relative units) were matched with those of co-locating reference data to obtain absolute values in $m^3\ m^{-3}$, following [85].

Accounting for the temporal frequency of S-1 observations of 1.5 to 4 days over the study area, the correlation analysis of the SM2RAIN-derived rainfall products is carried out for five-day accumulated rainfall.

5. Evaluation Results and Discussion

We evaluated SCATSAR-SWI data over Italy against reference soil moisture data from the model-, satellite-, and in situ-datasets, as well as rainfall data, following Section 4.2.

5.1. The SCATSAR-SWI Dataset

For the evaluation presented here, all available ASCAT and S-1 SSM observations over Italy and its neighbours were processed with the algorithm in Section 3 and the setup outlined in Section 4.1. The obtained data cube consists of SWI daily imagery over five Equi7Grid-tiles (each 600 km wide). The SCATSAR-SWI data is built up of 180 to 420 single observations per 500 m pixel from S-1 SSM (average 292), and of 1850 to 2450 from ASCAT SSM (average 2060), depending on the S-1 and ASCAT orbit coverages (see minimaps in Figure 4a/d). As one can see, the S-1 coverage is very inhomogeneous, which is directly resulting from the Sentinel observation scenario [31]. For the first two years during the period (October 2014 to October 2016), ASCAT features, on average, a seven times higher frequency than S-1, as the ASCAT satellites overpassed an average location 2.2 times a day and S-1 only 0.3 times a day. From October 2016 onwards, the S-1 mission is in full deployment using two satellites. With the inclusion of S-1B, twice as much S-1 SSM measurements are available over Europe, with ASCAT featuring then a 3.5 times higher frequency than S-1. Overall, the fused SSM dataset consist of ca. 80% ASCAT observations.

To give a first impression of the data, Figure 5 compares the SM patterns of the input SSM from ASCAT and S-1, and the SCATSAR-SWI with T-value = 5 for 23 July 2017. Figure 5d adds the porosity of the soil layer at 0–5 cm depth from ISRIC SoilGrids [63] to the comparison. On that day, ASCAT achieves almost full coverage over the area with evening overpasses alone, while S-1 morning and evening overpasses cover less area (and has no coverage on the next day (not shown)).

As obvious from Figure 5, the S-1 SSM and SCATSAR-SWI feature a much higher degree of spatial detail than the ASCAT SSM. Accordingly, the comparison with ISRIC data illustrates that the two kilometric products are much closer to the scale of soil characteristics. However, both SSM signals seem to be contained in the SWI, as the smooth patterns from ASCAT and also the detailed patterns from S-1 are reflected in the SWI (e.g., in southeastern Austria or in the border area of Slovenia and Croatia). Ideally, the SCATSAR-SWI provides a resolution of 1 km as a result from the SCAT data matching and the SAR signal contribution. However, the actual resolution of the present SCATSAR-SWI appears to be lying between the resolution of S-1 and the ASCAT input, owing also to the large share of ASCAT observations in the fused data cube.

Interestingly, typical wet biases over large cities are apparent as blue spots in the ASCAT data, e.g., over Vienna (northeast in Figure 5a) or Munich (northwest), which are completely rectified in the SCATSAR-SWI. It is also noticeable that the SCATSAR-SWI contains linear artefacts following the edges of ASCAT and S-1 orbit footprints. These artefacts are due to different timings of the inherent SSM observations in adjacent image sections. Such linear artefacts are already known from the 12.5 km ASCAT SWI (not shown, e.g., as available from Copernicus Global Land Service (CGLS) [86]) at lower T-values, but appear to be more pronounced in the SCATSAR-SWI. In a sense, their perceptibility result from migrating the SWI method from the coarse scale, which is usually displayed in a time-series domain, to a finer scale, which is usually displayed in the image domain.

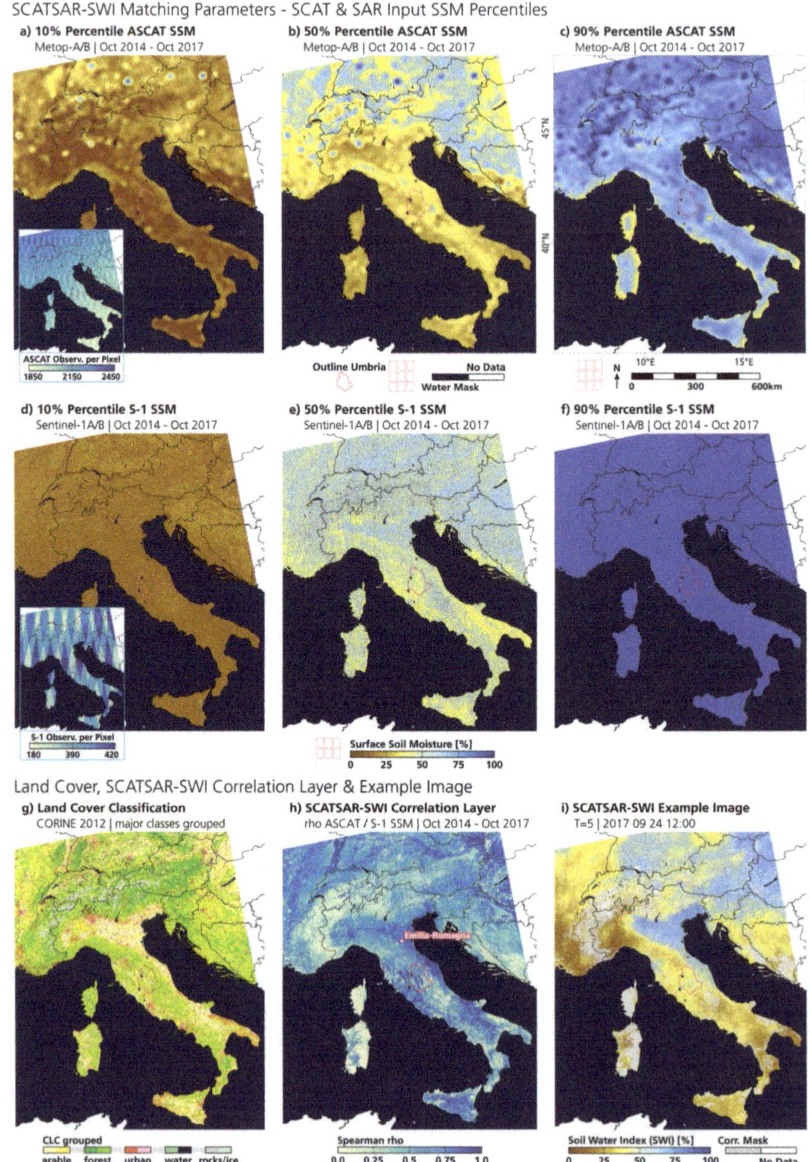

Figure 4. Collection of SCATSAR data fusion parameters and example SWI image over Italy, with land cover data for comparison. The Umbria region (acting as evaluation area) is outlined in red. No-data areas are white. (**a–c**) selected SCAT Matching Parameters; displaying here the 10%, 50%, and 90% percentiles of the ASCAT SSM distribution. The mini-map in (**a**) displays the number of ASCAT SSM observations in the period October 2014–October 2017; (**d–f**) as above, but for SAR from S-1 SSM; (**g**) land cover from CORINE 2012, grouped to major types; (**h**) SCATSAR-SWI Correlation Layer, holding the Spearman Rho between ASCAT and S-1 SSM; (**i**) example SCATSAR-SWI image at 24 September 2017, at *T*-value = 5 and with Correlation Mask applied.

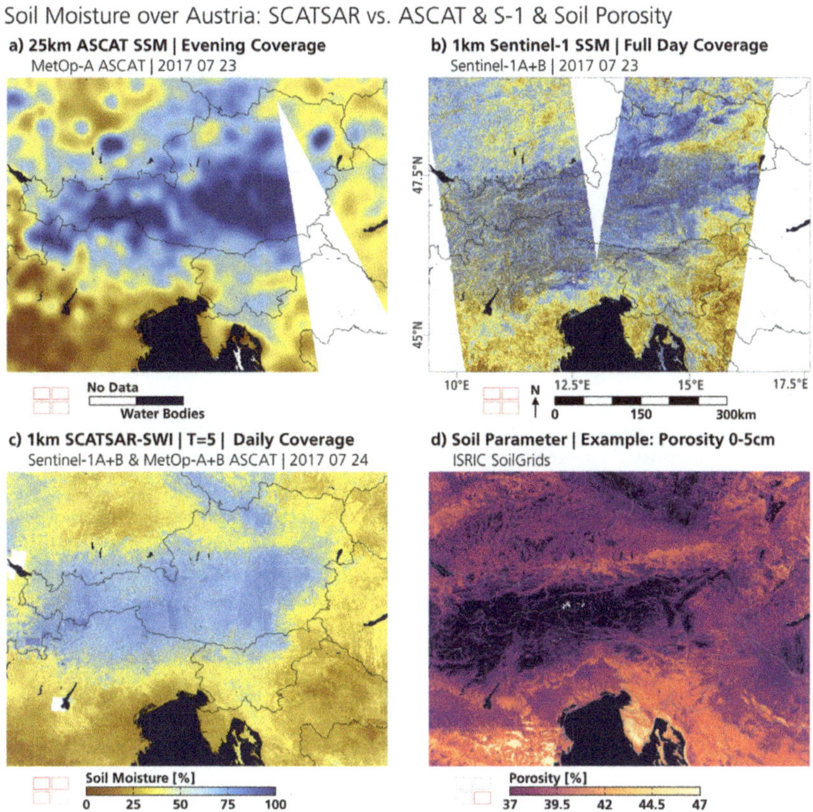

Figure 5. Example images for SCATSAR-SWI and its inputs, and soil porosity for the Austrian area. (**a**) 25 km Metop ASCAT SSM data from two overpasses on the evening of 23 July 2017, where brown colours indicate dry surface conditions, and blue wet conditions. White areas are not covered by the sensors. (**b**) as (**a**), but 1 km SSM from S-1 on the same day; (**c**) the 1 km SCATSAR-SWI at *T*-value = 5 on the same day; (**d**) for comparison, the porosity of the soil layer down to 5 cm depth.

The data fusion parameters are an important part of the SCATSAR-SWI retrieval. Figure 4 displays the SCATSAR-SWI Matching Parameters and the Correlation Layer, together with CORINE Land Cover and an example SCATSAR-SWI image over Italy and its neighbours, with the Umbria region in central Italy marked out. In addition, the ASCAT and S-1 satellite coverages are plotted as mini-maps that show total SSM observations during the evaluation period.

Figure 4a–c show the 10%-, 50%-, and 90%-percentiles of the ASCAT SSM input dataset, representing statistically low, average and high SSM during the period October 2014 to October 2017. The coarse resolution of this product is eminent, as no detailed features, e.g., in the scale of the land cover in Figure 4g, can be resolved with the ASCAT data. Moreover, along the coasts, drier SSM values occur that are potentially wrong, commonly owed to mixed sea/land backscattering effects in the large ASCAT footprints (exceedingly over Sardinia and Corsica). In addition, the wet biases over cities are obvious in all ASCAT SSM percentiles, especially in the northern section of the maps. One explanation for this could be the growth of urban areas during the ASCAT period from 2007 onwards, troubling the parametrisation of the ASCAT model in areas with soil turning into artificial surfaces, or with increasing built-up density. The 50%-percentile (median) map indicates that the period 2014–2017 was drier than normal in Italy, southern France and the Adria region as here the median SSM is largely around

30–40%. With the years 2014–2017 being reported drier than the previous years, this deviation from 50% SSM stems from the different observation periods and resulting parametrisation baselines of ASCAT (2007–2017) and SCATSAR (2014–2017). As the latter is identical with the S-1 observation period, S-1 features largely uniform SSM values in the 10%-, 50%-, and 90%-percentile maps in Figure 4d–f.

The Correlation Layer (CL) in Figure 4h holds the Spearman correlation coefficients between input SSM from ASCAT and S-1 for each pixel. The parameter is essential to the SCATSAR-SWI, as the SWI should be used only at locations with a reasonable good agreement between SCAT and SAR. The CL acts as a descriptor of how well the unlike-scaled signals of ASCAT and S-1 data match and thus indicates where the temporal stability concept for soil moisture is applicable. Indeed, the spatial patterns of the CL reflect many features in the CORINE land cover (as in Figure 4g), featuring, for example, rivers and cities. The CL shows generally high values over arable land, where large forests are absent, e.g., in the Po Valley in central-northern Italy, most of Sicily, the Rhine Valley at the French-German border area, or north and east of Vienna. Contrary, areas with a rough topography (e.g., the western Alps, the Apennines in central Italy, or Corsica and Sardinia) show very low values, due to a generally large geographical heterogeneity there. Similarly, areas with diverse land cover, with forests and agriculture mixed at the kilometre scale as, for example, Calabria in southern Italy, Tuscany west of Umbria, or southern Croatia and Bosnia show low values too. Large contiguous forests, however, like over the Apennines in northern Italy or over the eastern Alps in southern Austria, show medium correlation, further suggesting that spatial complexity governs the temporal stability.

Finally, Figure 4i shows an example SCATSAR-SWI image for 24 September 2017 at 12 h, with the Correlation Mask (CM) applied. The SWI (at T-value = 5) shows wet top soil layer conditions in the Po Valley and Venetia, over eastern Austria, western Hungary and Slovakia, and dry conditions over southeastern France, Piedmont, and southern Italy. Neither dry biases along the coasts nor wet biases around cities seem to be present, suggesting a successful data matching of the ASCAT signal. In addition, the spatial detail of the SAR input is preserved to some extent, while its noise is largely removed. The SWI images features well-defined SM gradients, for instance at the southern fringe of the Alps in north-eastern Italy. The CM (with $\rho = 0.3$ set as threshold) takes out pixels over the western and central Alps, large sections in Sardinia and Corsica, and more scattered, over central and southern Italy, and southern Croatia and Bosnia. Two square-like gaps in the Alps result from the ASCAT's SSF-flagging (done at the 12.5 km grid) that masks these locations (which in this case have a large share in water-bodies and rocks) as not enough valid input SSM data were available.

5.2. SCATSAR-SWI Signal Quality: Umbria Model Domain

Main results from the evaluation of the SCATSAR-SWI and its inputs against the Umbria model SM are displayed in Figure 3: Figure 3c is a map of the Pearson correlation coefficients (R) between the input ASCAT SSM and the model data, plotting 8991 1 km grid pixels. Obviously, ASCAT is uniformly agreeing well with the model, with a median R = 0.72. In contrast, the S-1 SSM input has a much lower average agreement with the model (Figure 3d, median R = 0.38) and shows a much more differentiated result, with low values over forests and high values over arable land as a comparison with the land cover in Figure 3b suggests. This behaviours of the S-1 SSM was already found in the recent study [31] and was related to the insufficient modelling of seasonal dynamics in vegetation density and its water content. However, the SCATSAR-SWI appears to be not much affected by this deficit of the SAR data, since it shows similar performance (Figure 3e, median R = 0.71) to the ASCAT product. The SCATSAR-SWI shows a less uniform pattern, with improved values over arable land (especially in the southwest) and decreased values over mountainous, forested areas in the southeast of Umbria. Not surprisingly, the correlation with the ASCAT SSM is (except for the mountainous southeastern area) very high and spatially consistent (Figure 3f, median R = 0.83), confirming that the ASCAT signal is successfully downscaled to the SCATSAR-SWI, devoid of quality degradation through, for instance, the data matching.

The results in Figure 3 discuss the SWI with T-value = 1, which is closest to the soil surface. The same analysis was done for T-value = 5, which represents SM just below the surface. Those results show as expected the same patterns (not shown) but slightly lower R-values (vs. model: median R = 0.64; vs. ASCAT: median R = 0.76) with the surface model data.

A separate statistical assessment was done for the area subset where the S-1 SSM input has a correlation higher than 0.5, mostly coinciding with zones in Umbria classified as arable (comparing Figure 3b,d). For this subset, the correlation is marginally better (vs. model: median R = 0.73; vs. ASCAT: median R = 0.84). ASCAT SSM shows a marginally lower agreement with the model over this subset (not shown, median R = 0.72), suggesting that the SAR component may not only add spatial detail but may also improve the temporal signal the SCATSAR-SWI over arable and flat areas.

The evaluation summary in Figure 6 plots the distributions of Pearson R and average root mean square difference (RMSD) for comparisons with in situ SM, ASCAT SSM, and model SM for both levels, at T-value = 1 and T-value = 5.

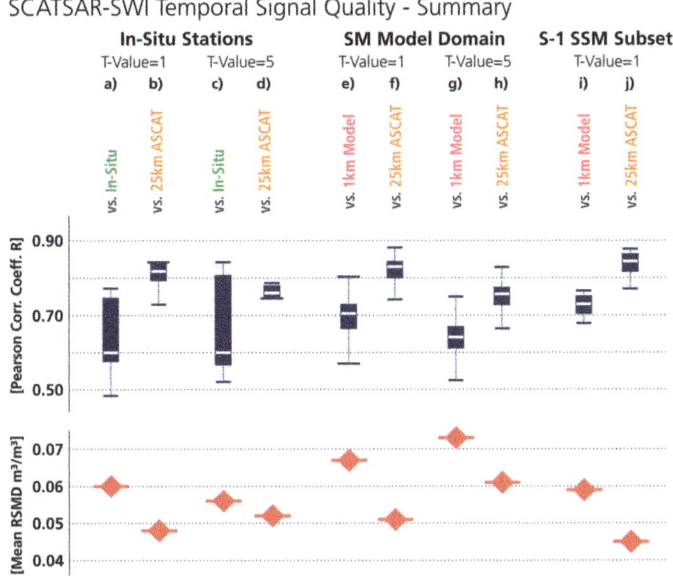

Figure 6. Compilation of statistics on Pearson R-values and average RMSD from the time series analyses between SCATSAR-SWI and SM references (Model SM, ASCAT SSM, In Situ SM) over in situ stations (**a–d**), Umbria SM model (**e–h**), and Umbria SM model subset area with good S-1 SSM performance ("S-1 SSM Subset", (**i,j**)).

5.3. SCATSAR-SWI Signal Quality: In Situ Stations

Figure 6a–d collects results from the evaluation of the SCATSAR-SWI against the reference data at the in situ stations. Overall, the agreement in SM dynamics is good (T-value = 1: median R = 0.60; T-value = 5: median R = 0.61), and has an acceptable average accuracy (T-value = 1: mean RMSD = 0.060 m^3/m^3; T-value = 5: mean RMSD = 0.056 m^3/m^3). Comparisons against ASCAT SSM at the in situ locations show very high correlations (T-value = 1: median R = 0.82; T-value = 5: median R = 0.77) as expected, verifying the measured performance over the Umbria model domain. ASCAT SSM itself shows comparable, slightly better agreement with in situ observations and identical average accuracy (not displayed, median R = 0.63, mean RMSD = 0.056 m^3/m^3), indicating that the temporal dynamics of the ASCAT signal is forwarded intact to the SCATSAR-SWI data.

The SCATSAR-SWI time series at the locations of the ground stations are plotted in Figure 7 against in situ and ASCAT reference, illustrating the high agreement of the data. We chose SWI at *T*-value = 5 for the comparison with the in situ measurements, which represent subsurface SM, and SWI at *T*-value = 1 for the reference from ASCAT (Surface SM).

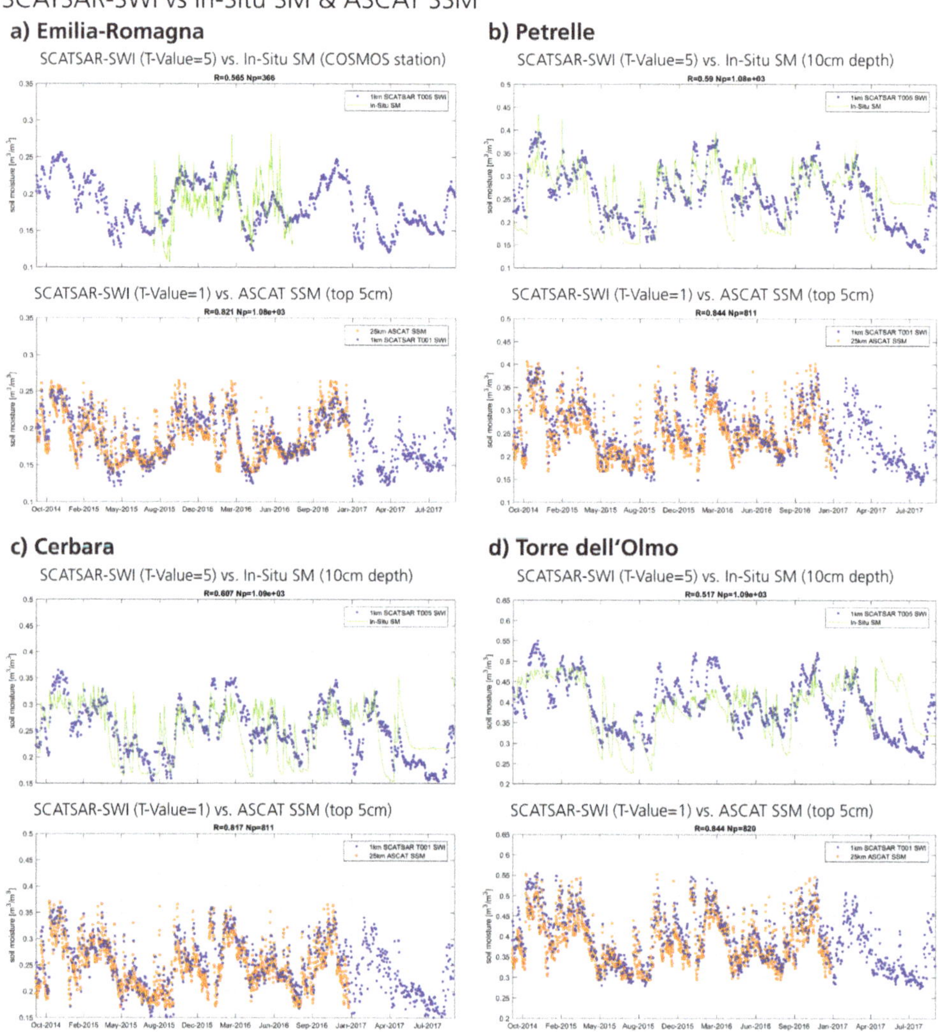

Figure 7. Comparisons of time series of SCATSAR-SWI against references (In situ SM and ASCAT SSM) at selected in situ stations. Values in volumetric units ($m^3\ m^{-3}$). All available data for the period October 2014–August 2017 is displayed. With Pearson Rho (R) and associated sample number of temporally matched data (Np).

The signal from the COSMOS-station in Emilia-Romagna from July 2015–August 2016 is well reproduced by the SCATSAR-SWI (Figure 7a). The COSMOS data is not a point measurement but represents an area of some hundreds of metres radius and is hence of particular interest. While the SCATSAR-SWI shows some wet bias during December 2015, and a dry bias during the subsequent spring, which both may be the consequence from an uncertain conversion to absolute SM due to the short temporal intersect of the two datasets, it accurately reproduces the timing of rises and drops in SM. The amplitude of the SM signal is much larger in the COSMOS data, underlining that the satellite product is not capable of fully capturing SM dynamics at the sub-kilometre scale. However, the SM signal of the ASCAT SSM (Figure 7a bottom), which describes the much larger 25 km SM signal, is well reflected in the SCATSAR-SWI.

Similar observations can be made from the time series plots at the other in situ stations (Figure 7b–d). The amplitudes of the Petrelle and Cerbara ground data show a good match with the SCATSAR-SWI, while some peaks are less pronounced, and many drops are much sharper in the in situ time series. Here the questions remains to what extent this is due to insufficient modelling of the satellite data or because of the scale mismatch between pointwise in situ and extensive remote sensing data.

Interestingly, at Petrelle, Cerbara, and most pronounced at Torre dell'Olmo, the SCATSAR-SWI contradicts the ground measurements when indicating very dry conditions during summer 2017, which was reported to be a very hot and dry season in Italy. The constant values at Petrelle and Cerbara from May to August 2017, clearly above the dry level, are peculiar and leaving doubt if irrigation measures or sensor issues may play a role here.

The above findings are widely supported by the time series analyses at the remaining two stations, which are included in the statistics in Figure 6 but not shown in the time series plots in Figure 7.

5.4. SCATSAR-SWI Rainfall Estimates over Italy

The experiments where we used the SCATSAR-SWI as input for the rainfall retrieval using the SM2RAIN method yielded very promising results (Figure 8). The first experiment, generating daily rainfall estimates at a 3 km sampling over entire Italy during the period April 2015–April 2016, reproduced the rainfall observations very well, as the average maps for this period almost coincide (Figure 8a,b). The correlation map between the five-day-accumulated rainfall from observations and SM2RAIN report values higher than 0.75 for a large part of Italy (median R = 0.72). Most notable regions with lower agreement are Alto Adige in the north and the Abruzzi in the centre, which are both of mountainous character.

The temporal evolution of the 5 day-accumulated areal mean for Italy (Figure 8d) also shows a very high agreement between observations and SM2RAIN estimates (R = 0.89, RMSE = 1.15 mm/day). Only during the winter period in December 2015 and January 2016, and to a lower degree during July 2015, one can see the five-day rainfall sums differ. The deviations during the cold period may result from incorrect detection of frozen/snow conditions in the SWI, supporting above findings as such conditions are commonly much more prevalent in high mountain areas.

Additionally, we tested the one-day-accumulated SM2RAIN data which potentially detects smaller rainfall events (Figure 8e). The analysis shows lower but again high agreement between the remote sensing and reference data (R = 0.79, RMSE = 2.22 mm/day). It reveals that almost all rainfall events reported in the observations are recorded by the SCATSAR-SWI-SM2RAIN, and, vice versa, only few small events are estimated by satellite product that are not in the observations. These events appear during the July 2015 and the winter season, suggesting that the overestimation in the five-day-time-series rather stems from a different number of light rainfall events than from different magnitude. Most interestingly, during strong rainfall events the SM2RAIN generally produces a smaller magnitude (except during winter), but a longer rainfall duration. We conclude that the SCATSAR-SWI based product, due to its daily resolution, cannot quantify the intensity of short-time rainfalls but can reliably estimate the accumulated precipitation.

Results from SM2RAIN from ASCAT SSM show similar performance, with slightly lower values in the five-day-accumulated correlation map (not displayed, median R = 0.71). When regarding the daily rainfall mean mini-map in Figure 8b, the SCATSAR-SWI data appears to reproduce rainfall with much more detail and shows less overall difference to the observation data than the ASCAT-based product, as over Calabria and Sicily for instance. However, concerning the one-day-accumulated areal means, the ASCAT performs significantly better (not displayed, R = 0.87, RMSE = 1.88 mm/day), which may also be attributed to the sub-daily observation frequency of the ASCAT product.

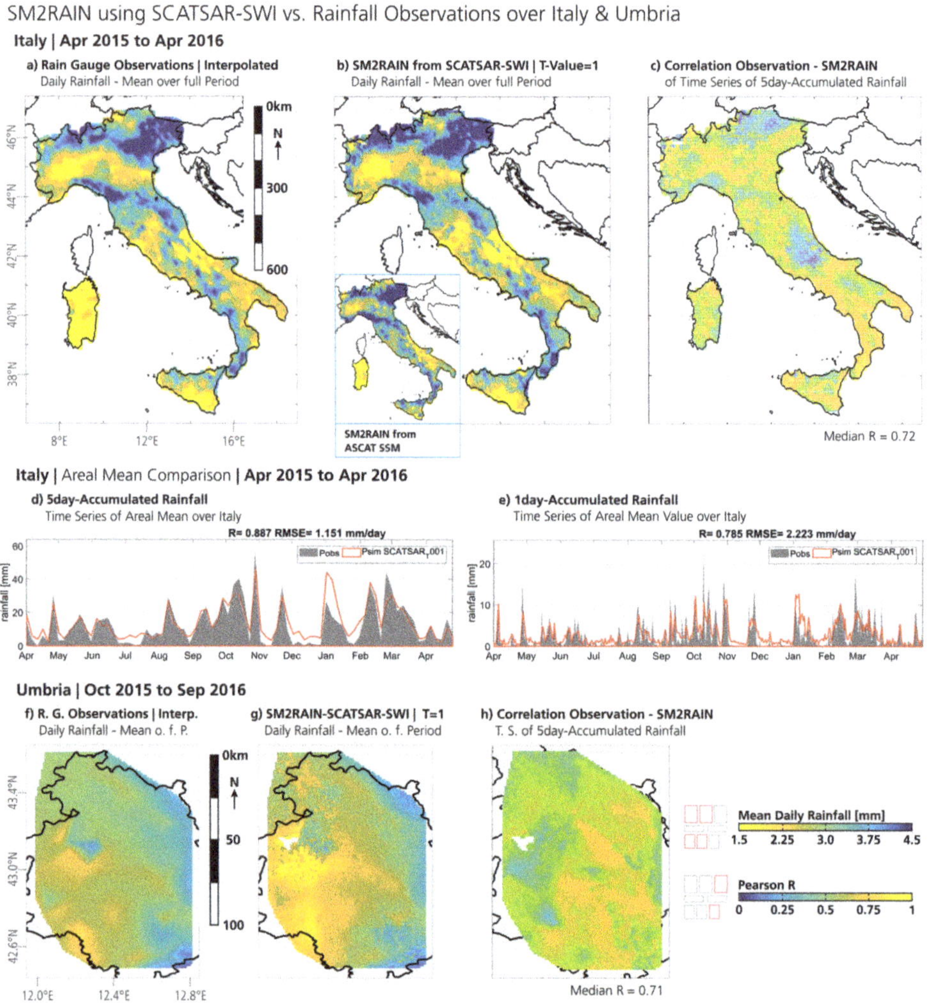

Figure 8. SM2RAIN estimates using SCATSAR-SWI data as input, compared to observed rainfall. (**a**) observed daily mean rainfall over Italy, averaged over April 2015–April 2016; (**b**) as (**a**) but from SM2RAIN using SCATSAR-SWI with *T*-value = 1, and with mini-map showing the result from SM2RAIN using ASCAT SSM as input; (**c**) correlation between 5 day-accumulated rainfall from observations and SM2RAIN-SCATSAR-SWI; (**d**) time series of the areal mean over Italy of five-day-accumulated rainfall from observations and SM2RAIN; (**e**) as (**d**) but for one-day-accumulated rainfall; (**f–h**) as (**a–c**) but for Umbria during the period October 2015–September 2016.

The second experiment over Umbria (with rainfall data at a 1 km sampling) during the period October 2015 to September 2016 shows very similar results as over Italy, but allows a closer look on the performance of the SCATSAR-SWI-SM2RAIN data (Figure 8f–h). Here as well, the two average rainfall maps are very similar, but with more noise-like artefacts and also more structure in the SM2RAIN map.

Interestingly, with regard to the land cover (as in Figure 3b), the highest correlations are detected over agricultural areas of Umbria (Figure 8h; with median R = 0.71 over the full domain), reproducing the already known patterns from the comparisons of S-1 SSM with land cover data. This suggests that the inclusion of S-1 SSM adds valuable information to the SCATSAR-SWI signal, enhancing the skill to detect rainfall-induced soil moisture changes where dense vegetation is absent. Comparisons with SM2RAIN from ASCAT SSM were done as above (not shown), and, as the results are very similar, confirm the findings above.

6. Conclusions

In this paper, we demonstrate how temporal filtering can bridge the scale gap in remote sensing of soil moisture. Absorbing the spadework on the SWI, our approach on the data fusion of SSM retrievals from scatterometer- and SAR- sensors allows the generation of a daily SM product at a kilometric scale, effectively combining the high temporal frequency of the coarse SCAT input with the fine spatial detail of the SAR input. By virtue of the temporal stability concept and the physical affinity of the (well-established) Metop ASCAT and the (novel) S-1 SAR sensors, the fusion of the SSM signals was proven not only possible but also successful in terms of preserving and mutually enhancing the input signals. Both input products contribute their individual assets to the SCATSAR-SWI as the evaluation yielded that (1) it can reproduce the temporal dynamics of in situ- and model- reference SM data at the same level as well-proven ASCAT SSM, and that (2) it supports the resolution of SM signals at the scale of land cover and soil characteristics and is hence much closer to the comprehension of typical users. Moreover, common caveats of the coarse-scaled SCAT SSM data like biases along coasts or around cities are pleasantly rectified through the data matching onto the SAR SSM patterns.

The SCATSAR-SWI method is at the forefront of current research on remote sensing of soil moisture. As built on an apt, multi-scale data cube that supports parallelisation, the implemented algorithm is capable of daily NRT-processing. It is, however, designed in a robust but plain fashion, also to account for the limited maturity of the SAR component, which is built on a data archive of a rather short baseline of currently only three years. The algorithm is yet designed in a way to be extensible for foreseen improvements. Currently, simple static weights for SCAT and SAR SSM inputs are used since the error model for the S-1 SSM is not fully developed. Once reliable error estimates—or as a more elaborate quality measure—Signal-to-Noise-Ratios (SNR)—are available, dynamic local weighting functions can be applied with the methodology presented here. This would facilitate a stronger and beneficial impact of the SAR signal over areas where it performs well, e.g., over agriculture, pasture, or in the vicinity of urban areas, which would be of particular interest for many users.

For now, in the present setup, the SCATSAR-SWI temporal dynamics are heavily relying on the SCAT component, simply due to the much larger observation number, but also accounting for the higher maturity of ASCAT. Consequently, impacts on SM from small-scaled events like local rainfall or irrigation activities that are potentially captured by the SAR component (e.g., as showed by [31]) are contained in the product only in an attenuated fashion. Nevertheless, the experiment on estimating rainfall from the SCATSAR-SWI data using the SM2RAIN approach yielded very satisfying results, raising high prospects on future realisations of the SCATSAR-SWI towards estimating precipitation.

A further boost in the SCATSAR-SWI's quality can be expected from future improvements in the downstream S-1 SSM algorithm, aiming for the mitigation of its deficits in modelling SM over areas with dense vegetation. Plans include the integration of a dynamic vegetation correction to model the contribution of plant growth and vigour, and a masking for frozen- and snow-conditions. Both are currently realised in the SCATSAR-SWI via the H-SAF's ASCAT SSM product, but only at its 12.5 grid, mingling small-scaled processes and leading to block-shaped SSF-flaggings in the fused product.

The Copernicus Global Land Service (CGLS, [86]) of the European Commission disseminates soil moisture products operationally and freely together with other bio-geophysical variables to enable the monitoring of the global vegetation, water, and energy budget. In recognition of the findings presented here, our method for generating the 1 km SCATSAR-SWI is currently in preparation towards operational product dissemination in the CGLS, featuring data in an initial phase over Europe, and subsequently on a global scale.

Author Contributions: B.B.-M. and W.W. conceived and designed the research on the algorithm. B.B.-M. performed the production, analysis, and interpretation of the data and wrote the article. C.P., T.M., S.H., and B.B.-M. designed and implemented the algorithm. L.B., C.M., S.M., and B.B.M. designed the evaluation experiments. S.M., C.M. L.C., and L.B. collected the reference data and carried out the evaluation experiments.

Funding: This research received funding from the Copernicus Global Land Service under the Framework Service Contract N°199494 (JRC).

Acknowledgments: We would like to thank our colleagues from the TU Wien who supported us with the Sentinel-1 data management and processing, most notably Stefano Elefante, Le Tuan, Vahid Freeman, and Senmao Cao. We also acknowledge the H-SAF services for provding the ASCAT data, most notably Sebastian Hahn, the Italian Civil Protection Department for providing the data from the national meteorological observation networks, and the CIMA Research Foundation for the elaboration of gauge-based rainfall data.

Conflicts of Interest: The authors declare no conflicts of interest. The funding sponsors had no role in the design of the study; in the collection, analyses, or interpretation of data; in the writing of the manuscript, or in the decision to publish the results.

References

1. Taylor, C.M.; de Jeu, R.A.M.; Guichard, F.; Harris, P.P.; Dorigo, W.A. Afternoon rain more likely over drier soils. *Nature* **2012**, *489*, 423–426. [CrossRef] [PubMed]

2. Legates, D.R.; Mahmood, R.; Levia, D.F.; DeLiberty, T.L.; Quiring, S.M.; Houser, C.; Nelson, F.E. Soil moisture: A central and unifying theme in physical geography. *Prog. Phys. Geogr.* **2011**, *35*, 65–86. [CrossRef]

3. Seneviratne, S.I.; Wilhelm, M.; Stanelle, T.; Hurk, B.; Hagemann, S.; Berg, A.; Cheruy, F.; Higgins, M.E.; Mcier, A.; Brovkin, V.; et al. Impact of soil moisture-climate feedback on CMIP5 projections: First results from the GLACE-CMIP5 experiment. *Geophys. Res. Lett.* **2013**, *40*, 5212–5217. [CrossRef]

4. Seneviratne, S.I.; Corti, T.; Davin, E.L.; Hirschi, M.; Jaeger, E.B.; Lehner, I.; Orlowsky, B.; Teuling, A.J. Investigating soil moisture—Climate interactions in a changing climate: A review. *Earth-Sci. Rev.* **2010**, *99*, 125–161. [CrossRef]

5. Koster, R.D.; Dirmeyer, P.A.; Guo, Z.; Bonan, G.; Chan, E.; Cox, P.; Gordon, C.T.; Kanae, S.; Kowalczyk, E.; Lawrence, D.; et al. Regions of strong coupling between soil moisture and precipitation. *Science* **2004**, *305*, 1138–1140. [CrossRef] [PubMed]

6. Bisselink, B.; Van Meijgaard, E.; Dolman, A.; De Jeu, R. Initializing a regional climate model with satellite-derived soil moisture. *J. Geophys. Res. Atmos.* **2011**, *116*. [CrossRef]

7. Albergel, C.; Balsamo, G.; de Rosnay, P.; Muñoz Sabater, J.; Boussetta, S. A bare ground evaporation revision in the ECMWF land-surface scheme: Evaluation of its impact using ground soil moisture and satellite microwave data. *Hydrol. Earth Syst. Sci. Discuss.* **2012**, *9*, 6715–6752. [CrossRef]

8. Brocca, L.; Ciabatta, L.; Massari, C.; Moramarco, T.; Hahn, S.; Hasenauer, S.; Kidd, R.; Dorigo, W.; Wagner, W.; Levizzani, V. Soil as a natural rain gauge: Estimating global rainfall from satellite soil moisture data. *J. Geophys. Res. Atmos.* **2014**, *119*, 5128–5141. [CrossRef]

9. Ciabatta, L.; Brocca, L.; Massari, C.; Moramarco, T.; Gabellani, S.; Puca, S.; Wagner, W. Rainfall-runoff modelling by using SM2RAIN-derived and state-of-the-art satellite rainfall products over Italy. *Int. J. Appl. Earth Obs. Geoinf.* **2016**, *48*, 163–173. [CrossRef]

10. Tramblay, Y.; Bouaicha, R.; Brocca, L.; Dorigo, W.; Bouvier, C.; Camici, S.; Servat, E. Estimation of antecedent wetness conditions for flood modelling in northern Morocco. *Hydrol. Earth Syst. Sci.* **2012**, *16*, 4375. [CrossRef]

11. Wanders, N.; Karssenberg, D.; Roo, A.D.; De Jong, S.; Bierkens, M. The suitability of remotely sensed soil moisture for improving operational flood forecasting. *Hydrol. Earth Syst. Sci.* **2014**, *18*, 2343–2357. [CrossRef]

12. Laiolo, P.; Gabellani, S.; Campo, L.; Silvestro, F.; Delogu, F.; Rudari, R.; Pulvirenti, L.; Boni, G.; Fascetti, F.; Pierdicca, N.; et al. Impact of different satellite soil moisture products on the predictions of a continuous distributed hydrological model. *Int. J. Appl. Earth Obs. Geoinf.* **2016**, *48*, 131–145. [CrossRef]

13. Sutanudjaja, E.; Van Beek, L.; De Jong, S.; Van Geer, F.; Bierkens, M. Calibrating a large-extent high-resolution coupled groundwater-land surface model using soil moisture and discharge data. *Water Resour. Res.* **2014**, *50*, 687–705. [CrossRef]

14. Qiu, J.; Gao, Q.; Wang, S.; Su, Z. Comparison of temporal trends from multiple soil moisture data sets and precipitation: The implication of irrigation on regional soil moisture trend. *Int. J. Appl. Earth Obs. Geoinf.* **2016**, *48*, 17–27. [CrossRef]

15. Miralles, D.G.; Teuling, A.J.; Van Heerwaarden, C.C.; Vilà-guerau De Arellano, J. Mega-heatwave temperatures due to combined soil desiccation and atmospheric heat accumulation. *Nat. Geosci.* **2014**, *7*, 345–349. [CrossRef]

16. Carrão, H.; Russo, S.; Sepulcre-Canto, G.; Barbosa, P. An empirical standardized soil moisture index for agricultural drought assessment from remotely sensed data. *Int. J. Appl. Earth Obs. Geoinf.* **2016**, *48*, 74–84. [CrossRef]

17. Bauer-Marschallinger, B.; Dorigo, W.A.; Wagner, W.; van Dijk, A.I. How oceanic oscillation drives soil moisture variations over mainland Australia: An analysis of 32 years of satellite observations. *J. Clim.* **2013**, *26*, 10159–10173. [CrossRef]

18. Dorigo, W.; de Jeu, R.; Chung, D.; Parinussa, R.; Liu, Y.; Wagner, W.; Fernández-Prieto, D. Evaluating global trends (1988–2010) in harmonized multi-satellite surface soil moisture. *Geophys. Res. Lett.* **2012**, *39*, L18405. [CrossRef]

19. Blunden, J.; Arndt, D.S. State of the climate in 2011. *Bull. Am. Meteorol. Soc.* **2012**, *93*, S1–S282. [CrossRef]

20. Walker, J.P.; Willgoose, G.R.; Kalma, J.D. In situ measurement of soil moisture: A comparison of techniques. *J. Hydrol.* **2004**, *293*, 85–99. [CrossRef]

21. De Jeu, R.; Dorigo, W. On the importance of satellite observed soil moisture. *Int. J. Appl. Earth Obs. Geoinf.* **2016**, *45*, 107–109. [CrossRef]

22. Wagner, W.; Hahn, S.; Kidd, R.; Melzer, T.; Bartalis, Z.; Hasenauer, S.; Figa-Saldaña, J.; de Rosnay, P.; Jann, A.; Schneider, S.; et al. The ASCAT soil moisture product: A review of its specifications, validation results, and emerging applications. *Meteorol. Z.* **2013**, *22*, 5–33. [CrossRef]

23. Li, L.; Gaiser, P.W.; Gao, B.C.; Bevilacqua, R.M.; Jackson, T.J.; Njoku, E.G.; Rudiger, C.; Calvet, J.C.; Bindlish, R. WindSat global soil moisture retrieval and validation. *IEEE Trans. Geosci. Remote Sens.* **2010**, *48*, 2224–2241. [CrossRef]

24. Kerr, Y.H.; Waldteufel, P.; Richaume, P.; Wigneron, J.P.; Ferrazzoli, P.; Mahmoodi, A.; Al Bitar, A.; Cabot, F.; Gruhier, C.; Juglea, S.E.; et al. The SMOS soil moisture retrieval algorithm. *IEEE Trans. Geosci. Remote Sens.* **2012**, *50*, 1384–1403. [CrossRef]

25. Chan, S.K.; Bindlish, R.; O'Neill, P.E.; Njoku, E.; Jackson, T.; Colliander, A.; Chen, F.; Burgin, M.; Dunbar, S.; Piepmeier, J.; et al. Assessment of the SMAP passive soil moisture product. *IEEE Trans. Geosci. Remote Sens.* **2016**, *54*, 4994–5007. [CrossRef]

26. Miyaoka, K.; Gruber, A.; Ticconi, F.; Hahn, S.; Wagner, W.; Figa-Saldaña, J.; Anderson, C. Triple collocation analysis of soil moisture from Metop-A ASCAT and SMOS against JRA-55 and ERA-Interim. *IEEE J. Sel. Top. Appl. Earth Obs. Remote Sens.* **2017**, *10*, 2274–2284. [CrossRef]

27. Burgin, M.S.; Colliander, A.; Njoku, E.G.; Chan, S.K.; Cabot, F.; Kerr, Y.H.; Bindlish, R.; Jackson, T.J.; Entekhabi, D.; Yueh, S.H. A comparative study of the SMAP passive soil moisture product with existing satellite-based soil moisture products. *IEEE Trans. Geosci. Remote Sens.* **2017**, *55*, 2959–2971. [CrossRef]

28. Woodhouse, I. *Introduction to Microwave Remote Sensing*; CRC Press: Boca Raton, FL, USA, 2006.

29. Pathe, C.; Wagner, W. Using ENVISAT ASAR global mode data for surface soil moisture retrieval over Oklahoma, USA. *IEEE Trans. Geosci. Remote Sens.* **2009**, *47*, 468–480. [CrossRef]

30. Dostálová, A.; Doubková, M.; Sabel, D.; Bauer-Marschallinger, B.; Wagner, W. Seven Years of Advanced Synthetic Aperture Radar (ASAR) Global Monitoring (GM) of Surface Soil Moisture over Africa. *Remote Sens.* **2014**, *6*, 7683–7707. [CrossRef]

31. Bauer-Marschallinger, B.; Naeimi, V.; Cao, S.; Paulik, C.; Schaufler, S.; Stachl, T.; Modanesi, S.; Ciabatta, L.; Brocca, L.; Wagner, W. Towards Global Soil Moisture Monitoring with Sentinel-1: Harnessing Assets and Overcoming Obstacles. *IEEE Trans. Geosci. Remote Sens.* in review.

32. Kaheil, Y.H.; Gill, M.K.; McKee, M.; Bastidas, L.A.; Rosero, E. Downscaling and assimilation of surface soil moisture using ground truth measurements. *IEEE Trans. Geosci. Remote Sens.* **2008**, *46*, 1375–1384. [CrossRef]

33. Verhoest, N.E.; van den Berg, M.J.; Martens, B.; Lievens, H.; Wood, E.F.; Pan, M.; Kerr, Y.H.; Al Bitar, A.; Tomer, S.K.; Drusch, M.; et al. Copula-based downscaling of coarse-scale soil moisture observations with implicit bias correction. *IEEE Trans. Geosci. Remote Sens.* **2015**, *53*, 3507–3521. [CrossRef]

34. Tomer, S.K.; Al Bitar, A.; Sekhar, M.; Zribi, M.; Bandyopadhyay, S.; Sreelash, K.; Sharma, A.; Corgne, S.; Kerr, Y. Retrieval and multi-scale validation of soil moisture from multi-temporal SAR data in a semi-arid tropical region. *Remote Sens.* **2015**, *7*, 8128–8153. [CrossRef]

35. Qu, W.; Bogena, H.; Huisman, J.A.; Vanderborght, J.; Schuh, M.; Priesack, E.; Vereecken, H. Predicting subgrid variability of soil water content from basic soil information. *Geophys. Res. Lett.* **2015**, *42*, 789–796. [CrossRef]

36. Gevaert, A.; Parinussa, R.M.; Renzullo, L.J.; van Dijk, A.I.; de Jeu, R.A. Spatio-temporal evaluation of resolution enhancement for passive microwave soil moisture and vegetation optical depth. *Int. J. Appl. Earth Obs. Geoinf.* **2016**, *45*, 235–244. [CrossRef]

37. Das, N.N.; Entekhabi, D.; Kim, S.B.; Jagdhuber, T.; Dunbar, S.; Yueh, S.; Colliander, A.; Lopez-baeza, E.; Martinez-Fernandez, J. High-Resolution Enhanced Product based on SMAP Active-Passive Approach using Sentinel-1A and 1B SAR Data. In Proceedings of the IEEE International Geoscience and Remote Sensing Symposium (IGARSS), Worth, TX, USA, 23–28 July 2017.

38. Peng, J.; Loew, A.; Merlin, O.; Verhoest, N.E. A review of spatial downscaling of satellite remotely sensed soil moisture. *Rev. Geophys.* **2017**, *55*, 341–366. [CrossRef]

39. Wagner, W.; Pathe, C.; Doubkova, M.; Sabel, D.; Bartsch, A.; Hasenauer, S.; Blöschl, G.; Scipal, K.; Martínez-Fernández, J.; Löw, A. Temporal Stability of Soil Moisture and Radar Backscatter Observed by the Advanced Synthetic Aperture Radar (ASAR). *Sensors* **2008**, *8*, 1174–1197. [CrossRef] [PubMed]

40. Wagner, W.; Lemoine, G.; Rott, H. A method for estimating soil moisture from ERS scatterometer and soil data. *Remote Sens. Environ.* **1999**, *70*, 191–207. [CrossRef]

41. Paulik, C.; Dorigo, W.; Wagner, W.; Kidd, R. Validation of the ASCAT Soil Water Index using in situ data from the International Soil Moisture Network. *Int. J. Appl. Earth Obs. Geoinf.* **2014**, *30*, 1–8. [CrossRef]

42. Wagner, W.; Noll, J.; Borgeaud, M.; Rott, H. Monitoring soil moisture over the Canadian prairies with the ERS scatterometer. *IEEE Trans. Geosci. Remote Sens.* **1999**, *37*, 206–216. [CrossRef]

43. Scipal, K.; Drusch, M.; Wagner, W. Assimilation of a ERS scatterometer derived soil moisture index in the ECMWF numerical weather prediction system. *Adv. Water Resour.* **2008**, *31*, 1101–1112. [CrossRef]

44. Naeimi, V.; Bartalis, Z.; Wagner, W. ASCAT Soil Moisture: An Assessment of the Data Quality and Consistency with the ERS Scatterometer Heritage. *J. Hydrometeorol.* **2009**, *10*, 555–563. [CrossRef]

45. Brocca, L.; Melone, F.; Moramarco, T.; Wagner, W.; Hasenauer, S. ASCAT soil wetness index validation through in situ and modeled soil moisture data in central Italy. *Remote Sens. Environ.* **2010**, *114*, 2745–2755. [CrossRef]

46. Schneider, S.; Wang, Y.; Wagner, W.; Mahfouf, J.F. Impact of ASCAT soil moisture assimilation on regional precipitation forecasts: A case study for Austria. *Mon. Weather Rev.* **2014**, *142*, 1525–1541. [CrossRef]

47. Doubková, M.; Van Dijk, A.I.; Sabel, D.; Wagner, W.; Blöschl, G. Evaluation of the predicted error of the soil moisture retrieval from C-band SAR by comparison against modelled soil moisture estimates over Australia. *Remote Sens. Environ.* **2012**, *120*, 188–196. [CrossRef] [PubMed]

48. Naeimi, V. Model Improvements and Error Characterization for Global ERS and METOP Scatterometer Soil Moisture Data. Ph.D. Thesis, TU Wien, Vienna, Austria, 2009.

49. Bartalis, Z.; Wagner, W.; Naeimi, V.; Hasenauer, S.; Scipal, K.; Bonekamp, H.; Figa, J.; Anderson, C. Initial soil moisture retrievals from the METOP-A Advanced Scatterometer (ASCAT). *Geophys. Res. Lett.* **2007**, *34*, 5–9. [CrossRef]

50. Brocca, L.; Hasenauer, S.; Lacava, T.; Melone, F.; Moramarco, T.; Wagner, W.; Dorigo, W.; Matgen, P.; Martínez-Fernández, J.; Llorens, P.; et al. Soil moisture estimation through ASCAT and AMSR-E sensors: An intercomparison and validation study across Europe. *Remote Sens. Environ.* **2011**, *115*, 3390–3408. [CrossRef]

51. Albergel, C.; de Rosnay, P.; Gruhier, C.; Muñoz Sabater, J.; Hasenauer, S.; Isaksen, L.; Kerr, Y.; Wagner, W. Evaluation of remotely sensed and modelled soil moisture products using global ground-based in situ observations. *Remote Sens. Environ.* **2012**, *118*, 215–226. [CrossRef]

52. Vreugdenhil, M.; Dorigo, W.A.; Wagner, W.; de Jeu, R.A.; Hahn, S.; van Marle, M.J. Analyzing the vegetation parameterization in the TU-Wien ASCAT soil moisture retrieval. *IEEE Trans. Geosci. Remote Sens.* **2016**, *54*, 3513–3531. [CrossRef]

53. Hahn, S.; Reimer, C.; Vreugdenhil, M.; Melzer, T.; Wagner, W. Dynamic Characterization of the Incidence Angle Dependence of Backscatter Using Metop ASCAT. *IEEE J. Sel. Top. Appl. Earth Obs. Remote Sens.* **2017**, *10*, 2348–2359. [CrossRef]

54. Eumetsat H-SAF Service Website. Available online: http://hsaf.meteoam.it (accessed on 20 February 2018).

55. Naeimi, V.; Paulik, C.; Bartsch, A.; Wagner, W.; Kidd, R.; Park, S.E.; Elger, K.; Boike, J. ASCAT Surface State Flag (SSF): Extracting information on surface freeze/thaw conditions from backscatter data using an empirical threshold-analysis algorithm. *IEEE Trans. Geosci. Remote Sens.* **2012**, *50*, 2566–2582. [CrossRef]

56. Torres, R.; Snoeij, P.; Geudtner, D.; Bibby, D.; Davidson, M.; Attema, E.; Potin, P.; Rommen, B.; Floury, N.; Brown, M.; et al. GMES Sentinel-1 mission. *Remote Sens. Environ.* **2012**, *120*, 9–24. [CrossRef]

57. The Remote Sensing Research Group of the Department of Geodesy and Geoinformation at the TU Wien. Available online: http://rs.geo.tuwien.ac.at (accessed on 20 February 2018).

58. Bauer-Marschallinger, B.; Cao, S.; Schaufler, S.; Paulik, C.; Naeimi, V.; Wagner, W. 1 km Soil Moisture from Downsampled Sentinel-1 SAR Data: Harnessing Assets and Overcoming Obstacles. *Geophys. Res. Abstr.* **2017**, *19*, 17330.

59. Naeimi, V.; Elefante, S.; Cao, S.; Wagner, W.; Dostalova, A.; Bauer-Marschallinger, B. Geophysical parameters retrieval from Sentinel-1 SAR data: A case study for high performance computing at EODC. In Proceedings of the 24th High Performance Computing Symposium, Pasadena, CA, USA, 3–6 April 2016; p. 10.

60. Hornacek, M.; Wagner, W.; Sabel, D.; Truong, H.L.; Snoeij, P.; Hahmann, T.; Diedrich, E.; Doubková, M. Potential for high resolution systematic global surface soil moisture retrieval via change detection using Sentinel-1. *IEEE J. Sel. Top. Appl. Earth Obs. Remote Sens.* **2012**, *5*, 1303–1311. [CrossRef]

61. Ali, I.; Naeimi, V.; Cao, S.; Elefante, S.; Bauer-Marschallinger, B.; Wagner, W. Sentinel-1 data cube exploitation: Tools, products, services and quality control. In Proceedings of the 2017 Conference on Big Data Space, Toulouse, France, 28–30 November 2017; pp. 40–43.

62. Bauer-Marschallinger, B.; Sabel, D.; Wagner, W. Optimisation of global grids for high-resolution remote sensing data. *Comput. Geosci.* **2014**, *72*, 84–93. [CrossRef]

63. Hengl, T.; de Jesus, J.M.; Heuvelink, G.B.; Gonzalez, M.R.; Kilibarda, M.; Blagotić, A.; Shangguan, W.; Wright, M.N.; Geng, X.; Bauer-Marschallinger, B.; et al. SoilGrids250m: Global gridded soil information based on machine learning. *PLoS ONE* **2017**, *12*, e0169748. [CrossRef] [PubMed]

64. The Equi7Grid definition and python tool package. Available online: https://github.com/TUW-GEO/Equi7Grid (accessed on 20 June 2018).

65. The scipy package. Available online: https://docs.scipy.org/doc/ (accessed on 20 March 2018).

66. Su, C.H.; Narsey, S.Y.; Gruber, A.; Xaver, A.; Chung, D.; Ryu, D.; Wagner, W. Evaluation of post-retrieval de-noising of active and passive microwave satellite soil moisture. *Remote Sens. Environ.* **2015**, *163*, 127–139. [CrossRef]

67. Reichle, R.H. Bias reduction in short records of satellite soil moisture. *Geophys. Res. Lett.* **2004**, *31*. [CrossRef]

68. Albergel, C.; Rüdiger, C.; Pellarin, T.; Calvet, J.C.; Fritz, N.; Froissard, F.; Suquia, D.; Petitpa, A.; Piguet, B.; Martin, E. From near-surface to root-zone soil moisture using an exponential filter: An assessment of the method based on in situ observations and model simulations. *Hydrol. Earth Syst. Sci. Discuss.* **2008**, *12*, 1323–1337. [CrossRef]

69. Gruber, A.; Crow, W.; Dorigo, W. Assimilation of spatially sparse in situ soil moisture networks into a continuous model domain. *Water Resour. Res.* **2018**, *54*, 1353–1367. [CrossRef]

70. Dirmeyer, P.A.; Wu, J.; Norton, H.E.; Dorigo, W.A.; Quiring, S.M.; Ford, T.W.; Santanello, J.A., Jr.; Bosilovich, M.G.; Ek, M.B.; Koster, R.D.; et al. Confronting weather and climate models with observational data from soil moisture networks over the United States. *J. Hydrometeorol.* **2016**, *17*, 1049–1067. [CrossRef] [PubMed]

71. Brocca, L.; Melone, F.; Moramarco, T. On the estimation of antecedent wetness conditions in rainfall–runoff modelling. *Hydrol. Process.* **2008**, *22*, 629–642. [CrossRef]

72. Büttner, G.; Soukup, T.; Kosztra, B. *CLC2012 Addendum to CLC2006 Technical Guidelines*; Final Draft; EEA: Copenhagen, Denmark, 2014.

73. Brocca, L.; Ciabatta, L.; Moramarco, T.; Ponziani, F.; Berni, N.; Wagner, W. Use of satellite soil moisture products for the operational mitigation of landslides risk in central Italy. In *Satellite Soil Moisture Retrieval*; Elsevier: New York, NY, USA, 2016; pp. 231–247.

74. Brocca, L.; Camici, S.; Melone, F.; Moramarco, T.; Martínez-Fernández, J.; Didon-Lescot, J.F.; Morbidelli, R. Improving the representation of soil moisture by using a semi-analytical infiltration model. *Hydrol. Process.* **2014**, *28*, 2103–2115. [CrossRef]

75. Brocca, L.; Melone, F.; Moramarco, T.; Wagner, W.; Naeimi, V.; Bartalis, Z.; Hasenauer, S. Improving runoff prediction through the assimilation of the ASCAT soil moisture product. *Hydrol. Earth Syst. Sci.* **2010**, *14*, 1881–1893. [CrossRef]

76. Lacava, T.; Matgen, P.; Brocca, L.; Bittelli, M.; Pergola, N.; Moramarco, T.; Tramutoli, V. A first assessment of the SMOS soil moisture product with in situ and modeled data in Italy and Luxembourg. *IEEE Trans. Geosci. Remote Sens.* **2012**, *50*, 1612–1622. [CrossRef]

77. Dorigo, W.A.; Wagner, W.; Hohensinn, R.; Hahn, S.; Paulik, C.; Xaver, A.; Gruber, A.; Drusch, M.; Mecklenburg, S.; van Oevelen, P.; et al. The International Soil Moisture Network: A data hosting facility for global in situ soil moisture measurements. *Hydrol. Earth Syst. Sci.* **2011**, *15*, 1675–1698. [CrossRef]

78. Dorigo, W.; Gruber, A.; De Jeu, R.; Wagner, W.; Stacke, T.; Loew, A.; Albergel, C.; Brocca, L.; Chung, D.; Parinussa, R.; et al. Evaluation of the ESA CCI soil moisture product using ground-based observations. *Remote Sens. Environ.* **2015**, *162*, 380–395. [CrossRef]

79. Wu, Q.; Liu, H.; Wang, L.; Deng, C. Evaluation of AMSR2 soil moisture products over the contiguous United States using in situ data from the International Soil Moisture Network. *Int. J. Appl. Earth Obs. Geoinf.* **2016**, *45*, 187–199. [CrossRef]

80. Zreda, M.; Shuttleworth, W.; Zeng, X.; Zweck, C.; Desilets, D.; Franz, T.; Rosolem, R. COSMOS: The cosmic-ray soil moisture observing system. *Hydrol. Earth Syst. Sci.* **2012**, *16*, 4079–4099. [CrossRef]

81. The Italian Civil Protection Department. Available online: http://www.protezionecivile.gov.it/jcms/en/home.wp (accessed on 20 March 2018).

82. Pignone, F.; Rebora, N.; Silvestro, F.; Castelli, F. GRISO—Rain. In *Operational Agreement 778/2009 DPC-CIMA, Year-1 Activity Report 272/2010*; CIMA Research Foundation: Savona, Italy, 2010.

83. Brocca, L.; Crow, W.T.; Ciabatta, L.; Massari, C.; de Rosnay, P.; Enenkel, M.; Hahn, S.; Amarnath, G.; Camici, S.; Tarpanelli, A.; et al. A review of the applications of ASCAT soil moisture products. *IEEE J. Sel. Top. Appl. Earth Obs. Remote Sens.* **2017**, *10*, 2285–2306. [CrossRef]

84. Ciabatta, L.; Massari, C.; Brocca, L.; Gruber, A.; Reimer, C.; Hahn, S.; Paulik, C.; Dorigo, W.; Kidd, R.; Wagner, W. SM2RAIN-CCI: A new global long-term rainfall data set derived from ESA CCI soil moisture. *Earth Syst. Sci. Data* **2018**, *10*, 267–280. [CrossRef]

85. Massari, C.; Brocca, L.; Tarpanelli, A.; Moramarco, T. Data assimilation of satellite soil moisture into rainfall-runoff modelling: A complex recipe? *Remote Sens.* **2015**, *7*, 11403–11433. [CrossRef]

86. The Copernicus Global Land Service. Available online: https://land.copernicus.eu/global/ (accessed on 20 June 2018).

 *remote sensing*

Article

Using SAR-Derived Vegetation Descriptors in a Water Cloud Model to Improve Soil Moisture Retrieval

Junhua Li * and Shusen Wang

Canada Centre for Mapping and Earth Observation, Natural Resources Canada, Ottawa, ON K1A 0E4, Canada; Shusen.Wang@Canada.ca

* Correspondence: Junhua.Li@Canada.ca; Tel.: +1-613-759-7205

Received: 26 June 2018; Accepted: 24 August 2018; Published: 29 August 2018

Abstract: The water cloud model (WCM) is a widely used radar backscatter model applied to SAR images to retrieve soil moisture over vegetated areas. The WCM needs vegetation descriptors to account for the impact of vegetation on SAR backscatter. The commonly used vegetation descriptors in WCM, such as Leaf Area Index (LAI) and Normalized Difference Vegetation Index (NDVI), are sometimes difficult to obtain due to the constraints in data availability in in-situ measurements or weather dependency in optical remote sensing. To improve soil moisture retrieval, this study investigates the feasibility of using all-weather SAR derived vegetation descriptors in WCM. The in-situ data observed at an agricultural crop region south of Winnipeg in Canada, RapidEye optical images and dual-polarized Radarsat-2 SAR images acquired in growing season were used for WCM model calibration and test. Vegetation descriptors studied include HV polarization backscattering coefficient (σ°_{HV}) and Radar Vegetation Index (RVI) derived from SAR imagery, and NDVI derived from optical imagery. The results show that σ°_{HV} achieved similar results as NDVI but slightly better than RVI, with a root mean square error of 0.069 m^3/m^3 and a correlation coefficient of 0.59 between the retrieved and observed soil moisture. The use of σ°_{HV} can overcome the constraints of the commonly used vegetation descriptors and reduce additional data requirements (e.g., NDVI from optical sensors) in WCM, thus improving soil moisture retrieval and making WCM feasible for operational use.

Keywords: soil moisture; Radarsat-2; SAR; water-cloud model; vegetation descriptor

---

1. Introduction

Soil moisture plays a key role in the terrestrial water cycle. The retrieval of soil moisture over a large area is important in the modeling and assessing drought impact [1], evapotranspiration [2], and water budget [3,4]. With high temporal and spatial variations, soil moisture data over large areas is difficult to obtain from in situ networks. Radar has a high backscattering sensitivity to soil moisture due to the high contrast of the microwave dielectric constant (ε) between dry soil (ε = 2~3) and water (ε = 80) [5]. It also has the advantage of observing the earth's surface day and night in all weather conditions. Therefore, radar remote sensing has the potential to measure soil moisture on a large scale at regular temporal intervals from space [6–10]. Over the past 30 years, considerable effort has been spent on using Synthetic Aperture Radar (SAR) imagery to retrieve soil moisture.

SAR incidence angle and polarization are important factors that affect soil moisture retrieval [5,8,11–14]. Radar backscatters normally decrease with increasing incidence angle [15,16]. The rate of decrease depends mostly on roughness conditions and land cover. Previous studies showed that SAR at low incidence angle is less sensitive to surface roughness [5,11–13,17] and vegetation [18,19] than at high incidence angle. Low incidence angles are thus optimal for soil moisture retrieval. The choice of polarizations also plays an important role in SAR-based soil moisture retrieval.

Single-polarization data is often used for soil moisture retrieval [5,11]. Several studies showed that HH polarization is more sensitive than HV to soil moisture but less sensitive than VV [13,20,21]. However some other studies disagreed. For example, Beaudoin et al. [22], Le Loan [23], and Baghdadi et al. [14] found that HH polarization is most relevant to soil moisture estimates. Moreover, the study by McNarin and Brisco [24] demonstrated that additional polarizations can provide more information content in a SAR dataset. For example, cross-polarized SAR images (HV or VH) are sensitive to crop structure within the total canopy volume and thus provide information that can be complementary to HH and VV imagery. Using multiple polarizations should, in theory, improve soil moisture estimation [25]. Improved soil moisture retrieval using multi-polarization SAR data was reported even under dense vegetation canopy [26,27]. It must be noted that Baghdadi et al. [28] found that the accuracy of the soil moisture estimation did not improve significantly (<0.01 cm^3/cm^3) for two bare soil sites when two polarizations (HH and HV) were used instead of only one polarization. There have been great interests to use full-polarized SAR data in soil moisture retrieval after the data is available [1,25,29] as it contains more information on the scattering objects than the single- and dual-polarization data. However, full-polarized SAR data has the disadvantage of narrow swath, which limits its applications in soil moisture retrieval over large areas. As such, dual-polarized SAR data is recommended for retrieving soil moisture over large areas because of its large swath width of up to 500 km (e.g., Radarsat-2 ScanSAR wide beam mode).

For the case without vegetation or with low vegetation cover, shorter wavelength SAR (especially X-band) is more sensitive to soil moisture and least sensitive to surface roughness than longer wavelength SAR [30–32]. For the presence of vegetation (especially dense vegetation), however, longer wavelength L-band, compared with the SAR X-band and C-band, is more suitable for soil moisture retrieval since it has stronger vegetation/soil penetration power and is less sensitive to vegetation canopy [9,33]. Unfortunately, currently, ALOS-2/PALSAR is the only long wavelength L-band SAR sensor currently in orbit. Since its data is limited, especially for regions outside Japan, the ALOS-2/PALSAR is significantly hampered for operational use in soil moisture retrieval. In contrast, Sentinel-1A, Sentinel-1B and Radarsat-2, which are C-band SAR satellites currently in orbit, can provide routine observations of Earth's surface over large areas. Radarsat Constellation Mission (RCM) with 4-days revisit interval will further expand current C-band SAR satellites' revisit capabilities (6-days for Sentinel-1A/B and 24-days for Radarsat-2). These capacities are promoting new scientific and operational perspectives, e.g., downscaling passive microwave soil moisture [34,35], in the retrieval of soil moisture continuously over large areas using C-band SAR imagery.

Soil moisture retrievals from C-band SAR are significantly affected by vegetation cover and surface roughness. A number of SAR backscatter models have been proposed to separate the backscattering contributions of soil and vegetation [6,14,36–39]. These models are generally categorised into three groups: theoretical, empirical, and semi-empirical. The theoretical models such as the Integral Equation Model (IEM) and the advanced IEM model [17,40,41] are complicated and require a large number of parameters. Baghdadi et al. [42–44] modified the IEM to reduce the IEM's input soil parameters from three to two [45]. On the other hand, the empirical models, e.g., Dubois model [46], are simple to develop but may have limitations in applicability for other sites due to their data and site dependency [6,47]. Recently Baghdadi et al. [31] improved the Dubois model for a reliable estimate of soil moisture. The semi-empirical models, e.g., Oh model [15] and Water Cloud Model (WCM) [48], start from a physical background and then use simulated or experimental datasets to simplify the theoretical backscattering models [8]. The WCM is often used in retrieving soil moisture and modelling of the scattering of vegetated areas for its simplicity [27]. In addition, the inversion techniques, including the Neural Network (NN) approach and the Change Detection (CD) method, are also used widely for soil moisture retrieval.

The neural network approach consists of a number of hidden neurons or nodes that work in parallel to convert data from an input vector to an output vector [9,48]. For soil moisture retrieval, the neural network is often trained using a synthetic database generated from SAR backscattering model such as IEM, Oh, and WCM models [39,49,50] or WCM model combined with the IEM model [39,51]. Based on the NN approach, Santi et al. [52] retrieved soil moisture from ENVISAT/ASAR data with an Root Mean Square Error (RMSE) as low as 0.023 m^3/m^3. Paloscia et al. [39] trained a neural network by using a synthetic database of backscattering coefficients simulated from WCM model combined with IEM model for a wide range of soil moisture, surface roughness, and vegetation index. The inputs to the neural networks were the SAR data and NDVI. The approach doesn't need roughness measurements. They achieved results with RMSE between 1.67 m^3/m^3 and 6.68 m^3/m^3 over several areas in Italy, Australia, and Spain using Sentinel-1 SAR data, which were very much in line with GMES requirements (with RMSE generally <5 m^3/m^3). The change detection method is based on the near linear relationship between SAR backscatter and soil moisture. It assumes the effect of vegetation and surface roughness on observed backscatter are consistent between acquisitions. The advantage of this method is that it can retrieve soil moisture in the absence of prior information of the study area when multi-temporal SAR data is available. Based on this method, Gao et al. [48] retrieved soil moisture over a site in Urgell (Catalunya, Spain) from multi-temporal Sentinel-1/SAR data, which combined with optical Sentinel-2 data, with an RMSE as low as 0.059 m^3/m^3. Zribi et al. [53] achieved a better result with a RMSE of about 0.035 m^3/m^3 in retrieving soil moisture over a semi-arid area using ASAR data. The Water Cloud Model (WCM), a simple semi-empirical backscatter model, can relate the backscattering coefficient (HH or VV polarizations) to soil properties (moisture and roughness) and vegetation properties (e.g., biomass, leaf area index) and thus it can be used to retrieve soil moisture from SAR imagery over densely vegetated areas [48]. Based on the WCM, Gherboudj et al. [27] retrieved soil moisture with an RMSE of 5.9% and 6.6% for two sites in crop fields from multi-polarized and multi-angular Radarsat-2 SAR data. Zribi et al. [54] achieved similar result (RMSE: ~0.06 m^3/m^3) in a semi-arid region from C-band ASAR data using the WCM. Kumar et al. [55] obtained a better soil moisture retrieval with an RMSE as low as 0.0419 m^3/m^3 from C-band ASAR data by the use of LAI in the WCM as the vegetation descriptor. The WCM is used in this study for its simplicity.

The performance of the WCM for soil moisture retrieval depends on the characterization of surface roughness and vegetation. Obtaining accurate information about surface roughness is difficult. First, the in-situ measurement of the surface roughness is quite challenging [56]. Second, the SAR-based surface roughness retrievals (e.g., depolarization ratio method) are mainly developed for bare soil and are less efficient in the presence of a vegetation canopy [27]. In this study, the effect of surface roughness is accounted for by using multi-temporal data over the same field, as surface roughness has little change over a short time period especially for crop fields during the growing time period [57].

Various vegetation descriptors such as plant height, leaf area index (LAI), vegetation water content, and normalized difference vegetation index (NDVI) have been used in WCM [14,27,38,55]. These vegetation descriptors are calculated using data either from in-situ measurements or from optical satellite remote sensing. The use of in-situ data is difficult in operational applications because of the high cost and time consuming in data collections, especially in remote areas. The use of optical satellite data is often limited by weather conditions such as cloud and haze.

Recently, some advances have been made toward developing SAR parameters for characterizing vegetation canopies. For example, the radar vegetation index (RVI), which is computed as a ratio of the cross-polarization scattering to the total scattering, has been used to estimate the biomass and the water content of a wheat crop [58]. Kumar et al. [59] showed RVI as a better alternative to NDVI for monitoring soybean and cotton. In other studies, the HV backscattering coefficient (σ°_{HV}) was found to be very sensitive to the vegetation biomass [60,61] and correlated with LAI [62,63]. Using SAR data to characterize vegetation in the WCM for soil moisture retrieval could be much more effective and beneficial than using in situ and optical sensor-based vegetation descriptors, but few studies are available.

The objective of this study was to assess the WCM performance in soil moisture retrieval by using SAR-derived vegetation descriptors. Specifically, the SAR-derived σ°_{HV} and RVI obtained from dual-polarized Radarsat-2 SAR imagery were tested using a simplified WCM. The NDVI derived from

RapidEye images was also used in the WCM for comparison. The goal of this study is to provide a practical tool for continuous mapping of soil moisture over large areas by future SAR missions such as RCM.

2. Methods

A simplified WCM was employed in this study to retrieve soil moisture. Assuming the effect of soil surface roughness on observed backscatter for a given site is consistent over a short period, in such a case, the temporal change in SAR backscattering only reflects the change of vegetation and soil moisture. Therefore the multi-temporal SAR data was used in the WCM for this study. Since the multi-temporal SAR imagery may be in different viewing geometries, an incidence angle normalization was conducted to make the SAR data radiometrically comparable. The RVI was then computed from these normalized SAR images. The accuracy of the retrieved soil moisture was evaluated by root mean square error (RMSE) and the Pearson correlation coefficient (R). The detailed WCM description and data processing methods are given below.

2.1. A Simplified Water-Cloud Model (WCM)

The water-cloud model, initially developed by Attema and Ulaby [36], considers the vegetation canopy as a cloud containing water droplets randomly distributed within the canopy. It provides solutions for the backscattering coefficients for the vegetation canopy as well as the underlying soil [23,28]. The WCM for a given co-polarization (pp) is formulated as:

$$\sigma^{\circ}{}_{pp} = \sigma^{\circ}{}_{veg} + \tau^2 \sigma^{\circ}{}_{soil}, \tag{1}$$

where

$$\sigma^{\circ}{}_{veg} = AV_1 cos\theta_i \left(1 - \tau^2\right), \tag{2}$$

$$\tau^2 = e^{-2BV_2 sec\theta_i}, \tag{3}$$

where $\sigma^{\circ}{}_{pp}$ (in power) is the observed canopy backscattering coefficient, which is represented as the sum of vegetation volume scattering $\sigma^{\circ}{}_{veg}$ and the bare soil scattering $\sigma^{\circ}{}_{soil}$, τ^2 is the two-way transmissivity of the vegetation, A and B are the model coefficients, and V_1 and V_2 are the vegetation descriptors. According to Bai and He [42], the WCM can be further simplified by expanding τ^2 through the Maclaurin series [64]:

$$\tau^2 = e^{-2BV_2 sec\theta_i} = 1 - \frac{2BV_2}{cos\theta_i} + \frac{2B^2V_2^2}{cos^2\theta_i} + \cdots\cdots, \tag{4}$$

Only the first two items are preserved in Equation (4). Combining Equations (1)–(4), the WCM is simplified as:

$$\sigma^{\circ}{}_{pp} = 2ABV_1{}^2 + \left(1 - \frac{2BV_2}{cos\theta_i}\right)\sigma^{\circ}{}_{soil}, \tag{5}$$

The V_1 and V_2 are often reduced to a single vegetation descriptor V ($V = V_1 = V_2$) in the WCM. As such, Equation (5) can be further simplified through variable substitution:

$$\sigma^{\circ}{}_{pp} = aV^2 + bV\sigma^{\circ}{}_{soil} + \sigma^{\circ}{}_{soil}, \tag{6}$$

in which $a = 2AB$ and $b = -2B/cos\theta_i$ are the coefficients of the simplified WCM. Comparing to Equation (1), Equation (6) can simplify the computation of the unknown model coefficients [38].

The bare soil scattering $\sigma^{\circ}{}_{soil}$ is represented by the function $f(R, M_s)$ of surface roughness (R) and soil moisture (M_s). For a field especially an agriculture field during the crop growing time period, R is considered constant for a short period. In such a case, when multi-temporal SAR data is used over the same field, the temporal change in bare soil scattering only reflects the change of soil moisture with time for a site, and the $\sigma^\circ{}_{soil}$ has a linear relationship with M_s, which can be expressed as:

$$\sigma^\circ{}_{soil}(db) = cM_s + d,\tag{7}$$

where c can be considered as the sensitivity of SAR to soil moisture and d indicates the backscatter due to surface roughness. The parameters a, b, and c for a specific vegetation are assumed to be constant for a specific time period and the parameter d varies with surface roughness. It must be noted that the backscattering coefficients in Equations (6) and (7) are given in *power* and *db* formats, respectively. When submitting Equation (7) to Equation (6), the unit conversion must be conducted to make the units of both equations are the same. As mentioned earlier in the Introduction, the V is often parameterized by LAI, vegetation water content, or NDVI estimated using in-situ or optical remote sensing data, and it is parameterized using HV backscattering and RVI from Radarsat-2 dual-polarization (HH + HV) imagery in this study.

2.2. Incidence Angle Normalization of SAR Imagery

Given a short time period, the amount of repeat pass SAR images may not be sufficient for fitting the WCM due to their low temporal resolutions. As a solution, all available SAR images were used to produce a time series dataset. These images may not be radiometrically comparable because they could be captured at different incidence angles (especially low incidence angles), which have a significant impact on radar backscatters [16]. For example, Oh et al.[15] showed that even as little as a 5° incidence angle can sometimes have ~3 db difference in radar backscatters, which may correspond to about 15% soil moisture change when incidence angle was lower than 35°. As such, incidence angle normalization toward a single reference angle is required for the time series SAR images. For this purpose, several studies [65,66] used empirical regression approaches, which assume linear function of SAR backscatter to incidence angles. This kind of approaches is site- and sensor-specific. In this study, we normalized the SAR backscatters to a reference angle based on the theoretical model of Lambert's law:

$$\sigma^\circ\left(\theta_{ref}\right) = \frac{cos^2\left(\theta_{ref}\right)}{cos^2(\theta_i)}\sigma^\circ(\theta_i),\tag{8}$$

where $\sigma^\circ(\theta_i)$ is the incidence angular dependent radar backscatter, θ_i and θ_{ref} represent the local incidence angle and the reference incidence angle, respectively, and $\sigma^\circ\left(\theta_{ref}\right)$ is the normalized radar backscatter to a reference incidence angle θ_{ref}. The Lambert's law assumes that the relationship between the incidence angle and amount of scattering per unit surface area follows the cosine law. This behaviour is typical for the middle range of incidence angles [67]. The model is simple but it was found to be reasonably representative for many types of terrains [65,66]. The model has been applied for agriculture land surfaces by many researchers [57,68–72].

2.3. Radar Vegetation Index (RVI)

The RVI is normally derived from quad-polarization SAR data using the equation proposed by Kim and Van Zyl [73]:

$$RVI_{quad} = \frac{8\sigma^\circ_{HV}}{\sigma^\circ_{HH} + \sigma^\circ_{VV} + 2\sigma^\circ_{HV}},\tag{9}$$

where σ°_{HH}, σ°_{HV}, and σ°_{VV} are backscattering coefficients of HH, HV and VV polarizations, respectively. Since we use dual-polarization (HH + HV) SAR data for soil moisture retrieval, VV polarization data is not available. Instead, we calculate the RVI from the dual-polarization SAR data for each Radarsat-2 image using the following equation as modified by Charbonneau et al. [74],

$$RVI = \frac{4\sigma^{\circ}_{HV}}{\sigma^{\circ}_{HH} + \sigma^{\circ}_{HV}}. \tag{10}$$

2.4. Normalized Difference Vegetation Index (NDVI)

The NDVI is computed as the ratio of the difference and sum of the reflectance measurements acquired in the near infrared (NIR) and red spectral regions. It can be written as:

$$NDVI = \frac{\rho_{NIR} - \rho_{RED}}{\rho_{NIR} + \rho_{RED}}, \tag{11}$$

where ρ_{NIR} and ρ_{RED} are the reflectance at NIR and red spectral wavebands respectively.

2.5. Model Evaluation

The model results for soil moisture are compared to in-situ measurements using the Pearson correlation coefficient (R) and the Root Mean Square Error (RMSE) as given below:

$$R = \frac{\sum_{i=1}^{n} \left(Mso_i - \overline{Mso}\right)\left(Msp_i - \overline{Msp}\right)}{\sqrt{\sum_{i=1}^{n}\left(Mso_i - \overline{Mso}\right)^2 \sum_{i=1}^{n}\left(Msp_i - \overline{Msp}\right)^2}}, \tag{12}$$

$$RMSE = \sqrt{\frac{\sum_{i=1}^{n}(Msp_i - Mso_i)^2}{n}}, \tag{13}$$

where Mso_i and Msp_i are the retrieved and in-situ soil moisture at site i, respectively. $\overline{Mso}$ and $\overline{Msp}$ are their corresponding mean values, and n is the total number of sample sites. In this study, the wide range of variability for in-situ soil moisture and LAI data and large time span over the growing season provided solid inputs for the model testing, validation and analysis. However, the difficulties and costs associated with field measurements at this scale largely constrained the size of the in situ datasets. To address this limitation, the Leave-One-Out-Cross-Validation (LOOCV) method was used for model evaluation [9]. In this method, one of the data samples was left out of each time for model evaluation and the remaining $n - 1$ data samples were used to train the model. This resulted in a total of n models trained using the $n - 1$ data samples. The results of the model evaluation can be obtained by computing R and RMSE using all unique sample estimation from each of the n models. One advantage of the LOOCV method over other traditional validation methods, e.g., splitting the dataset equally into one subset for training and one subset for validation, is that it can effectively reduce the impact of small amount of data samples. The result from this LOOCV evaluation is generally regarded as a more conservative estimate of the model performance than that trained on all samples [75].

3. Study Area and Datasets

3.1. Study Area

Our study area was located in the SMAP Validation Experiment 2012 (SMAPVEX12) site [31] (Figure 1). The background image shows the coverage (30 km by 50 km) of our study region. It is an agricultural region (90% are crop fields) located in south of Winnipeg in Manitoba, Canada. The crop types include corn, soybean, canola, wheat, and pasture. The soil texture varies greatly across the study region providing a large range of soil moisture levels. Soil moisture, soil temperature, and other surface characteristics (vegetation, roughness, soil density, etc.) data were collected during a six-week field campaign in 2012 (7 June–17 July). Accompanying the field campaign, remotely sensed satellite and airborne data were acquired at a time close to the in-situ data collections. The remote sensing data includes SMOS, AMSR-E, Radarsat-2, RapidEye, SPOT-4, DMC International Ltd. (DMCii) and Uninhabited Aerial Vehicle Synthetic Aperture Radar (UAVSAR). The crop map was produced from a supervised classification of imagery acquired by SPOT-4, DMCii, and Radarsat-2. The wealth of data collected during the intensive field campaign provided a good opportunity for developing and testing our soil moisture retrieval models. More detail information about the study region and the datasets can be found in McNarin et al. [76].

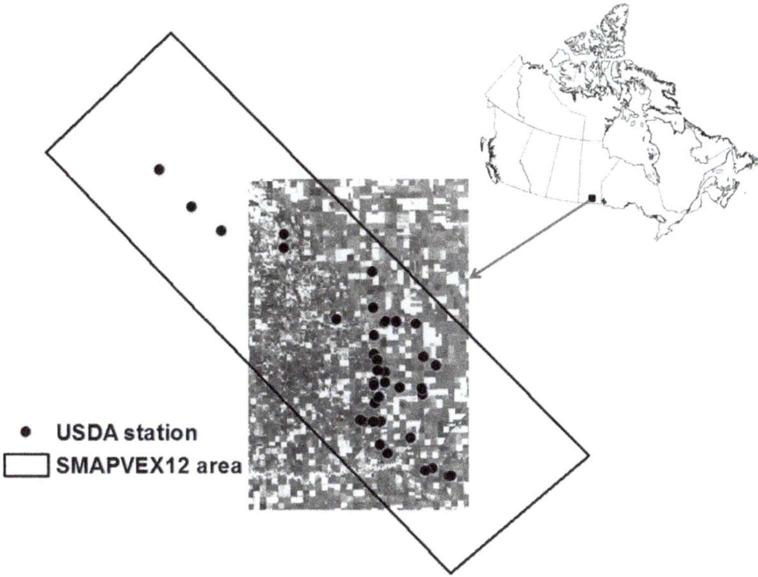

Figure 1. The location of the study area.

3.2. Data and Pre-Processing

Time series of Radarsat-2 (in single look complex (SLC) format) and RadpidEye (in L1B format) images acquired during the SMAPVEX12 field campaign were used in this study. RapidEye refers to the constellation of five satellites, which provide daily-revisit high resolution (5 m) images. A RapidEye image has five spectral bands: Blue (440–510 nm), Green (520–590 nm), Red (630–690 nm), Red-Edge (690–730 nm), and Near-Infrared (760–880 nm). All the satellite images were obtained through Canada's National Earth Observation Data Framework (NEODF). There are a total of 12 Radarsat-2 scenes and seven RapidEye scenes of which details are given in Table 1. In addition, a 30 m resolution Shuttle Radar Topography Mission (SRTM) DEM data covering the study area was downloaded from http://eros.usgs.gov for orthorectifying the satellite images.

The SLC (single look complex) format data for the time series of Radarsat-2 images was first multi-looked. The digital number (DN) values of HH and HV polarizations were then converted to backscattering coefficients (denoted as σ°_{HH} and σ°_{HV}, respectively) in power format. σ°_{HH} and σ°_{HV} were then orthorectified in UTM projection with 20 m resolution by using the SRTM DEM data. While orthorectifying the images, the local incidence angle for each pixel was computed scene by scene. A 5×5 enhanced Lee filter was applied to reduce speckle noise. All the processes were completed using the GAMMA Remote Sensing software. Table 1 shows that the incidence angles of the Radarsat-2 images vary from 20° to 37°. To make these images radiometrically comparable, we used Equation (8) to normalize the backscattering coefficients to a reference incidence angle of 25° since low incidence angle (<35°) is less sensitive to surface roughness [12,13]. The RVI images were then generated from these incidence angle normalized images by using Equation (10).

Table 1. The satellite images used in this study.

Radarsat-2					RapidEye
Acquisition Date (2012)	Flight Direction	Mode	Polarizations	Incidence Angle	Acquisition Date (2012)
5 June	Descending	FQ3W	HH, HV, VH,VV	20.0–23.6°	4 June
6 June	Ascending	S3	HH, HV	30.4–37°	12 June
13 June	Ascending	FQ10W	HH, HV, VH, VV	28.4–31.6°	28 June
19 June	Descending	S3	HH, HV	30.4–37°	5 July
20 June	Ascending	FQ6W	HH, HV, VH, VV	23.7–27.2°	14 July
27 June	Ascending	FQ2W	HH, HV, VH, VV	19.7–22.7°	21 July
29 June	Descending	FQ3W	HH, HV, VH, VV	20.0–23.6°	27 July
30 June	Ascending	S3	HH, HV	30.4–37°	
7 July	Ascending	FQ10W	HH, HV, VH, VV	28.4–31.6°	
14 July	Ascending	FQ6W	HH, HV, VH, VV	23.7–27.2°	
21 July	Ascending	FQ2W	HH, HV, VH, VV	19.7–22.7°	
24 July	Ascending	S3	HH, HV	30.4–37°	

For the time series L1B RapidEye images, a series of preprocessing procedures were conducted using the PCI Geomatica software including the conversion of DN values to Top of Atmosphere (TOA) reflectance, atmospheric corrections for converting TOA reflectance to surface reflectance, and orthorectifiction. Finally, all the images represented surface reflectance in UTM projection with 20 m resolution. The NDVI images were then generated from these processed images by using Equation (11). Due to the differences of acquisition dates between Radarsat-2 imagery and RapidEye imagery (see Table 1), we used the B-Spline interpolation method to interpolate the NDVI images to generate time series NDVI images corresponding to the acquisition dates of Radarsat-2.

Considering that the in situ soil moisture data collected in the field campaign was only coincident with flight overpasses of the UAVSAR rather than the Radarsat-2, we used soil moisture measurements (0–5 cm) from the U.S. Department of Agriculture's (USDA) stations, which were installed during the SMAPVEX12 campaign. The USDA datasets provided hourly calibrated soil moisture measurements from 4 June to 18 July 2012. For each station, only soil moisture recorded close (less than 30 min) to the acquisition time of Radarsat-2 imagery was used in this study. The soil surface roughness and LAI data measured during the SMAPVEX12 were also used for analysis. Each measurement of soil surface roughness and LAI represents a crop field where a USDA station was located. For most of the USDA stations, LAI were measured on 12, 13, 19, 20, 27, 29, and 30 June, and 6, 7, 13, and 14 July 2012. Since the dates of LAI measurements were different from the acquisition dates of Radarsat-2 imagery, the LAI measurements were also interpolated to match the acquisition dates of Radarsat-2 imagery using the B-Spline interpolation method, similar to that for NDVI. All the in-situ measurements were downloaded from https://smapvex12.espaceweb.usherbrooke.ca and were processed in ArcGIS shapefile format. More details about in-situ measurements can be found in McNairn et al. [75] or the above website.

As mentioned earlier, the effect of surface roughness can be eliminated by using multi-temporal data over the same field. However, soil surface roughness varies with the fields even having the same crop types. To make data as most as possible for analysis, we used the in-situ data with similar roughness in the WCM for each crop type. The soil moisture data were processed as follows. For each crop type, the USDA soil moisture stations with similar surface roughness are grouped. The group with the largest number of stations were selected for this study. For each selected soil moisture station, the time series of soil moisture (M_s) and LAI measurements at the acquisition times of Radarsat-2 images, as well as the co-located 3 × 3 pixel averaged Radarsat-2 backscatter coefficients ($\sigma°_{HH}$ and $\sigma°_{HV}$), RVI and RapidEye NDVI values, were extracted. Using a 3 × 3 pixel averaged value to replace a single pixel value can reduce errors caused by co-registration errors between satellite images and soil moisture measurement sites. It is noted that the study was based on data representing each specific in-situ measurement site and not the whole crop field. Table 2 lists the statistics of the selected soil moisture and LAI measurements for five crop types: soybean, canola, corn, wheat, and pasture. The root mean square (RMS) roughness values of the selected fields for each crop type are 0.31–0.42 cm (soybean), 1.31–1.33 cm (canola), 1.23–1.28 cm (corn), 1.27–1.29 cm (wheat), and 0.6–0.74 cm (pasture), which are also listed in Table 2.

Table 2. The in-situ soil moisture (0–5 cm) and Leaf Area Index (LAI) data used in this study.

Crop Types	Soil Moisture (m^3/m^3)			LAI (m^2/m^2)			RMS * (cm)
	Range	Average	n *	Range	Average	n	
Soybean	0.155–0.467	0.317	26	0.11–2.43	0.92	18	0.31–0.42
Canola	0.042–0.419	0.24	27	0.31–6.33	3.12	18	1.31–1.33
Corn	0.122–0.354	0.211	23	0.09–3.92	1.04	21	1.23–1.28
Wheat	0.123–0.37	0.245	27	0.59–5.15	2.38	20	1.27–1.29
Pasture	0.0398–0.217	0.14	25	1.3–7.19	3.36	18	0.6–0.74

* n is the number of in-situ measurements and *RMS* is the root mean square roughness.

4. Results and Discussion

SAR backscatters are generally related to vegetation properties such as shape, height, size, and density, which vary with vegetation types [77]. Different vegetation types may present different behavior in WCM's calibrations [27,78,79]. In this study, therefore, the simplified WCM (Equation (6)) was applied to five different crop types: soybean, canola, pasture, wheat, and corn where σ°_{HH} was used as the soil moisture retriever σ°_{pp}. The σ°_{HV}, RVI, and NDVI were separately used as the vegetation descriptor V. The WCM was calibrated for each crop type separately. The LOOCV method was used for the model evaluation. We calculated R values between the in-situ and modelled soil moisture for each crop type. Figure 2 show the plots of the R values for the five crop types. The results showed that the R values vary with crop types and vegetation descriptors.

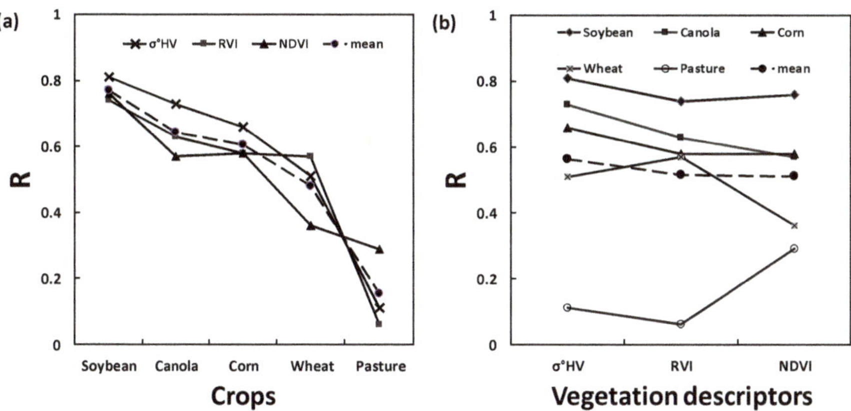

Figure 2. The plots of R values of the retrieved and measured soil moisture against (**a**) crop types and (**b**) vegetation descriptors.

The soil moisture retrieval for soybean, canola, and corn generally showed higher R values than those for wheat and pasture for all the three vegetation descriptors (Figure 2a). Among the five crop types, soybean and pasture have the best and the worst soil moisture retrieval, respectively. These results are confirmed by the mean R values of the three vegetation descriptors. The soybean, canola, and corn have mean R values ranging from 0.61 to 0.77 while the wheat and pasture have mean R

values of 0.48 and 0.15, respectively. Figure 2b shows the mean R values for σ°_{HV}, RVI, and NDVI are 0.57, 0.52, and 0.51. It also demonstrates that σ°_{HV} used as the vegetation descriptor in the WCM gave slightly better soil moisture retrieval results than RVI and NDVI. The results are also confirmed by Figure 3, which shows the scatter plots with R values between the in-situ measured and model retrieved moisture together for all crop types. The R values for σ°_{HV}, RVI, and NDVI are 0.59, 0.54, and 0.54, respectively. The values of RMSE between retrieved and in-situ soil moisture are 0.069 m^3/m^3, 0.085 m^3/m^3, and 0.071 m^3/m^3 for σ°_{HV}, RVI, and NDVI, respectively. The results are similar to those in the literature for soil moisture retrieval for other crops from SAR data using the WCM [12,28]. Overall, σ°_{HV} as the vegetation descriptor in the WCM shows similar accuracy as NDVI but slightly outperforms RVI for the retrieval of soil moisture. It worth mentioning that the in-situ LAI as the vegetation descriptor in the WCM achieved slightly better results ($R = 0.8$ and RMSE = 0.067 m^3/m^3) than σ°_{HV}. Nevertheless our study showed that the use of σ°_{HV} can overcome the constraints of the commonly used vegetation descriptors (e.g., NDVI) in WCM such as data availability and weather dependency, thus improving soil moisture retrieval.

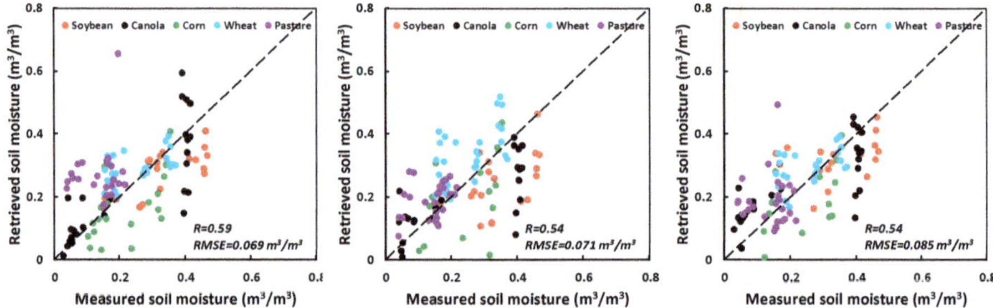

Figure 3. Plots of soil moisture measured from U.S. Department of Agriculture's (USDA) stations and soil moisture retrieved from Radarsat-2 using (**a**) backscatters of HV polarization σ°_{HV}, (**b**) RVI, and (**c**) NDVI in the WCM for all crops.

To further understand the results, we examined the relationships between the LAI and the vegetation descriptors for each crop type. Figure 4 shows the scatter plots of the LAI vs. the three vegetation descriptors of σ°_{HV}, RVI, and NDVI. The R values between the LAI and the three vegetation descriptors were given in Figure 5. The results show that soybean, canola, and corn generally have better correlations (R) between the LAI and three vegetation descriptors than wheat and pasture. Soybean, canola and corn have mean R values larger than 0.65 while wheat and pasture have mean R values around 0.3, which helps to explain the results shown in Figure 2a. The results indicate that the crops that have a stronger correlation between LAI and vegetation descriptors generally demonstrate better soil moisture retrievals when using the WCM.

Figure 4 show that for some crop types, e.g., wheat and pasture, the relationships between vegetation descriptors (σ°_{HV}, RVI, and NDVI) are not good. Moreover, for some crop types, vegetation descriptors do not saturate with LAI. More specifically, for soybean, the σ°_{HV} has a better sensitivity to LAI than RVI and NDVI. It could result in the more accurate soil moisture retrieval (with $R = 0.81$). Canola shows LAI values ranging from 0.3 m^2/m^2 to 6.3 m^2/m^2 with a mean value of 3.12 (Table 2). All three vegetation descriptors present a good relationship with LAI, but σ°_{HV} and RVI are more sensitive to LAI than NDVI , which is saturated at a LAI of around 2.0 m^2/m^2 as shown in Figure 4. This could be the reason why NDVI as the vegetation descriptor in the WCM achieves less accurate soil moisture retrievals ($R = 0.57$) than σ°_{HV} and RVI. For corn, the three vegetation descriptors show good linear relationships with LAI when LAI is less than 2.0 m^2/m^2, but they saturate quickly. Their similar relationships with LAI could lead to similar soil moisture estimations (Figure 2b). For wheat, all three vegetation descriptors saturate with LAI and have weak relationships with LAI, with NDVI showing slightly better correlations than σ°_{HV} and RVI. Some of the LAI having values between 3.0 m^2/m^2 and 5.0 m^2/m^2 (Figure 4) indicated some observations were in late growing stage of wheat with dense canopy, as a result, the penetration of SAR is in fact limited when vegetation canopy is dense. The three vegetation descriptors present similar soil moisture estimations (Figure 2b). We also tested LAI as the vegetation descriptor in the WCM. The LAI gives the values of 0.59 and 0.07 m^3/m^3 for R and RMSE, separately. It shows that even LAI as the vegetation descriptor cannot significantly improve the result at this growing stage. However the estimation accuracies are still acceptable. The partial reason for the results is that NDVI and LAI are often influenced by vegetation water content [80], in addition, Kim et al. [58] also showed that RVI is highly correlated with vegetation water content in wheat. The vegetation water content could correlate with the soil moisture underneath the canopy. For pasture of which the LAI values ranged from 2.0 m^2/m^2 to 7.0 m^2/m^2 (Table 1), all three vegetation descriptors were saturated and had weak relationships with LAI ($R = 0.47$ for σ°_{HV}, $R = 0.04$ for RVI and $R = 0.29$ for NDVI). However, NDVI leads to better soil moisture retrieval than σ°_{HV} and RVI which needs to be explained by further studies. In general, Figure 4 shows the σ°_{HV} became saturated at relatively higher LAI values and is more sensitive to LAI than RVI and NDVI for some crop types The results are consistent with the studies of Paloscia [62], Simoes et al. [63], and Jiao et al. [81] which showed that σ°_{HV} is related to LAI and vegetation biomass.

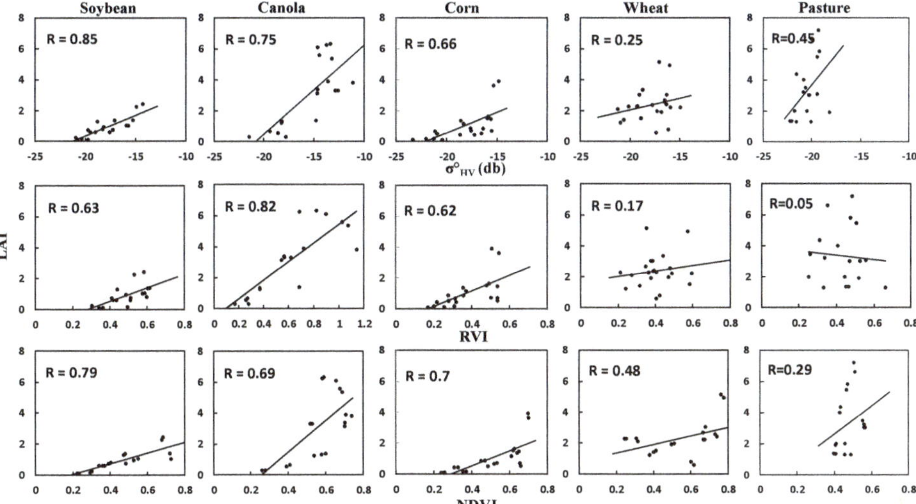

Figure 4. The plots of LAI against three vegetation descriptors: σ°_{HV} (top row), RVI (middle row) and NDVI (bottom row) for each crop type.

Although, σ°_{HV} is generally more sensitive to vegetation than NDVI and RVI, its backscatter still contains some soil moisture signals, which may introduce errors. However, this study shows, overall, σ°_{HV} as the vegetation descriptor presents similar soil moisture estimation accuracy as NDVI and slightly outperforms RVI (Figure 3). Actually Baghdadi et al. showed that cross polarization is more sensitive to vegetation cover than to soil moisture and the soil contribution in cross polarization quickly becomes lower than the vegetation contribution. The sensitivity of cross polarization to soil moisture strongly decreases when both incidence angle and vegetation density increases. Therefore σ°_{HV} only introduce errors when the WCM is applied to areas with less vegetation. Moreover, the sensitivity decreases when the soil moisture decreases. For example, the soil contribution becomes negligible for incidence angle higher than 25° from an NDVI of 0.27 and 0.39 for soil moisture with 0.05 m^3/m^3 and 0.10 m^3/m^3, respectively [16]. The soil contribution in cross polarization to the total signal is low in the case of well-developed vegetation cover. Moreover, in this study, 25% of in-situ data were collected when LAI was less than 1.0. Majority of the in-situ data has LAI larger than 1.0. The influence of soil contribution to HV backscatter on the soil moisture estimation is low in this study. The study also showed that σ°_{HV} ($R = 0.63$) achieved better soil moisture estimations than RVI ($R = 0.53$) and NDVI (0.52) for LAI larger than 2.0 and vice versa for LAI less than 1.0 ($R = 0.80$, 0.86, 0.83 for σ°_{HV}, RVI and NDVI, separately).

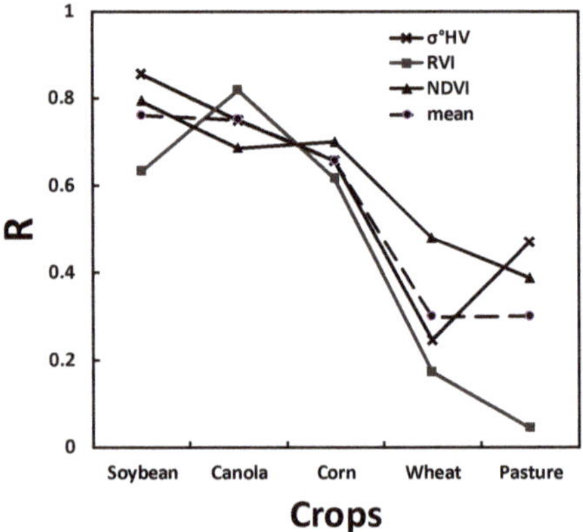

Figure 5. The plots of *R* values between LAI and vegetation descriptors for crop types calculated from Figure 4.

As stated above, NDVI has a saturation point at lower LAI than σ°_{HV} for some crop types. The reason could be that NDVI is a measure of vegetation greenness whereas σ°_{HV} is a measure of volumetric scattering. In general, vegetation reaches the maximum greenness before it becomes matured volumetrically due to radar's penetration. Normally σ°_{HV} and RVI have similar results in vegetation characterization [82]. The dispersion of σ°_{HV} and RVI in this study may be related to the effect of crop planting directions. RVI was calculated only from HH and HV polarization in this study. A previous study [83] reported that the planting row orientations in a sugarcane field had an influence on the HH polarization but not on the HV polarization. In addition, it can be also noticed from Figure 3 that the retrieved soil moisture is overestimated for relatively dry sites ($M_s < 0.25$ m^3 m^{-3}) and underestimated for wet sites ($M_s > 0.25$ m^3 m^{-3}) due to the difference of penetration of SAR in dry and wet soils. In addition, the data we used in this study is from the SMAPVEX12, which is a six-week field campaign in 2012 (7 June–17 July). The database for some crop types may cover a limited period. It could be useful to test WCM for a vegetation growth cycle in future study.

Ideally, the Radarsat-2 data should be in the same viewing geometry. Since Radarsat-2 has a revisit time of 24 days, it is difficult to obtain enough repeat pass images within a short period for the analysis. Instead, we used all available Radarsat-2 data, which are in different modes and possess different incidence angles. To reduce the effect of SAR incidence angle, we normalized all Radarsat-2 images to a reference angle theoretical model of Lambert's law. However, it should be noted that the theoretical approach may not be the most appropriate one and could introduce errors [40]. The assessments of errors and different approaches are beyond the scope of this paper. Therefore, although an incidence angle normalization process was applied to these images to minimize the effect of different viewing geometries on SAR backscatters, the residual errors may still exist. The SAR imagery limitations can be improved after the RCM data become available. The RCM has a 4-day revisit time, which makes it possible to acquire sufficient repeat pass SAR images in a short period.

5. Conclusions

This study investigated the capability of using SAR-derived vegetation descriptors in a WCM for improving soil moisture retrieval over a vegetated area. Two vegetation descriptors, σ°_{HV} and RVI, derived from Radarsat-2/SAR were studied. The results were compared to those obtained from using NDVI. The use of σ°_{HV} in WCM achieved similar soil moisture retrieval as the use of NDVI but slightly better than the use of RVI. The RMSE between retrieved and observed soil moisture were 0.069 m^3/m^3, 0.085 m^3/m^3 and 0.071 m^3/m^3 for σ°_{HV} , RVI and NDVI, respectively. The corresponding R were 0.59, 0.54, and 0.54, respectively. The results can be explained from the relationships of the vegetation descriptors with LAI, which shows σ°_{HV} saturates at relatively higher LAI values for some crop types and is generally more sensitive to vegetation than RVI and NDVI. Soil contribution in HV polarization brings an effect on the soil moisture estimation over areas with less vegetation but the effect is limit in this study. The small difference of RVI and σ°_{HV} (or NDVI) in soil moisture retrieval indicates that RVI is also a relevant vegetation descriptor in the WCM. Moreover, the use of σ°_{HV} or RVI in the WCM overcomes the dependency of WCM on in-situ or optical remote sensing data for deriving vegetation descriptors and further reduces additional data requirements for WCM, thus improves soil moisture retrieval. The study also indicates that the use of dual-polarized SAR images presents a practical way to retrieve soil moisture over a large area since the dual-polarized images can have a large swath width of up to 500 km (e.g., Radarsat-2 ScanSAR wide mode). The RCM will further enhance the capability with its rapid revisit and compact polarimetric (CP) configuration. The RCM CP is considered as a possible alternative of quad-polarized (QP) system but with wider image swath. Therefore the use of CP data in WCM could achieve a better soil moisture retrieval over a large area even under dense vegetation canopy.

Author Contributions: J.L. performed the experiment, analyzed the data, validated the results and wrote the manuscript. S.W. provided crucial guidance and support during the research, and revised the manuscript.

Funding: The research received no external funding.

Acknowledgments: This work was supported by the Groundwater Geoscience Program and Remote Sensing Program of Natural Resources Canada. We gratefully thank the investigators of the SMAP Validation Experiment 2012 (SMAPVEX12) for their sharing the SMAPVEX12 in-situ measurements with the scientific community. The editor and anonymous reviewers are also thanked for their valuable comments and suggestions, which improved an earlier version of this paper.

Conflicts of Interest: The authors declare no conflict of interest.

References

1. Hajnsek, I.; Jagdhuber, T.; Schön, H.; Papathanassiou, K.P. Potential of estimating soil moisture under vegetation cover by means of PolSAR. *IEEE Trans. Geosci. Remote Sens.* **2009**, *47*, 442–454. [CrossRef]
2. Wang, S. Simulation of evapotranspiration and its response to plant water and CO_2 transfer dynamics. *J. Hydrometeorol.* **2008**, *9*, 426–443. [CrossRef]
3. Wang, S.; McKenney, D.W.; Shang, J.; Li, J. A national-scale assessment of long-term water budget closures for Canada's watersheds. *J. Geophys. Res.* **2014**, *119*, 8712–8725. [CrossRef]
4. Wang, S.; Huang, J.; Li, J.; Rivera, A.; McKenney, D.W.; Sheffield, J. Assessment of water budget for sixteen large drainage basins in Canada. *J. Hydrol.* **2014**, *512*, 1–15. [CrossRef]
5. Ulaby, F.T.; Moore, R.K.; Fung, A.K. *Microwave Remote Sensing: Active and Passive*; Artech House: Dedham, MA, USA, 1986; Volume III.
6. Zribi, M.; Baghdadi, N.; Holah, N.; Fafin, O. New methodology for soil surface moisture estimation and its application to ENVISAT-ASAR multi incidence data inversion. *Remote Sens. Environ.* **2005**, *96*, 485–496. [CrossRef]
7. Srivastava, H.S.; Patel, P.; Sharma, Y.; Navalgund, R.R. Large-area soil moisture estimation using multiincidence-angle RADARSAT-1 SAR data. IEEE Trans. *Geosci. Remote Sens.* **2009**, *47*, 2528–2535. [CrossRef]

8. Petropoulos, G.P.; Ireland, G.; Barrett, B. Surface soil moisture retrievals from remote sensing: Current status, products & future trends. *Phys. Chem. Earth* **2015**, *83*, 36–56.

9. Alexakis, D.D.; Mexis, F.K.; Vozinaki, A.K.; Daliakopoulos, I.N.; Tsanis, I.K. Soil moisture content estimation based on sentinel-1 and auxiliary earth observation products. A Hydrological Approach. *Sensors* **2017**, *17*, 1455. [CrossRef] [PubMed]

10. Zhang, Y.; Gong, J.; Sun, K.; Yin, J.; Chen, X. Estimation of soil moisture index using multi-temporal Sentinel-1 images over Poyang Lake ungauged zone. *Remote Sens.* **2018**, *10*, 12. [CrossRef]

11. Autret, M.; Bernard, R.; Vidal-Madjar, D. Theoretical study of the sensitivity of the microwave backscattering coefficient to the soil surface parameters. *Int. J. Remote Sens.* **1989**, *10*, 171–179. [CrossRef]

12. Zribi, M.; Gorrab, A.; Baghdadi, N. A new soil roughness parameter for the modelling of radar backscattering over bare soil. *Remote Sens. Environ.* **2014**, *152*, 62–73. [CrossRef]

13. Holah, N.; Baghdadi, N.; Zribi, M.; Bruand, A.; King, C. Potential of ASAR/ENVISAT for the characterization of soil surface parameters over bare agricultural fields. *Remote Sens. Environ.* **2005**, *96*, 78–86. [CrossRef]

14. Baghdadi, N.; Hajj, M.E.; Zribi, M.; Bousbih, S. Calibration of the water cloud model at C-band for winter crop fields and grasslands. *Remote Sens.* **2017**, *9*, 969. [CrossRef]

15. Oh, Y.; Sarabandi, K.; Ulaby, F.T. An empirical model and an inversion technique for radar scattering from bare soil surfaces. *IEEE Trans. Geosci. Remote Sens.* **1992**, *30*, 370–381. [CrossRef]

16. Mladenova, I.E.; Jackson, T.J.; Bindlish, R.; Hensley, S. Incidence angle normalization of radar backscatter data. *IEEE Trans. Geosci. Remote Sens.* **2013**, *51*, 1791–1804. [CrossRef]

17. Fung, A.K.; Chen, K.S. Dependence of the surface backscattering coefficients on roughness, frequency and polarization stats. *Int. J. Remote Sens.* **1992**, *13*, 1663–1680. [CrossRef]

18. Sun, G.; Simonett, D.S.; Strahler, A.H. A radar backscatter model for discontinuous coniferous forests. *IEEE Trans. Geosci. Remote Sens.* **1991**, *29*, 639–650. [CrossRef]

19. Wang, Y.; Day, J.L.; Davis, F.W.; Melack, J.M. Modeling L-band radar backscatter of Alaskan boreal forest. *IEEE Trans. Geosci. Remote Sens.* **1993**, *31*, 1146–1154. [CrossRef]

20. Li, Y.Y.; Zhao, K.; Ren, J.H.; Ding, Y.L.; Wu, L.L. Analysis of the dielectric constant of saline-alkali soils and the effect on radar backscattering coefficient: A case study of soda alkaline saline soils in Western Jilin Province using RADARSAT-2 data. *Sci. World J.* **2014**, *2014*, 563015. [CrossRef] [PubMed]

21. Zhang, T.; Zeng, Q.; Li, Y.; Xiang, Y. Study on relation between InSAR coherence and soil moisture. In Proceedings of the ISPRS Congress, Beijing, China, 3–11 July 2008.

22. Beaudoin, A.; Le Toan, T.; Gwyn, Q.H.J. SAR observations and modeling of the C-band backscatter variability due to multiscale geometry and soil moisture. *IEEE Trans. Geosci. Remote Sens.* **1990**, *28*, 886–895. [CrossRef]

23. Le Toan, T. Active microwave signatures of soil and crops: Significant results of three years of experiments. In Proceedings of the Symposium IGARSS'82, Munich, Germany, 1–4 June 1982.

24. McNairn, H.; Brisco, B. The application of C-band polarimetric SAR for agriculture: A review. *Can. J. Remote Sens.* **2004**, *303*, 525–542. [CrossRef]

25. Barrett, B.W.; Dwyer, E.; Whelan, P. Soil moisture retrieval from active spaceborne microwave observations: An evaluation of current techniques. *Remote Sens.* **2009**, *1*, 210–242. [CrossRef]

26. Romshoo, S.A.; Koike, M.; Hironaka, S.; Oki, T.; Musiake, K. Influence of Surface and Vegetation Characteristics on C-band Radar Measurements for Soil Moisture Content. *J. Indian Soc. Remote Sens.* **2002**, *30*, 229. [CrossRef]

27. Gherboudj, I.; Maga gi, R.; Berg, A.A.; Toth, B. Soil moisture retrieval over agriculture fields from multi-polarized and multi-angular Radarsat-2 SAR data. *Remote Sens. Environ.* **2011**, *115*, 33–43. [CrossRef]

28. Baghdadi, N.; Holah, N.; Zribi, M. Soil moisture estimation using multi-incidence and multi-polarization ASAR data. *Int. J. Remote Sens.* **2006**, *27*, 1907–1920. [CrossRef]

29. Cloude, S.R.; Pottier, E. A review of target decomposition theorems in radar polarimetry. *IEEE Trans. Geosci. Remote Sens.* **1996**, *34*, 498–518. [CrossRef]

30. Baghdadi, N.; Zribi, M.; Loumagne, C.; Ansart, P.; Anguela, T.P. Analysis of TerraSAR-X data and their sensitivity to soil surface parameters over bare agriculture fields. *Rmote Sens. Environ.* **2008**, *112*, 4370–4379. [CrossRef]

31. Baghdadi, N.; Choker, M.; Zribi, M.; Hajj, M.E.; Paloscia, S.; Verhoest, N.E.C.; Lievens, H.; Baup, F.; Mattia, F. A new empirical model for radar scattering from bare soil surfaces. *Remote Sens.* **2016**, *8*, 920. [CrossRef]

32. Aubert, M.; Baghdadi, N.; Zribi, M.; Douaoui, A.; Loumagne, C.; Baup, F.; Garrigues, S. Analysis of TerraSAR-X data sensitivity to bare soil moisture, roughness, composition and soil crust. *Remote Sens. Environ.* **2011**, *115*, 1801–1810. [CrossRef]

33. Das, K.; Paul, P.K. Present status of soil moisture estimation by microwave remote sensing. *Cogent Geosci.* **2015**, *1*, 1084669. [CrossRef]

34. He, L.; Hong, Y.; Wu, X.; Ye, N.; Walker, J.P.; Chen, X. Investigation of SMAP active-passive downscaling algorithms using combined Sentinel-1 SAR and SMAP radiometer data. *IEEE Trans. Geosci. Remote Sens.* **2018**, *56*, 4906–4918. [CrossRef]

35. Li, J.; Wang, S.; Gunn, G.; Joosse, P.; Hazen, A.J. A model for downscaling SMOS soil moisture using Sentinel-1 SAR data. *Int. J. Appl. Earth Obs.* **2018**, *72*, 109–121. [CrossRef]

36. Attema, E.P.W.; Ulaby, F.T. Vegetation modeled as a water cloud. *Radio Sci.* **1978**, *13*, 357–364. [CrossRef]

37. Oh, Y. Quantitative retrieval of soil moisture content and surface roughness from multipolarized radar observations of bare soil surfaces. *IEEE Trans. Geosci. Remote Sens.* **2004**, *42*, 596–601. [CrossRef]

38. Bai, X.; He, B. Potential of Dubois model for soil moisture retrieval in prairie areas using SAR and optical data. *Int. J. Remote Sens.* **2015**, *36*, 5737–5753. [CrossRef]

39. Paloscia, S.; Pettinato, S.; Santi, E.; Notarnicola, C.; Pasolli, L.; Reppucci, A. Soil moisture mapping using Sentinel-1 images: Algorithm and preliminary validation. *Remote Sens. Environ.* **2013**, *134*, 234–248. [CrossRef]

40. Fung, A.K. *Microwave Scattering and Emission Models and Their Applications*; Artech House: Nordwood, MA, USA, 1994.

41. Chen, K.-S.; Wu, T.-D.; Tsang, L.; Li, Q.; Shi, J.; Fung, A.K. Emission of rough surfaces calculated by the integral equation method with comparison to three-dimensional moment method simulations. *IEEE Trans. Geosci. Remote Sens.* **2006**, *27*, 3831–3852. [CrossRef]

42. Baghdadi, N.; Saba, E.; Aubert, M.; Zribi, M.; Baup, F. Evaluation of radar backscattering models IEM, Oh, and Dubois for SAR data in X-band over bare soils. *IEEE Geosci. Remote Sens. Lett.* **2011**, *6*, 1160–1164. [CrossRef]

43. Baghdadi, N.; Chaaya, J.A.; Zribi, M. Semiempirical calibration of the integral equation model for SAR data in C-band and cross polarization using radar images and field measurements. *IEEE Geosci. Remote Sens. Lett.* **2011**, *8*, 14–18. [CrossRef]

44. Baghdadi, N.; Zribi, M.; Paloscia, S.; Verhoest, N.E.; Lievens, H.; Baup, F.; Mattia, F. Semi-empirical calibration of the integral equation model for co-polarized L-band backscattering. *Remote Sens.* **2015**, *7*, 13626–13640. [CrossRef]

45. Choker, M.; Baghdadi, N.; Zribi, M.; El Hajj, M.; Paloscia, S.; Verhoest, N.; Lievens, H.; Mattia, F. Evaluation of the Oh, Dubois and IEM models using large dataset of SAR signal and experimental soil measurements. *Water* **2017**, *9*, 38. [CrossRef]

46. Dubois, P.; van Zyl, J.; Engman, T. Measuring soil moisture with imaging radars. *IEEE Trans. Geosci. Remote Sens.* **1995**, *33*, 915–926. [CrossRef]

47. Gorrab, A.; Zribi, M.; Baghdadi, N.; Lili-Chabaane, Z.; Mougenot, B. Multi-frequency analysis of soil moisture vertical heterogeneity effect on radar backscatter. In Proceedings of the 2014 1st International Advanced Technologies for Signal and Image Processing (ATSIP), Sousse, Tunisia, 17–19 March 2014; pp. 379–384.

48. Gao, Q.; Zribi, M.; Escorihuela, M.J.; Baghdadi, N. Synergetic Use of Sentinel-1 and Sentinel-2 Data for Soil Moisture Mapping at 100 m Resolution. *Sensors* **2017**, *17*, 1966. [CrossRef] [PubMed]

49. Palosica, S.; Pampaloni, P.; Pettinato, S.; Santi, E. A comparison of algorithms for retrieving soil moisture from ENVISAT/ASAR images. *IEEE Trans. Geosci. Remote Sens.* **2008**, *46*, 3274–3284. [CrossRef]

50. Hajj, M.E.; Baghdadi, N.; Zribi, M.; Belaud, G.; Cheviron, B.; Courault, D.; Charron, F. Soil moisture retrieval over irrigated grassland using X-band SAR data. *Remote Sens. Environ.* **2016**, *176*, 202–218. [CrossRef]

51. Hajj, M.E.; Baghdadi, N.; Zribi, M.; Bazzi, H. Synergic use of Sentinel-1 and Sentinel-2 images for operational soil moisture mapping at high spatial resolution over agricultural areas. *Remote Sens.* **2017**, *9*, 1292. [CrossRef]

52. Santi, E.; Paloscia, S.; Pettinato, S.; Notarnicola, C.; Pasolli, E.; Pistocchi, A. Comparison between SAR Soil Moisture Estimates and Hydrological Model Simulations over the Scrivia Test Site. *Remote Sens.* **2013**, *5*, 4961–4976. [CrossRef]

53. Zribi, M.; Kotti, F.; Amri, R.; Wagner, W.; Shabou, M.; LiliChabaane, Z.; Baghdadi, N. Soil moisture mapping in a semiarid region, based on ASAR/wide swath satellite data. *Water Resour. Res.* **2014**, *50*, 823–835. [CrossRef]

54. Zribi, M.; Chahbi, A.; Shabou, M.; Lili-Chabaane, Z.; Duchemin, B.; Baghdadi, N.; Amri, R.; Chehbouni, A. Soil surface moisture estimation over a semi-arid region using ENVISAT ASAR radar data for soil evaporation evaluation. *Hydrol. Earth Syst. Sci.* **2011**, *15*, 345–358. [CrossRef]

55. Kumar, K.; Rao, H.P.S.; Arora, M.K. Study of water cloud model vegetation descriptors in estimating soil moisture in Solani catchment. *Hydrol. Process.* **2015**, *29*, 2137–2148. [CrossRef]

56. Borgeaud, M.; Attema, E.; Salgado-Gispert, G.; Bellini, A.; Noll, J. Analysis of bare soil surface roughness parameters with ERS-1 SAR data. In Proceedings of the Symposium on Retrieval of Bio- and Geophysical Parameters from SAR data for Land Applications, Toulouse, France, 17–20 October 1995.

57. Tomer, S.K.; Bitar, A.A.; Sekhar, M.; Zribi, M.; Bandyopadhyay, S.; Sreelash, K.; Sharma, A.K.; Corgne, S.; Kerr, Y. Retrieval and Multi-scale Validation of Soil Moisture from Multi-temporal SAR data in a Semi-Arid Tropical Region. *Remote Sens.* **2015**, *7*, 8128–8153. [CrossRef]

58. Kim, Y.; Jackson, T.; Bindlish, R.; Lee, K. Retrieval of wheat growth parameters with radar vegetation indices. *IEEE Geosci. Remote Sens.* **2013**, *11*, 808–812.

59. Kumar, S.D.; Rao, S.S.; Sharma, J.R. Radar vegetation index as an alternative to NDVI for monitoring of soybean and cotton. *Indian Cartogr.* **2013**, *33*, 91–96.

60. Kasischke, E.S.; Tanase, M.A.; Bourgeau-Chavez, L.L.; Borr, M. Soil moisture limitations on monitoring boreal forest regrowth using spaceborne L-band SAR data. *Remote Sens. Environ.* **2011**, *115*, 227–232. [CrossRef]

61. Rao, S.S.; Kumar, S.D.; Das, S.N.; Nagaraju, M.S.S.; Venugopal, M.V.; Rajankar, O.; Laghate, P.; Reddy, M.S.; Joshi, A.K.; Sharma, J.R. Modified Dubois model for estimating soil moisture with dual polarized SAR data. *J. Indian Soc. Remote Sens.* **2013**, *41*, 865–872.

62. Paloscia, S. An Empirical Approach to Estimating Leaf Area Index from Multifrequency SAR Data. *Int. J. Remote Sens.* **1998**, *19*, 359–364. [CrossRef]

63. Simões, M.S.; Rocha, J.V.; Lamparelli, R.A.C. Growth Indices and Productivity in Sugarcane. *Sci. Agric.* **2005**, *62*, 23–30. [CrossRef]

64. Kseneman, M.; Gleich, D. Soil Moisture Estimation from X-Band Data Using Tikhonov Regularization and Neural Net. *IEEE Trans. Geosci. Remote Sens.* **2013**, *51*, 3885–3898. [CrossRef]

65. Wagner, W.; Bloeschl, G.; Pampaloni, P.; Calvet, J.-C.; Bizzarri, B.; Wigneron, J.-P.; Kerr, Y. Operational readiness of microwave remote sensing of soil moisture for hydrologic applications. *Hydrol. Res.* **2007**, *38*, 1–20. [CrossRef]

66. Mason, D.C.; Garcia-Pintado, J.; Cloke, H.L.; Dance, S.L. Evidence of a topographic signal in surface soil moisture derived from ENVISAT ASAR wide swath data. *Int. J. Appl. Earth Obs.* **2016**, *45*, 178–186. [CrossRef]

67. Ulaby, F.T.; Moore, R.K.; Fung, A.K. *Microwave Remote Sensing, Active and Passive, Volume II: Radar Remote Sensing and Surface Scattering and Emission Theory*; Artech House: Boston, MA, USA, 1982.

68. Lievens, H.; Verhoest, E.C. On the retrieval of soil moisture in wheat fields from L-band SAR based on water cloud modelling, the IEM, and effective roughness parameters. *IEEE Geosci. Remote Sens.* **2011**, *8*, 740–744. [CrossRef]

69. Martinez-Agirre, A.; Alvarez-Mozos, J.; Lievens, H.; Verhoest, E.C.; Gimenez, R. Sensitivity of C-band backscatter to surface roughness parameters measured at differenct scales. In Proceedings of the 2015 IEEE International Geoscience and Remote Sensing Symposium, Milan, Italy, 26–31 July 2015.

70. Lievens, H.; Verhoest, E.C. Spatial and temporal soil moisture estimation from RADARSAT-2 imagery over Flevland, The Netherlands. *J. Hydrol.* **2012**, *456*, 44–56. [CrossRef]

71. Keyser, E.D.; Lievens, H.; Vernieuwe, H.; Alvarez-Mozos, J.; De Baets, B.; Verhoest, E.C. Assessment of the impact of uncertainty on modeled soil surface roughness on SAR-retrieved soil moisture uncertainty. In Proceedings of the ISPRS TC VII Symposium—100 Years ISPRS, Vienna, Austria, 5–7 July 2010; Volume XXXVIII.

72. Van der Velde, R.; Salama, M.S.; Eweys, O.A.; Wen, J.; Wang, Q. Soil moisture mapping using combined active/passive microwave observations over the east of the Netherlands. *IEEE J. Sel. Top. Appl.* **2014**, *8*, 4355–4372. [CrossRef]

73. Kim, Y.; van Zyl, J.J. A Time-Series Approach to Estimate Soil Moisture Using Polarimetric Radar Data. *IEEE Trans. Geosci. Remote Sens.* **2009**, *47*, 2519–2527. [CrossRef]

74. Charbonneau, F.; Trudel, M.; Fernandes, R. Use of Dual Polarization and Multi-Incidence SAR for soil permeability mapping. In Proceedings of the 2005 Advanced Synthetic Aperture Radar (ASAR) Workshop, St-Hubert, QC, Canada, 15–17 November 2005.

75. Wang, S.; Zhou, F.; Russell, H.A.J. Estimating snow mass and peak river flows for the Mackenzie river basin using GRACE satellite observations. *Remote Sens.* **2017**, *9*, 256. [CrossRef]

76. McNairn, H.; Jackson, T.; Wiseman, G.; Belair, S.; Berg, A.; Bullock, P.; Colliander, A.; Cosh, M.; Kim, S.; Magagi, R.; et al. The Soil Moisture Active Passive validation experiment 2012 (SMAPVEX12): Pre-launch calibration and validation of the SMAP Satellite. *IEEE Trans. Geosci. Remote Sens.* **2015**, *53*, 2784–2801. [CrossRef]

77. Macelloni, G.; Paloscia, S.; Pampaloni, P.; Marliani, F.; Gai, M. The relationship between the backscattering coefficient and the biomass of narrow and broad leaf crops. *IEEE Trans. Geosci. Remote Sens.* **2001**, *39*, 873–884. [CrossRef]

78. Van De Griend, A.A.; Wigneron, J.P. The b-factor as a function of frequency and canopy type at h-polarization. *IEEE Trans. Geosci. Remote Sens.* **2004**, *42*, 786–794. [CrossRef]

79. Ulaby, F.T.; Wilson, E.A. Microwave attenuation properties of vegetation canopies. *IEEE Trans. Geosci. Remote Sens.* **1985**, *23*, 746–753. [CrossRef]

80. Zarco-Tejada, P.J.; Rueda, C.A.; Ustin, S.L. Water content estimation in vegetation with MODIS reflectance data and model inversion methods. *Remote Sens. Environ.* **2003**, *85*, 109–124. [CrossRef]

81. Jiao, X.; McNairn, H.; Shang, J.; Pattey, E.; Liu, J.; Champagne, C. The sensitivity of RADARSAT-2 polarimetric SAR data to corn and soybean Leaf Area index. *Can. J. Remote Sens.* **2011**, *37*, 69–81. [CrossRef]

82. Moran, M.S.; Alonso, L.; Moreno, J.; Mateo, M.C.; de la Cruz, F.; Montoro, A. A radarsat-2 quad-polarized time series for monitoring crop and soil conditions in Barrax, Spain. *IEEE Trans. Geosci. Remote Sens.* **2012**, *50*, 1057–1070. [CrossRef]

83. Picoli, M.; Lamparelli, R.; Sano, E.; de Mello, J.; Rocha, J. Effect of sugarcane planting row directions on ALOS/PALSAR satellite images. *GISci. Remote Sens.* **2013**, *50*, 349–357.

remote sensing

Article

Sensitivity of Sentinel-1 Backscatter to Vegetation Dynamics: An Austrian Case Study

Mariette Vreugdenhil [1,*], Wolfgang Wagner [1], Bernhard Bauer-Marschallinger [1], Isabella Pfeil [1], Irene Teubner [1], Christoph Rüdiger [2] and Peter Strauss [3]

[1] Department of Geodesy and Geoinformation, TU Wien, 1040 Vienna, Austria;
    wolfgang.wagner@geo.tuwien.ac.at (W.W.); bernhard.bauer-marschallinger@geo.tuwien.ac.at (B.B.-M.);
    isabella.pfeil@geo.tuwien.ac.at (I.P.); irene.teubner@geo.tuwien.ac.at (I.T.)
[2] Department of Civil Engineering, Monash University, 3800 Clayton, Australia; chris.rudiger@monash.edu
[3] Institute for Land and Water Management Research (IKT), Federal Agency of Water Management,
    3252 Petzenkirchen, Austria; peter.strauss@baw.at
* Correspondence: mariette.vreugdenhil@geo.tuwien.ac.at; Tel.: +43-1-55801-12260

Received: 20 July 2018; Accepted: 29 August 2018; Published: 1 September 2018

Abstract: Crop monitoring is of great importance for e.g., yield prediction and increasing water use efficiency. The Copernicus Sentinel-1 mission operated by the European Space Agency provides the opportunity to monitor Earth's surface using radar at high spatial and temporal resolution. Sentinel-1's Synthetic Aperture Radar provides co- and cross-polarized backscatter, enabling the calculation of microwave indices. In this study, we assess the potential of Sentinel-1 VV and VH backscatter and their ratio VH/VV, the cross ratio (CR), to monitor crop conditions. A quantitative assessment is provided based on in situ reference data of vegetation variables for different crops under varying meteorological conditions. Vegetation Water Content (VWC), biomass, Leaf Area Index (LAI) and height are measured in situ for oilseed-rape, corn and winter cereals at different fields during two growing seasons. To quantify the sensitivity of backscatter and microwave indices to vegetation dynamics, linear and exponential models and machine learning methods have been applied to the Sentinel-1 data and in situ measurements. Using an exponential model, the CR can account for 87% and 63% of the variability in VWC for corn and winter cereals. In oilseed-rape, the coefficient of determination (R^2) is lower ($R^2 = 0.34$) due to the large difference in VWC between the two growing seasons and changes in vegetation structure that affect backscatter. Findings from the Random Forest analysis, which uses backscatter, microwave indices and soil moisture as input variables, show that CR is by and large the most important variable to estimate VWC. This study demonstrates, based on a quantitative analysis, the large potential of microwave indices for vegetation monitoring of VWC and phenology.

Keywords: Sentinel-1; vegetation water content; microwave indices; crops

---

1. Introduction

With increasing stress on food supply due to the growing world population and changing climate [1,2], vegetation monitoring and risk mitigation are essential for ensuring food security. By closely tracking crop conditions, agricultural droughts and subsequent crop losses could be better dealt with and yield predictions can be improved. In addition, crop monitoring can assist in more sustainable land management and reducing the use of pesticides and fertilizers.

Spaceborne microwave remote sensing provides the means to monitor vegetation and soil conditions on a range of scales. Synthetic Aperture Radars (SAR) provide observations at a high spatial resolution in the order of tens of meters, which can be used for agricultural crop monitoring [3].

However, until recently, high resolution SAR observations were not frequent enough to be used to monitor vegetation dynamics in a way to be useful to farmers. With the launch of the European Space Agency (ESA) Copernicus Sentinel-1 satellite series, backscatter observations are available at an unprecedented temporal and spatial resolution, with a revisit time of 1.5–4 days over Europe and a spatial resolution of 20 m. Since microwaves are sensitive to the water content in the soil and vegetation and other variables influencing backscatter, i.e., soil roughness and vegetation structure, the challenge in microwave remote sensing is to retrieve the vegetation signal.

At a large scale, many studies have demonstrated the use of microwave sensors for vegetation monitoring, like EUMETSATs Metop Advanced SCATterometers (ASCAT), JAXAs Advanced Microwave Scanning Radiometer 2 (AMSR2), ESAs Soil Moisture Ocean Salinity (SMOS) mission and NASAs Soil Moisture Active Passive (SMAP), [4–11]. The temporal sampling for these products is 1–2 days, but the spatial resolution is relatively coarse with pixels covering tens of kilometres. Often, Vegetation Optical Depth is derived, which is an indicator of the water content in the above ground biomass. At the field scale, many studies have used backscatter directly or indices thereof to find a relation to vegetation dynamics. Ferrazzoli et al. [12] found that HV-polarized backscatter at C-band correlated strongly ($R^2 = 0.75$) with crop biomass over colza, wheat and alfalfa, but that saturation occurred in corn, sunflower and sorghum. Paloscia et al. [13] found high correlations between vegetation biomass and HV-backscatter over broad leaf crops such as sunflower. In addition, Macelloni et al. [14] found an increase in VH backscatter with increasing Leaf Area Index (LAI) over rapeseed sites in Italy and Sweden. Ratios of co- and cross-polarized backscatter observations, i.e., the Radar Vegetation Index (RVI) [10,15] and Cross Ratio (CR) [16], were found to distinguish well between vegetation densities and high linear correlations were found to in situ measured Normalized Difference Vegetation Index (NDVI), LAI and Vegetation Water Content (VWC) over different crops [15]. Wiseman et al. [17] compared dry biomass to C-band RADARSAT backscatter for a six-week period in southern Manitoba, Canada. Significant correlations were found for corn, soybean and oilseed-rape, which increased when applying a logarithm to the observations. In addition, radar backscatter was also found to be sensitive to crop structure changes and phenology. This was also found by Mattia et al. [18] and Satalino et al. [19], where backscatter changed drastically with the emergence of heads in wheat. More recently, Veloso et al. [20] compared Sentinel-1 time series to NDVI time series for wheat, oilseed-rape, corn, soybean and sunflower over test sites in France. Good correspondence was found between SAR data and NDVI. Particularly, the VH/VV ratio could be used for monitoring crop growth cycles. A qualitative comparison was performed between VH/VV and in situ measured biomass for barley and corn and showed a good agreement. These studies demonstrate the potential of SAR and especially Sentinel-1 to monitor vegetation dynamics.

The aim of this study is to further quantify the potential of Sentinel-1 backscatter to monitor vegetation dynamics. We assess the sensitivity of Sentinel-1 VH and VV backscatter and ratio thereof to vegetation dynamics by comparing them to in situ measured vegetation variables, such as VWC, LAI, height and biomass. Destructive vegetation samples were taken for two consecutive years, with very different meteorological conditions, during the growing season of winter cereals, corn and oilseed-rape. A linear model, exponential model and random forest machine learning are used to understand the signal and assess the potential of combining microwave indices from Sentinel-1 to estimate VWC. Testing the use of freely available microwave indices and products in combination with machine learning approaches ensures applicability on a large scale and to ultimately develop predictive models for VWC. The advantage of the presented approach is that no a priori information on vegetation structure is needed, which was often the case in previous studies estimating VWC. This work advances from previous studies by providing for the first time a quantitative performance assessment of Sentinel-1 for monitoring vegetation dynamics over multiple crop types and years.

2. Data

2.1. Site Description

The study is performed in the Hydrological Open Air Laboratory (HOAL) [21], which is a 66 hectare large catchment located in Petzenkirchen, Austria (48°9 N, 15°9 E) and managed by the Austrian Federal Agency of Water Management and TU Wien. Elevation varies between 268 and 323 m and the average slope is 8%. The main land use in the catchment is agriculture and most common crops are winter wheat and corn. Dominant soil types are Cambisols and Planisols with medium to poor infiltration capacity. Average temperature in the HOAL is 9.5°C and mean annual precipitation is 823 mm per year. The peak of precipitation is usually in summer.

In 2016 and 2017, samples of vegetation variables were taken every 12 or 24 days during the growing season, coinciding with the same orbit of Sentinel-1. Only during one orbit of Sentinel-1 samples were taken, since the incidence angle is the same for this orbit. The overpass time of Sentinel-1 was at 5:09 a.m. and sampling was done as close to this time as possible. A total of six sampling units (SU) were selected for sampling of vegetation and soil moisture. Different crops were sampled: oilseed-rape, corn and winter cereals barley and wheat. Measurements were taken at random locations within a 20 m radius from the centre of the SU. The location of the SUs is depicted in Figure 1.

2.2. In Situ Data

2.2.1. Biomass and Vegetation Water Content

Wet biomass (BM) and VWC are determined by destructive sampling and oven drying of the samples. Samples were separated in stems and leafs, resulting in three VWC measurements, total VWC (VWC_t), leaf VWC (VWC_l) and stem VWC (VWC_s). Since the croplands are owned by farmers and sampling is frequent, per sampling day a number of plants or rows were cut and BM and VWC_t are determined per plant or row. In corn and oilseed-rape, three plants per sampling day were taken. In winter cereal, two or three rows were cut with a total length of 50 centimeters. To acquire an estimate of BM and VWC per m^2, the row or plant density on 1 m^2 was counted for every croptype and subsequently BM and VWC at 1 m^2 were calculated. In the laboratory, samples were separated for stems and leafs and separately weighted before and after oven-drying. Oven-drying was done at 70 °C for 24 to 48 h until the dry weight was stable.

2.2.2. Leaf Area Index

Leaf Area Index (LAI) is measured using a Licor LAI-2000 Plant Canopy Analyzer (Lincoln, NE, USA). The LAI-2000 measures light intensity with a fish-eye lense in five concentric field of views, with zenith angles of 7, 23, 38, 53 and 68 degrees. For every measurement, one above-canopy and six below-canopy readings are acquired. The ratio between the above- and below-canopy reading is used to calculate gap fraction and subsequently LAI using the built-in software C2000 LI-COR software (Lincoln, NE, USA). For each SU and sampling day, a total of five repeat LAI measurements were taken, which were distributed randomly around the SU centre. The five LAI measurements were then averaged.

2.2.3. Vegetation Height and Status

Vegetation height was measured at five locations for every sampling unit and sampling day and averaged. Height was noted in the field. In addition, photographs were taken of the vegetation to document vegetation status, e.g., seeding, stem extension, flowering, heading, harvest and corn growth stages. An overview of vegetation status is given in Table 1.

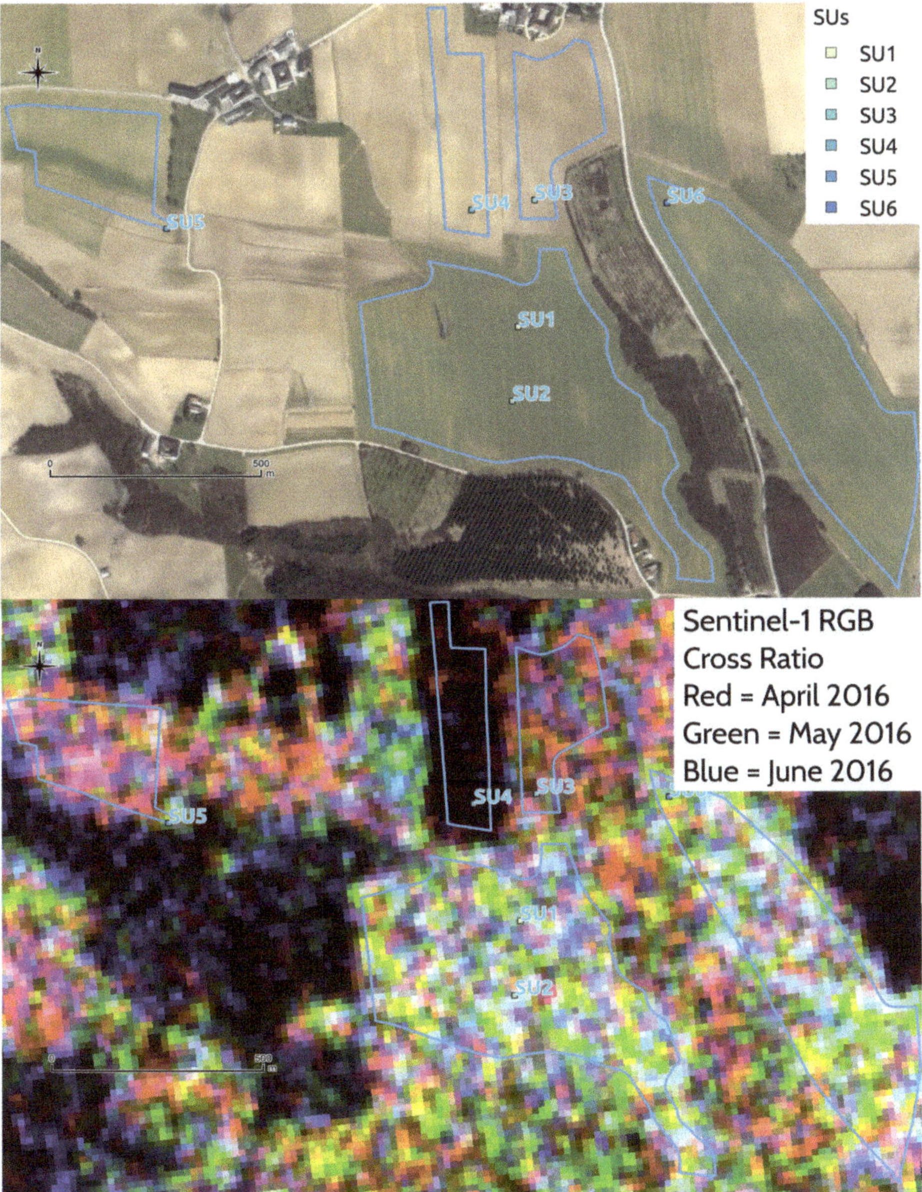

Figure 1. Map of the HOAL indicating the Sampling Unit locations (blue squares) and the area per field used to calculate Sentinel-1 indices (blue polygons). The top graph shows a Google Earth Image as the background, the lower figure depicts monthly averaged Sentinel-1 CR for the months April (Red), May (Green) and June (Blue) 2016.

Table 1. Crop type, number of samples (NS), seeding and harvest dates (year/month/day and day of year) per year and SU. In addition, timing of phenological stages stem extension (St.), flowering (Fl.), heading (He.) and ripening (Ri.) are given in day of year for oilseed-rape (Rape) and winter cereal (Cereal) and growth stages' three leaves (V3), 12 leaves and cob development (V12), flowering (R1) and maturity (R6) are given for corn.

SU	Area	Crop	NS	Seeding	Harvest	St./V3	Fl./V12	He./R1	Ri./R6
1	5 ha	Rape	3	'15/08/29 (241)	'16/07/23 (205)		~104		~130
		Cereal	11	'16/11/29 (334)	'17/07/19 (200)	~135		~150	~170
2	6.1 ha	Rape	11	'15/08/29 (241)	'16/07/23 (205)		~104		~130
		Cereal	11	'16/11/29 (334)	'17/07/19 (200)	~135		~150	~170
3	2.6 ha	Cereal	9	'15/10/05 (278)	'16/07/23 (205)	~100		~145	~165
		Corn	11	'17/04/22 (112)	'17/10/25 (298)	~135	~180	~205	~252
4	2.3 ha	Corn	10	'16/04/27 (118)	'16/09/30 (274)	~150	~188	~203	~253
		Cereal	11	'16/10/31 (305)	'17/07/19 (200)	~135		~150	~170
5	3.2 ha	Cereal	9	'15/10/02 (275)	'16/07/01 (183)	~100		~130	~150
		Rape	11	'16/08/25 (238)	'17/07/19 (200)		~100		~135
6	9.4 ha	Rape	0	'15/08/28 (240)	'16/07/20 (202)		~104		~130
		Cereal	11	'16/11/01 (306)	'17/07/19 (200)	~135		~150	~170

2.2.4. Soil Moisture and Precipitation

Soil moisture is monitored continuously with the in situ soil moisture network, which is installed in the HOAL. Soil moisture and soil temperature are measured using low-current Time Domain Transmission (TDT) probes called SPADE (sceme.de GmbH i.G., Horn-Bad Meinberg, Germany). In the summer of 2013, soil moisture stations were installed at 30 locations throughout the HOAL. Per station, four sensors are horizontally installed at the following depths: 0.05, 0.10, 0.20 and 0.50 m. The network is calibrated in natural soil, by creating a test bed in the field with homogenized soil and TDT sensors and laboratory calibrated TDR sensors were installed at 0.10 m depth and two at 0.20 m depth. The four TDT sensors were calibrated with the TDR as the reference measurement of soil moisture. For this study, we used soil moisture from one station at 5 cm, which is representative for the rest of the catchment. Precipitation is recorded every minute at four OTT Pluvio rain gauges distributed throughout the catchment. In this study, we used daily sums of precipitation, averaged for the four rain gauges.

2.3. Sentinel-1

Sentinel-1 is part of Europe's Copernicus programme and at the moment has two satellites in orbit, Sentinel-1A and Sentinel-1B launched in April 2014 and 2016, respectively. The Sentinel-1 satellites carry Synthetic Aperture Radars (SAR), providing backscatter at C-band (5.405 GHz). The acquisition mode over (non-polar) land is Interferometric Wide (IW) swath mode. The SAR instruments are designed to provide co- and cross-polarized backscatter over a 250 km swath at a 20 m spatial resolution in single look. The temporal revisit time of one Sentinel-1 satellite is 12 days, and temporal coverage is 1.5–4 days over Europe using both Sentinel-1A and Sentinel-1B.

3. Methods

This study investigates the potential of microwave indices for crop monitoring in two ways. First, the temporal evolution of vegetation variables, soil moisture and Sentinel-1 VH and VV backscatter and CR are investigated and discussed in detail. Secondly, quantitative analyses are performed by calculating correlation statistics between backscatter data fitted to in situ observations using a linear model, exponential model and Random Forest (RF) modelling.

3.1. Microwave Indices from Sentinel-1

Sentinel-1 data processed at the Earth Observation Data Centre for Water Resources monitoring GmbH (EODC) and TU Wien using the Sentinel Application Platform (SNAP) toolbox provided by ESA. The first step is the radiometric calibration that converts the intensity into normalized backscatter (σ°). This is followed by the terrain correction using Range Doppler Terrain correction with the Shuttle Radar Topography Mission digital elevation model provided with SNAP and georeferencing. The resulting σ° is available at a 10 m sampling.

To bypass incidence angle effects and other observation geometry effects, only one orbit is used, providing Sentinel-1A observations every 12 days. For every crop field that contains an SU, σ° is averaged over the whole field in the linear domain. To rule out border effects, a 50 m buffer from the field borders is excluded from the calculation. The areas used and an RGB image of S1 Cross Ratio between VH and VV backscatter for April, May and June are shown in Figure 1. First, both VV and VH backscatter are averaged and then CR is calculated as VH/VV in the linear domain. As a last step, all values are converted to the logarithmic domain.

Total backscatter from a vegetated surface does not only comprise scattering from the vegetation itself. It also includes the backscatter originating from the underlying surface attenuated by the vegetation, and interaction between soil and vegetation. To simplify backscatter from vegetated surfaces, vegetation is often regarded as a cloud of randomly distributed water droplets, which are structurally held in place by dry matter. The predominant mechanism responsible for backscatter from vegetation is volume scattering. Often, cross-polarized backscatter is most indicative for these scattering mechanisms and high correlations have been found between cross-polarized backscatter at C-Band and crop condition indicators such as LAI and biomass [12,13,22]. Since cross-polarized backscatter increases more strongly with volume scattering than co-polarized backscatter, CR increases with vegetation. By using the CR, the effect of soil moisture is reduced as well as soil-vegetation interaction effects [20]. However, a challenge in using cross-polarized backscatter or CR is the effect of soil roughness and vegetation structure. Soil roughness also causes depolarisation when soils are rough and have the same backscatter or CR values as a vegetated surface. Vegetation structure significantly impacts backscatter and, for some crops, structure can change throughout the growing season. Hence, monitoring VWC_t using SAR is complex and more research is needed quantifying the potential of SAR for crop monitoring.

3.2. Linear and Exponential Model

For all quantitative analyses, in situ and remote sensing observations are aggregated per crop type over the different fields and years. In situ measurements of vegetation variables were only taken when crops are present on the field and not during intercropping periods. The Sentinel-1 observations are temporally matched to the in situ measurements and analyses are thus only performed on dates when crops are present. To relate Sentinel-1 backscatter and CR to in situ observations, two models are used. First, Sentinel-1 observations are directly compared to in situ observations using a linear model. Secondly, as done by Wiseman et al. [17], an exponential model is developed, calculating the logarithm of the dependent variables. Temporal correlation coefficients (R^2) are calculated between results from the linear and exponential model results and the in situ data of VWC_t, VWC_l, VWC_s, BM, LAI and height. Results of the temporal correlation analyses are discussed per crop type and per model, i.e., linear or exponential.

3.3. Random Forest Modeling

The sensitivity of Sentinel-1 backscatter to VWC_t is further quantified using a supervised random forest (RF) machine learning approach. RF has the advantage that it does not make any assumptions about the relation between input variables and the response variable and can identify nonlinear relationships. RF uses multiple regression trees that are trained on the response variable, in this study VWC_t. RF selects a random set of input variables and data points. Four RF trees all with a maximum of 500 decision trees and a depth of 30 are trained using all in situ observations; a separate model per crop and a model using all data and crop type as categorical variable. As inputs, the following EO based microwave indices and products are used: Sentinel-1 VH, VV and CR. Since soil moisture could be an important indicator for VWC_t, surface soil moisture from in situ observations is used as additional input variable. Two soil moisture variables are used, the soil moisture at the moment of the Sentinel-1 overpass (SSM) and antecedent soil moisture (ASSM), which represents soil moisture averaged over the previous three days.

The advantage of RF is the analysis of feature importance, quantifying the relative importance of the different variables for estimating VWC_t. The variable importance quantifies the decrease in performance of the model when the variable is left out of the regressor. The Out-Of-Bag (OOB) R^2 score is calculated to assess the performance of the models. Bootstrapping is used to train the RF and the OOB R^2 score is the average R^2 calculated from the trees that do not contain a certain value for VWC_t in the respective bootstrap sample. We emphasize that due to the sample size RF is not used to predict VWC_t but as a tool to analyse the sensitivity of the different microwave indices to VWC_t and their importance to represent variability in VWC_t.

4. Results and Discussion

4.1. Time Series Analysis

Two complete growing seasons have been sampled and are covered by Sentinel-1. In Austria, these two years have differed significantly from each other in terms of water availability, temperature and radiation. In addition, 2016 received more than average precipitation especially in January, February, May and June, whereas 2017 received less rainfall than average. These significantly different meteorological conditions between the years provide the opportunity to investigate the sensitivity of backscatter to vegetation dynamics under varying meteorological conditions.

4.1.1. Oilseed-Rape

Oilseed-rape is a broadleaf plant and has a different vegetation structure than cereals. In the HOAL, oilseed-rape is seeded at 28, 29 and 25 of August 2015 and 2016 (doy 240–241 and 238) and harvested at 23, 20 and 19 of July (doy 205, 202 and 200) for 2016 (Figure 2) and 2017 (Figure 3), respectively. As for planting, flowering and ripening also happened around the same time for both years, with flowering starting at doy 100–104 and development of pods and ripening at doy 130–135. VWC_t however differs greatly between the two years, with values around 15 kg/m^2 in 2016 and 4 kg/m^2 in 2017. This is likely caused by the different meteorological conditions between the two years. Where 2016 was an exceptionally wet year, 2017 was drier especially in May, June and July. The difference in water content originates mainly from the stem water content, which is higher in 2016 than in 2017. After a rainfall event in June 2017, water content increases with 4 kg/m^2 and is more comparable to 2016. This indicates rapid changes of stem water content in oilseed-rape when soil moisture is available.

For both years, the CR increases until the end of May (doy 150), apart from a peak in January 2016 and a dip in February and April 2017, which are caused by frozen soils and snow cover. VH also increases throughout the growing season. Starting around doy 110, a decrease is observed in both CR and VH backscatter, followed by a strong increase of several dB. The small decrease coincides with the start of flowering at doy 100–104. The strong increase is observed from doy 120 to doy 132, which is at a similar time as the start of ripening. Both the decrease and strong increase related to flowering and ripening were also observed by Wiseman et al. [17] in Canada with C-band Radarsat and Veloso et al. [20] over France with Sentinel-1. The increase of VH with increasing vegetation was also observed by Macelloni et al. [14], who found that cross-polarized backscatter increased with increasing LAI over test sites in Montespertoli, Italy, and Fjardhundra, Sweden. VV backscatter varies over the year simultaneously with soil moisture, with lowest VV backscatter on doy 120 in 2016, which coincides with a drop in soil moisture. However, the decrease in soil moisture starting from doy 125 in both years is not reflected in VV backscatter. Instead, VV backscatter increases together with CR and VH backscatter.

Even though there are clear differences in VWC_t, as a result of a large difference in VWC_s, and soil moisture, the CR, VV and VH are very similar for 2016 and 2017. This suggests that both VV and VH backscatter are mostly sensitive to the leaf water content but not the stem water content.

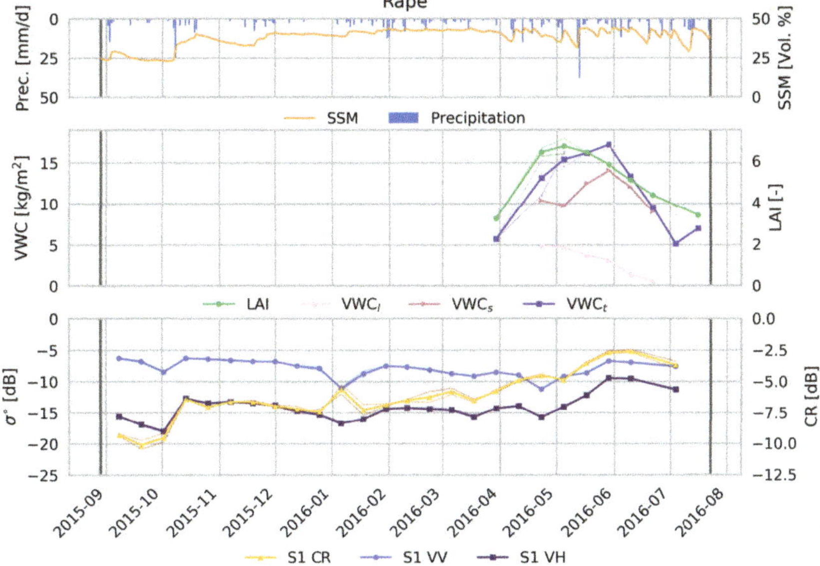

Figure 2. Time series of precipitation (blue) and soil moisture at 5 cm depth (orange) (top). Total VWC (VWC_t, blue), stems (VWC_s, pink) and leafs (VWC_l, light pink), LAI (green) (middle). S-1 VH (dark blue) and VV (light blue) and CR (yellow) (bottom). The separate fields are indicated with transparent lines. Grey vertical lines are planting and harvesting dates.

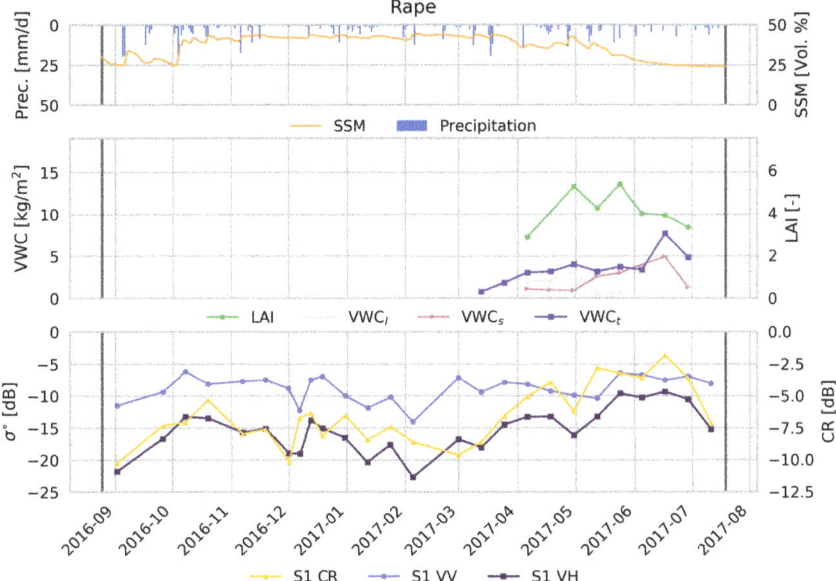

Figure 3. Time series of Sentinel-1 and in situ variables as in Figure 2 for 2017.

4.1.2. Corn

Time series of in situ data and Sentinel-1 for Corn are shown in Figures 4 and 5. Like oilseed-rape, corn is a broadleaf plant. Corn is planted on 27 (doy 118) and 22 (doy 112) of April and harvested on 30 September 2016 (doy 274) and 25 October 2017 (doy 298). Around doy 135 in 2017 and doy 150 in 2016, growth stage V3 is reached, when three leaves are visible and the plant is around 20 cm high. Stage V10–V12, when 10–12 leafs have emerged, is reached at doy 180 in 2017 and 188 in 2016. The pollination stage (VT/R1) is reached at doy 203 in 2016 and doy 192 in 2017. VWC_t increases rapidly until doy 225 in 2016 and doy 205 in 2017. In 2016, VWC_t increases steadily and starts to decrease when soil moisture starts to decrease. In 2017, the decrease occurs earlier and VWC_t then varies until harvesting.

The CR starts to increase strongly from doy 135 in 2017 and 150 in 2016, 3–4 weeks after planting at growth stage V3 (Figures 4 and 5 and Table 1). In both years, CR increases strongly until doy 180, from -12 dB to -6 dB. The end of the strong increase coincides with growth phase V10–V12. At this time, VWC_t also exceeds 2 kg/m^2 and LAI exceeds 3. Similar thresholds were found by Ferrazzoli et al. [12] and Jiao et al. [23] where backscatter and indices saturated at LAI of 2–3. In both years, CR keeps increasing slightly until doy 192 in 2017 and 205 in 203 in 2016. At this time, leaves have fully developed, dry matter has accumulated the most mass and pollination starts (VT/R1). In 2017, CR varies with VWC_t, decreasing from doy 204 to 216 and then increasing to doy 240 and decreasing again during the ripening phase until harvest. This shows that, when the crop is fully grown, CR responds to small changes in VWC_t. In general, VV backscatter decreases from doy 150, simultaneously with SSM. Between doy 150 and 200, VV backscatter is lower in 2017, as is soil moisture. The sensitivity of VV to soil moisture in corn can be explained by the row distance between corn. With a row spacing of 70 cm, bare soil is still visible until a late stage, hence VV backscatter is still sensitive to soil moisture and not yet strongly attenuated by vegetation. In the same period, VH backscatter increases, with increasing VWC_t and with increasing LAI and closing crop cover. The same was found by Macelloni et al. [14] and Ferrazzoli et al. [12] who found an increase in C-band backscatter with increasing LAI in corn. After harvest of the corn at doy 290 in 2016 and 273 in 2017, backscatter and CR drop.

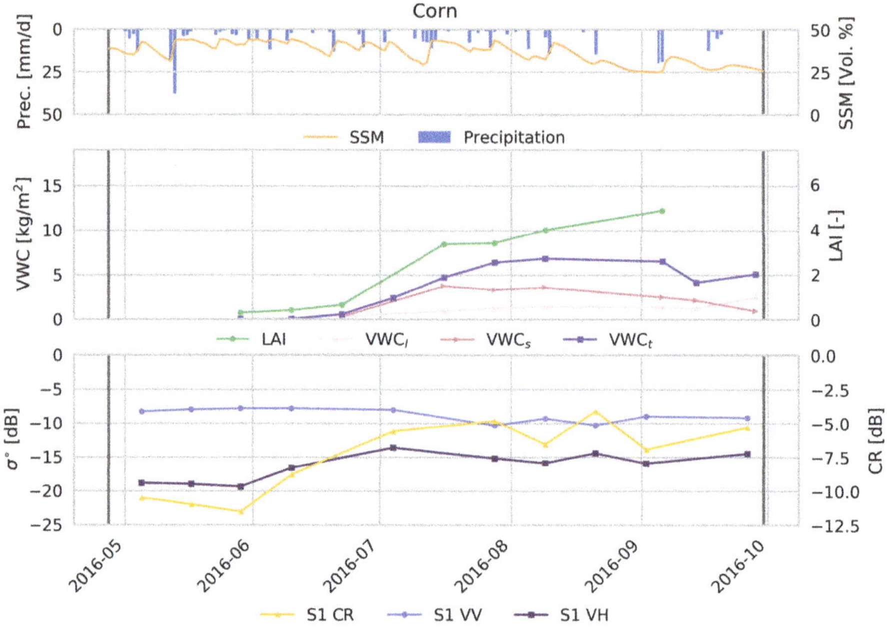

Figure 4. Time series of Sentinel-1 and in situ variables as in Figure 2 for corn 2016.

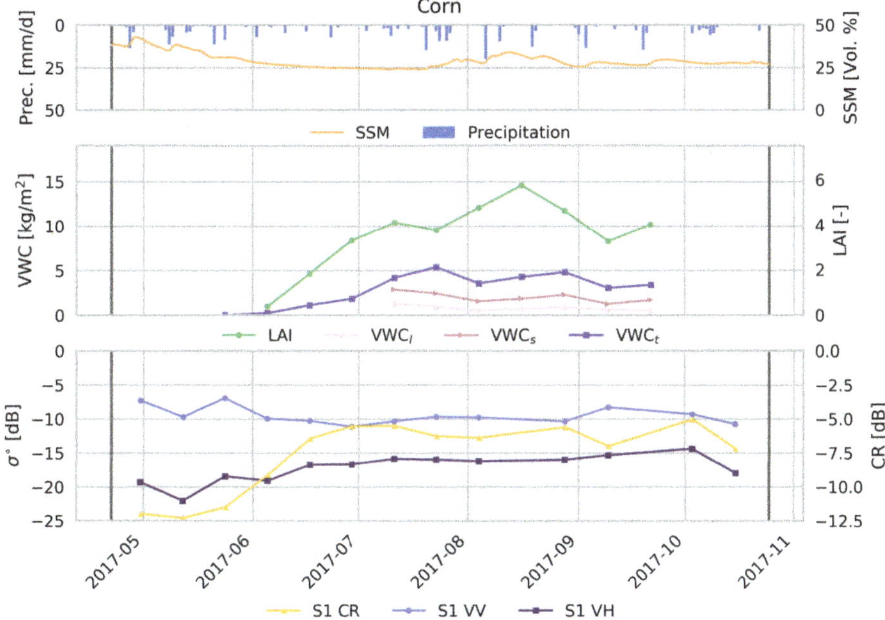

Figure 5. Time series of Sentinel-1 and in situ variables as in Figure 2 for corn 2017.

For corn, monitoring of growth stages V10–R1 is of great importance to farmers. Moisture and heat stress during the stages V10–R1 have the largest effect on final yield. In addition, management practices during these stages, e.g., application of fertilizers, can have a positive effect on yields. Hence, information of the timing of these phases and water content of the plant is pivotal and can improve farming practices.

4.1.3. Winter Cereals

In the HOAL, winter cereals comprise barley and wheat, both narrow-leaf crops. The growing season, phenology and structure of the two crops is very similar. The time series of the in situ data and Sentinel-1 data is shown for the growing season of 2016 and 2017 in Figures 6 and 7. Winter cereals were planted in October 2015 and November 2016 (doy 183, 200, 205). The earlier planting date and mild winter in 2016 led to an earlier start of the stem extension phase in 2016, at the middle of April (doy 100). In 2017, January and April were colder than usual and this in combination with the later planting date led to the stem extension phase to start half of May (doy 135). Heading occurred in the last weeks of May and beginning of June (doy 130–150) for both years. The shift in growth phases between 2016 and 2017 is also observed in VWC_t (Figures 6 and 7), which increases at an earlier stage in 2016 (doy 90) than in 2017 (doy 120). VWC_t keeps increasing during the grain filling period until its peak around doy 150, which is the same for both years. During the ripening stage, VWC_t decreases steadily until harvest.

Strong variations in CR in winter, especially January 2016 and February 2017, are caused by frozen soils and snow cover as was also observed in the CR for oilseed-rape. A clear steady increase in CR can be seen from doy 0, which is during the tillering phase of winter cereals, which was also observed for winter barley by Veloso et al. [20]. The shift in growth phases between years, as was observed in in situ measurements, can also be observed in CR. CR starts to increase earlier and is overall higher in 2016 (Figure 6). This is also visible in the maxima of CR, which is at doy 102 in 2016 (Figure 6) and doy 126 in 2017 (Figure 7). The maximum in CR coincides with the start of the stem extension phase. During the stem extension phase, CR starts to decrease. For all fields, a dip in CR is observed around doy 150, coinciding with the maximum in VWC_t and the start of heading and flowering. During the grain fill period, CR starts to rise again and then decreases or stays constant. The changes in CR are most likely driven by phenology and large changes in vegetation structure during stem extension and heading. Similar variations related to structure were also found by Mattia et al. [18] and Satalino et al. [24]. VV backscatter decreases until the beginning of the stem extension phase of the winter cereal. Although soil moisture decreases too, small variations in soil moisture are not reflected in the VV signal. When VWC_t in the crop decreases (from day 150), a slight increase in VV backscatter can be observed even though soil moisture is still decreasing (especially in 2017). In addition, VV backscatter is lower between doy 85 and 115 in 2016 than in 2017, even though soil moisture was higher in 2016. Since vegetation water content in winter cereal was higher during this time in 2016, the lower VV backscatter in 2016 is likely caused by the increased attenuation. VH backscatter decreases during the last period of tillering. The decrease occurs earlier in 2016 than in 2017. As soon as stem extension starts, VH starts to increase again, likely as a result of volume scattering mechanisms.

The sensitivity of CR to structural changes related to growth stages of winter cereals provides pivotal information on the growth stage to farmers. Winter cereals, especially wheat, are very sensitive to diseases during the flowering stage when soils are wet and temperatures are high, i.e., Fusarium a fungal disease leading to harmful mycotoxins. Infection risk can be reduced by targeted application of pesticides during flowering Chala et al. [25]. Hence, additional information on timing of flowering can assist in disease prevention and limit the use of chemicals.

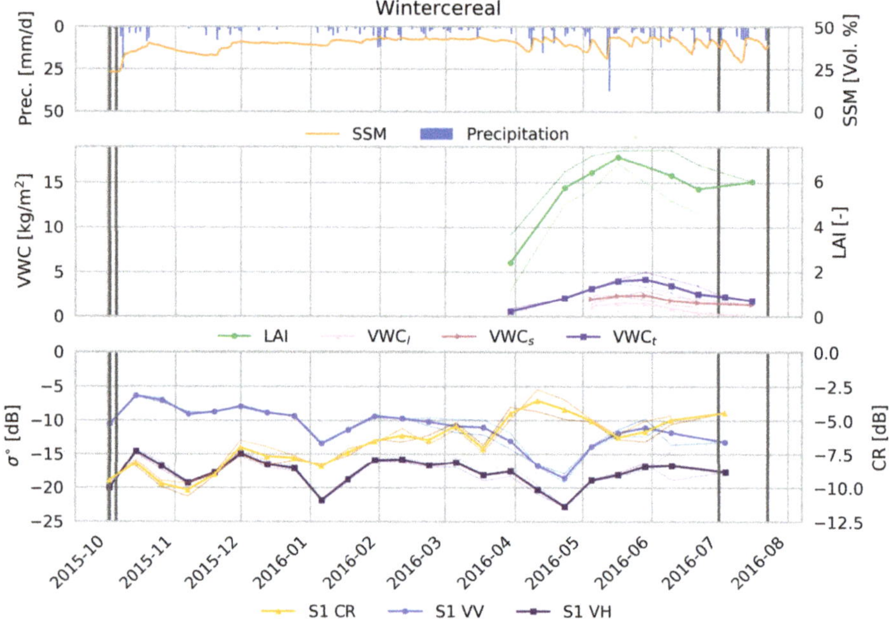

Figure 6. Time series of Sentinel-1 and in situ variables as in Figure 2 for winter cereal 2016.

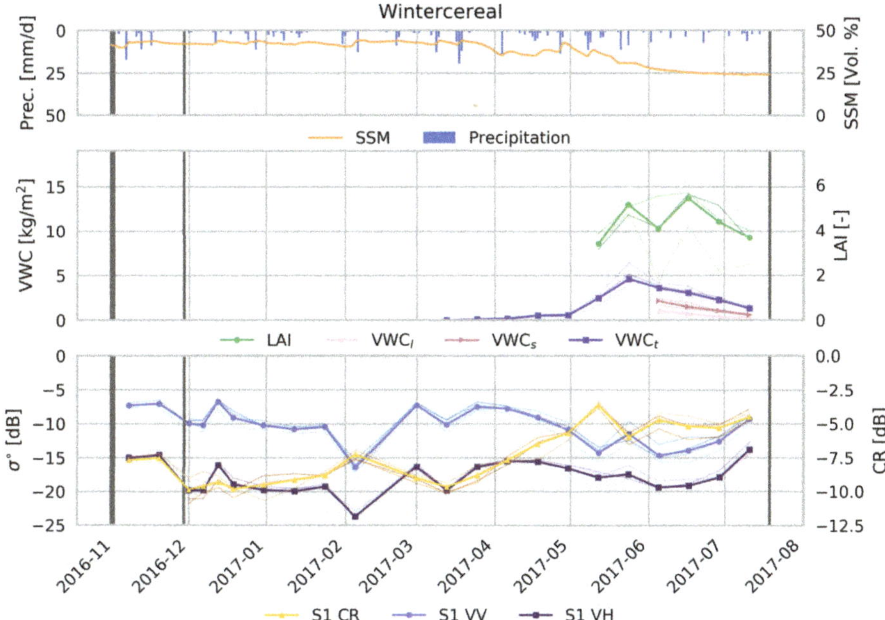

Figure 7. Time series of Sentinel-1 and in situ variables as in Figure 2 for winter cereal 2017.

4.2. Temporal Evolution of CR

Figure 8 shows the CR averaged per crop type for all fields within the HOAL for the years 2015–2017 and averaged over all years. The CR for oilseed-rape and winter cereal, which are planted in the previous autumn, starts to rise in spring (doy 70–150) and decreases in early summer (doy 170–200). CR for oilseed-rape increases steadily although some variations are visible around doy 110, which is related to flowering, as discussed in Section 4.1.1. In winter cereal, CR shows a clear dip around doy 130–150, associated with the heading as discussed in Section 4.1.3. Between years, a clear difference between the CR can be seen. In 2016, CR starts to rise earlier, from January onward. This is most likely caused by the mild winter of 2016 compared to 2015 and 2017. Corn is usually planted in April and starts to rise later in the year (doy 150–170) and decreases with harvest in autumn (doy 250–300). The CR increases continuously until the ripening stage commences. In 2017, CR in corn has its peak in June and decreases steadily until October. This could be due to the dry conditions of 2017 compared to 2015 and 2016, especially in June, July and August.

For all crops, the variability is much smaller during the growing period of the crop. Around planting time and harvest, the variability is higher since data is not masked for the planting and harvest dates per field. Due to differences in planting and harvesting dates and different field management practices, e.g., some fields are left bare while others have inter-cropping, variability of CR increases.

In the last window, the CR shows a consistent and distinct behaviour for every crop type even though different years are averaged with very different meteorological conditions and crops rotate on a yearly basis between fields. This emphasizes the sensitivity of CR to vegetation dynamics.

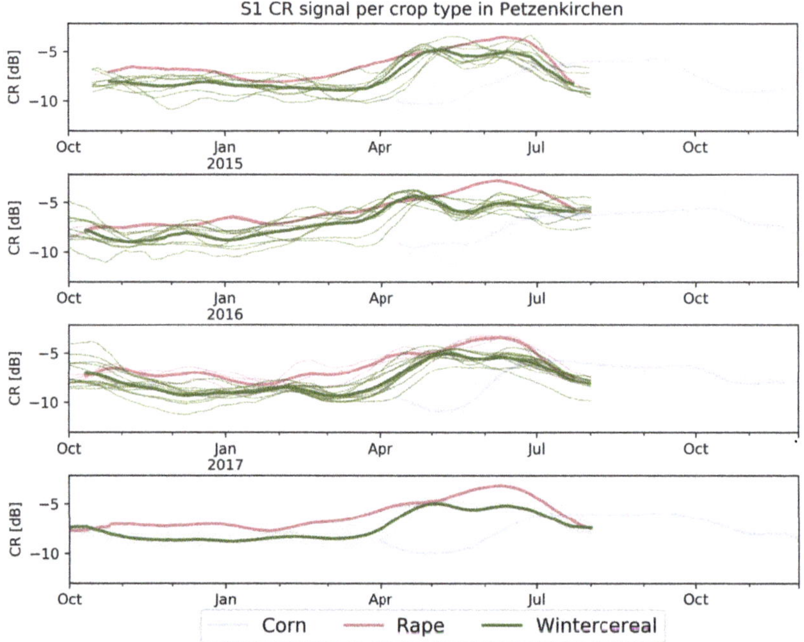

Figure 8. Cross Ratio of Sentinel-1 for different crop types per year (2015–2017) and averaged over all fields and years (lower window). The shaded areas represent the 25th and 75th percentile.

4.3. Quantitative Comparison

4.3.1. Linear and Exponential Model Results

The scatterplots in Figure 9 and R^2 values in Table 2 illustrate the relation between Sentinel-1 parameters and in situ measured variables. From the three microwave indices, i.e., VH, VV and CR, CR shows the highest correlation to VWC_t for all crops, with R^2 of 0.16, 0.48 and 0.22 for oilseed-rape (not-significant), corn and winter cereals, respectively, and $R^2 = 0.21$ for all crops together. CR correlates strongest to LAI in corn ($R^2 = 0.62$), and to height for oilseed-rape ($R^2 = 0.51$) and winter cereals ($R^2 = 0.50$). In corn, VH backscatter is most sensitive to LAI ($R^2 = 0.61$) and VWC_l ($R^2 = 0.49$). This was also found by Macelloni et al. [14], who, using a simple first order radiative transfer model, showed that, at the C-band, leaves make a significant contribution to scattering in broad leaf crops. Contrary to oilseed-rape and winter cereal, in corn, the VV is moderately sensitive to soil moisture ($R^2 = 0.24$), which was also observed in the time series evolution. This can be explained by the fact that, for the other crops, sampling started in March, when the canopy was already closed, and attenuation already affects the VV signal. The planting date in April and the row spacing of corn leaves sufficient bare soil to still observe variations in soil moisture.

The scatterplots (Figure 9) clearly indicate the nonlinear relationship between CR and VWC_t with a large sensitivity to low VWC_t values and saturation at high VWC_t. This behaviour is observed for all vegetation variables and for all crop types. At denser vegetation, with higher VWC_t, BM, height and LAI, the CR appears to saturate and small uncertainties in CR can lead to large variations in VWC_t. This was also seen in the time series analysis, where CR was more sensitive to vegetation variables especially in the early growing season. Saturation at higher crop density was also observed by Ferrazzoli et al. [26] for HV backscatter, Paloscia et al. [27] for VH/VV and Wiseman et al. [17] for HV backscatter and the Cloude–Pottier decomposition measure entropy. Consequently, the nonlinear relationship negatively affects the correlation of the linear model. As done by Wiseman et al. [17], an exponential model is developed, calculating the logarithm of the dependent variables. The R^2 for all crops increases with 0.42 to $R^2 = 0.63$. Values for R^2 increase dramatically for winter cereals and corn to $R^2 = 0.87$ and $R^2 = 0.63$. In winter cereals CR is still a strong indicator for height ($R^2 = 0.68$); however, using an exponential model, CR is almost as sensitive to VWC_t as it is to height ($R^2 = 0.63$). For oilseed-rape, CR is still most sensitive to height and to less extent to VWC_t ($R^2 = 0.34$).

The low explained variability of VWC_t by CR in oilseed-rape can be explained by the large difference in VWC_t between 2016 and 2017. The large difference in total VWC_t originates from the difference in stem water (Figures 2 and 3). Correlation to VWC_s is also low ($R^2 = 0.36$). With the wavelength of C-band, it is expected that, for broad leaf crops, backscatter is more sensitive to the leaf water content. However, there is also no strong correlation between VWC_l and CR ($R^2 = 0.31$). Similar results were found by Wiseman et al. [17] and attributed the low explained variance in dry biomass to the effect of crop phenology and structure due to flowering and seeding rather than accumulation of biomass as discussed in Section 4.1.1.

4.3.2. Random Forest Model Results

The RF modelling is used to elucidate the potential of different microwave indices to describe variability in VWC. The RF is applied to the logarithm of the VWC_t and results of OOB R^2 score and feature importance are illustrated in Table 3.

For oilseed-rape, the OOB R^2 score of 0.31 indicates that the RF model is not able to estimate VWC_t better than a single input feature, as, for example, an exponential model using only CR. The feature importance does show that CR is the most important input variable to predict VWC_t with a relative importance of 0.26. ASSM has a similar relative importance followed by SSM. This can be explained by the large difference in VWC_t between 2016 and 2017, where also soil moisture was higher in 2016 than 2017 (Figures 2 ande 3). The VWC_t in oilseed-rape is strongly related to soil moisture availability and hence soil moisture is an important predictive variable for VWC_t in oilseed-rape.

For winter cereals, the RF approach provides better results than the linear and exponential model, $R^2 = 0.81$ for RF versus $R^2 = 0.63$ for the exponential model. For winter cereals, CR varied with phenological stages and vegetation structure after the tillering phase. As such, more complex machine learning methods, which include more input variables, can better explain variability in narrow-leaf crops like winter cereals. Still, the most important variable to estimate VWC_t is CR with a relative importance of 0.31, followed by ASSM and S1VV, demonstrating the potential of CR for monitoring VWC_t in winter cereal.

For corn, the exponential model performs better than the RF approach with $R^2 = 0.87$ for the exponential model and $R^2 = 0.74$ for RF. This indicates that simple exponential model between CR and VWC_t performs better for corn, and that introducing SSM and ASSM decreases the predictive performance. This is further emphasized by the variable importance for corn, where the Sentinel-1 input variables CR and VH are most important with relative importances of 0.30 and 0.25. This can be explained by the fact that the VV backscatter for corn is sensitive to soil moisture, so that additional information on soil moisture is not of added value.

The advantage of RF is that it can also use categorical variables for building the regression model. This is done for the last RF model where all data from all crops are grouped and categorical values are given for the different crop types. The OOB R^2 score of 0.80 demonstrates the ability to describe the variability in VWC_t based on the microwave indices and soil moisture. Looking at the feature importance, it is evident that the CR is the variable best capturing variability in VWC_t with a relative importance of 0.35, followed by, although at a much lower importance, VV (0.17) and VH (0.16).

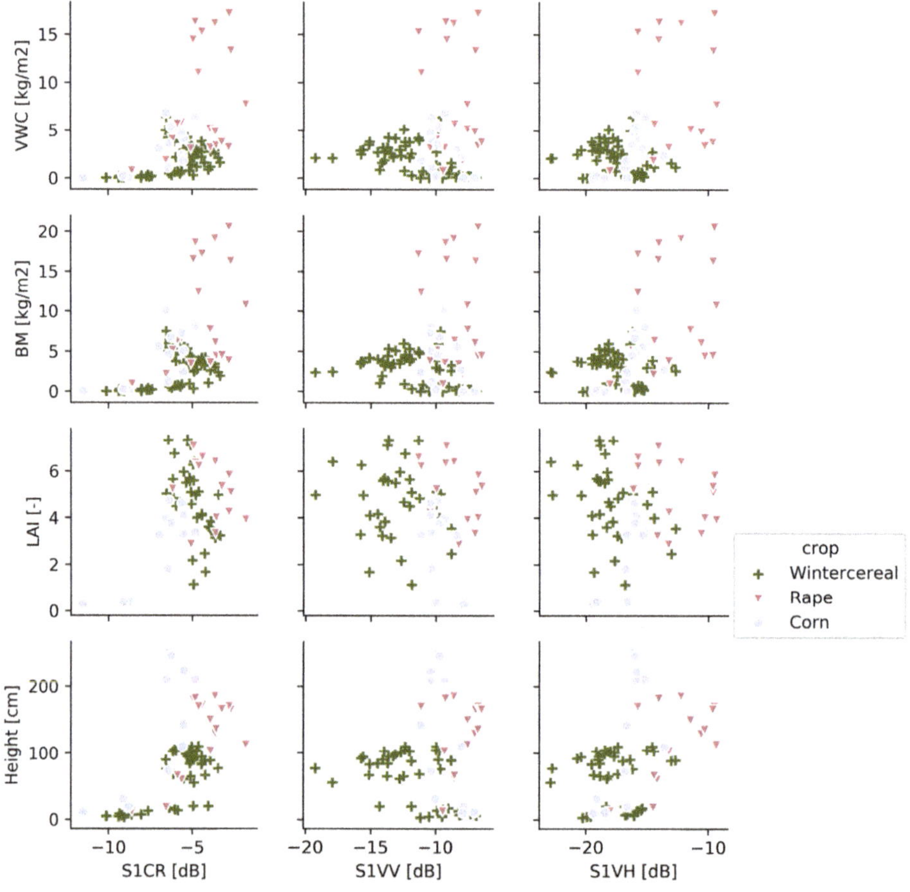

Figure 9. Scatter plot of S1 CR (left), VV (middle) and VH (right) backscatter with in situ measured vegetation variables VWC, BM and Height, and in situ measured soil moisture for the different crop types Oilseed-rape (red), Corn (purple) and winter cereal (green).

Table 2. R^2 per crop type (Oilseed-rape, Corn and Winter cereal) between Sentinel-1 CR, VH and VV for variables total VWC (VWC_t), VWC from leafs (VWC_l), VWC from stems (VWC_s), biomass (BM), Leaf Area Index (LAI), plant height (H) and soil moisture (SM) for both the linear and exponential model. In black are significant R^2 values ($p < 0.05$).

Crop	Model	Var	VWC_t	VWC_l	VWC_s	BM	LAI	H	SM
Oilseed-rape	linear	CR	0.16	0.27	0.14	0.19	0.03	**0.39**	0.07
		VH	0.03	0.29	0.06	0.05	0.03	0.15	0.16
		VV	0.02	0.19	0.00	0.01	0.15	0.00	0.15
	exponential	CR	**0.34**	**0.31**	**0.36**	**0.34**	0.08	**0.51**	0.06
		VH	0.10	0.23	0.12	0.12	0.01	0.23	0.16
		VV	0.01	0.11	0.00	0.00	0.13	0.01	0.16

Table 2. *Cont.*

Crop	Model	Var	VWC_t	VWC_l	VWC_s	BM	LAI	H	SM
Corn	linear	CR	0.48	0.16	0.02	0.42	0.62	0.55	0.07
		VH	0.44	0.49	0.16	0.43	0.61	0.48	0.00
		VV	0.15	0.01	0.23	0.10	0.19	0.20	0.24
	exponential	CR	0.87	0.18	0.11	0.85	0.78	0.83	0.09
		VH	0.62	0.35	0.27	0.63	0.73	0.61	0.00
		VV	0.42	0.00	0.28	0.39	0.27	0.40	0.27
Winter cereal	linear	CR	0.22	0.34	0.22	0.26	0.30	0.50	0.16
		VH	0.08	0.14	0.25	0.04	0.13	0.00	0.01
		VV	0.25	0.04	0.12	0.22	0.02	0.21	0.04
	exponential	CR	0.63	0.27	0.19	0.64	0.22	0.68	0.15
		VH	0.02	0.37	0.35	0.01	0.10	0.00	0.01
		VV	0.35	0.19	0.38	0.32	0.01	0.28	0.04

Table 3. OOB R^2 score and variable importance of Sentinel-1 CR, VH and VV, and surface soil moisture (SSM) and antecedent soil moisture averaged over three days prior (ASSM) from the RF modelling for VWC_t per crop type.

Crop	OOB R^2 Score	f1	f2	f3	f4	f5
Oilseed-rape	0.31	S1CR 0.26	ASSM 0.26	SSM 0.20	S1VH 0.15	S1VV 0.13
Corn	0.74	S1CR 0.30	S1VH 0.25	ASSM 0.18	S1VV 0.15	SSM 0.12
Winter cereal	0.81	S1CR 0.31	ASSM 0.20	S1VV 0.18	SSM 0.16	S1VH 0.16
All	0.80	S1CR 0.35	S1VV 0.17	S1VH 0.16	ASSM 0.12	SSM 0.11

5. Conclusions

This study assessed the sensitivity of Sentinel-1 SAR backscatter and a ratio derived thereof to vegetation dynamics. In situ data on VWC, biomass, height and LAI was collected over two growing seasons for oilseed-rape, corn and winter cereals. Using three approaches, a linear model, exponential model and random forest modelling, a quantitative analysis between SAR parameters and in situ data was performed.

Time series analysis demonstrated the sensitivity of backscatter and CR to vegetation variables such as total vegetation water content, but also the sensitivity to changes in vegetation structure. In particular, the flowering and seeding in oilseed-rape and the stem extension and head development in winter cereals affected backscatter and CR, diminishing its sensitivity to VWC_t. These observations correspond to previous studies that observed the same behaviour in the backscatter signals.

The modelling demonstrated the nonlinear relationship between CR and vegetation variables. Initially, low R^2 values were found between CR and VWC for oilseed-rape and winter cereals when using a linear model. Moderate R^2 values were found for corn. Using an exponential model, CR was able to account for 34%, 63% and 87% of the variability in VWC in oilseed-rape, corn and winter cereal. In oilseed-rape, the CR cannot account for the large differences in VWC_t which were observed between the two growing seasons. In addition, at high vegetation density and high CR, CR saturates and small uncertainties in CR can lead to large errors in VWC_t. The random forest modelling demonstrated the relative importance of CR for estimating VWC_t dynamics in crops, where the variable importance over all crops was 0.35 for CR. This was followed by VV and VH backscatter. It also showed that, for corn and oilseed-rape, a simple exponential model accounts better for variation in VWC_t, but, for winter

Remote Sens. **2018**, 10, 1396

cereals, the RF model performs better when including additional soil moisture data. Since CR in winter cereals is strongly affected by changes in structure, using these additional input variables as is done in the RF approach increase the performance of the model.

The presented research demonstrates the high sensitivity of microwave indices to vegetation dynamics especially for corn and winter cereals and at low VWC_t. It also provides valuable insights to ultimately predict VWC based on Earth Observation data using either simple models or machine learning approaches. Knowledge on VWC in crops can aid farmers in monitoring crop health and help decide when to irrigate and as a result could increase water use efficiency. In addition, this study confirms previous findings on changes in backscatter which are indicative for flowering and seeding of oilseed-rape. Furthermore, timing of heading and flowering in winter cereals can be observed. Since the risk of disease in oilseed-rape and winter cereals is high during flowering, information on the timing of these stages can assist in disease prevention and limit the use of chemicals. Consequently, although SAR data on itself is still affected by structural effects, valuable information can be gained on growth phases of crops. Although this study focused on a small region in Austria, it does show that information on vegetation dynamics can be obtained from Sentinel-1 observations. Further studies should focus on applying the developed method to other regions of interest. Furthermore, the added value of SAR data in combination with, for example, visible near infra-red based data could lead to improved vegetation monitoring capabilities and needs further exploration.

Author Contributions: M.V. and W.W. conceived and designed the research; M.V. performed the research, analyzed and interpreted the data and wrote the paper; B.B.-M. and C.R. interpreted the data; P.S. contributed analysis tools for the in situ validation and provided interpretation of results; I.P. collected in situ samples and provided input on analysis of results, I.T. provided input and interpretation of results on the random forest modelling. All authors have contributed to writing of the manuscript.

Funding: This research is funded by the Austrian Space Application Programme 13 (ASAP13) of the Austrian Research Promotion Agency (FFG)and the Doctoral Programme on Water Resource Systems of the Austrian Science Fund (FWF) (DK-plus W1219-N22).

Acknowledgments: Many thanks go out to Sebastian Hahn for reviewing the manuscript and contributions during discussions.

References

1. Schmidhuber, J.; Tubiello, F.N. Global food security under climate change. *Proc. Natl. Acad. Sci. USA* **2007**, *104*, 19703–19708. [CrossRef] [PubMed]
2. Godfray, H.C.J.; Beddington, J.R.; Crute, I.R.; Haddad, L.; Lawrence, D.; Muir, J.F.; Pretty, J.; Robinson, S.; Thomas, S.M.; Toulmin, C. Food security: The challenge of feeding 9 billion people. *Science* **2010**, *327*, 812–818. [CrossRef] [PubMed]
3. Steele-Dunne, S.C.; McNairn, H.; Monsivais-Huertero, A.; Judge, J.; Liu, P.W.; Papathanassiou, K. Radar remote sensing of agricultural canopies: A review. *IEEE J. Sel. Top. Appl. Earth Observ. Remote Sens.* **2017**, *10*, 2249–2273. [CrossRef]
4. Wagner, W.; Lemoine, G.; Rott, H. A method for estimating soil moisture from ERS scatterometer and soil data. *Remote Sens. Environ.* **1999**, *70*, 191–207. [CrossRef]
5. Kerr, Y.; Waldteufel, P.; Richaume, P.; Wigneron, J.P.; Ferrazzoli, P.; Mahmoodi, A.; Al Bitar, A.; Cabot, F.; Gruhier, C.; Juglea, S.; et al. The SMOS Soil Moisture Retrieval Algorithm. *IEEE Trans. Geosci. Remote Sens.* **2012**, *50*, 1384–1403. [CrossRef]
6. Parinussa, R.M.; Holmes, T.R.H.; Wanders, N.; Dorigo, W.A.; de Jeu, R.A.M. A Preliminary Study toward Consistent Soil Moisture from AMSR2. *J. Hydrometeorol.* **2014**, *16*, 932–947. [CrossRef]
7. Van der Schalie, R.; Parinussa, R.M.; Renzullo, L.J.; van Dijk, A.I.J.M.; Su, C.H.; de Jeu, R.A.M. SMOS soil moisture retrievals using the land parameter retrieval model: Evaluation over the Murrumbidgee Catchment, southeast Australia. *Remote Sens. Environ.* **2015**, *163*, 70–79. [CrossRef]

8.   Liu, Y.Y.; de Jeu, R.A.M.; McCabe, M.F.; Evans, J.P.; van Dijk, A.I.J.M. Global long-term passive microwave satellite-based retrievals of vegetation optical depth. *Geophys. Res. Lett.* **2011**, *38*. [CrossRef]

9.   Jones, M.O.; Jones, L.A.; Kimball, J.S.; McDonald, K.C. Satellite passive microwave remote sensing for monitoring global land surface phenology. *Remote Sens. Environ.* **2011**, *115*, 1102–1114. [CrossRef]

10. McColl, K.A.; Entekhabi, D.; Piles, M. Uncertainty analysis of soil moisture and vegetation indices using Aquarius scatterometer observations. *IEEE Trans. Geosci. Remote Sens.* **2014**, *52*, 4259–4272. [CrossRef]

11. Vreugdenhil, M.; Dorigo, W.A.; Wagner, W.; Jeu, R.A.M.d.; Hahn, S.; Marle, M.J.E.v. Analyzing the Vegetation Parameterization in the TU-Wien ASCAT Soil Moisture Retrieval. *IEEE Trans. Geosci. Remote Sens.* **2016**, *54*, 3513–3531. [CrossRef]

12. Ferrazzoli, P.; Paloscia, S.; Pampaloni, P.; Schiavon, G.; Solimini, D.; Coppo, P. Sensitivity of microwave measurements to vegetation biomass and soil moisture content: A case study. *IEEE Trans. Geosci. Remote Sens.* **1992**, *30*, 750–756. [CrossRef]

13. Paloscia, S.; Macelloni, G.; Pampaloni, P. The relations between backscattering coefficient and biomass of narrow and wide leaf crops. In Proceedings of the 1998 IEEE International Geoscience and Remote Sensing Symposium Proceedings, Seattle, WA, USA, 6–10 July 1998; Volume 1, pp. 100–102.

14. Macelloni, G.; Paloscia, S.; Pampaloni, P.; Marliani, F.; Gai, M. The relationship between the backscattering coefficient and the biomass of narrow and broad leaf crops. *IEEE Trans. Geosci. Remote Sens.* **2001**, *39*, 873–884. [CrossRef]

15. Kim, Y.; Jackson, T.; Bindlish, R.; Lee, H.; Hong, S. Radar vegetation index for estimating the vegetation water content of rice and soybean. *IEEE Geoscie. Remote Sensi. Lett.* **2012**, *9*, 564–568.

16. Paloscia, S.; Macelloni, G.; Pampaloni, P.; Sigismondi, S. The potential of C- and L-band SAR in estimating vegetation biomass: the ERS-1 and JERS-1 experiments. *IEEE Trans. Geosci. Remote Sens.* **1999**, *37*, 2107–2110. [CrossRef]

17. Wiseman, G.; McNairn, H.; Homayouni, S.; Shang, J. RADARSAT-2 polarimetric SAR response to crop biomass for agricultural production monitoring. *IEEE J. Sel. Top. Appl. Earth Observ. Remote Sens.* **2014**, *7*, 4461–4471. [CrossRef]

18. Mattia, F.; Le Toan, T.; Picard, G.; Posa, F.I.; D'Alessio, A.; Notarnicola, C.; Gatti, A.M.; Rinaldi, M.; Satalino, G.; Pasquariello, G. Multitemporal C-band radar measurements on wheat fields. *IEEE Trans. Geosci. Remote Sens.* **2003**, *41*, 1551–1560. [CrossRef]

19. Satalino, G.; Balenzano, A.; Mattia, F.; Davidson, M.W. C-band SAR data for mapping crops dominated by surface or volume scattering. *IEEE Geosci. Remote Sensi. Lett.* **2014**, *11*, 384–388. [CrossRef]

20. Veloso, A.; Mermoz, S.; Bouvet, A.; Le Toan, T.; Planells, M.; Dejoux, J.F.; Ceschia, E. Understanding the temporal behavior of crops using Sentinel-1 and Sentinel-2-like data for agricultural applications. *Remote Sens. Environ.* **2017**, *199*, 415–426. [CrossRef]

21. Blöschl, G.; Blaschke, A.P.; Broer, M.; Bucher, C.; Carr, G.; Chen, X.; Eder, A.; Exner-Kittridge, M.; Farnleitner, A.; Flores-Orozco, A.; et al. The Hydrological Open Air Laboratory (HOAL) in Petzenkirchen: A hypothesis-driven observatory. *Hydrol. Earth Syst. Sci.* **2016**, *20*. [CrossRef]

22. McNairn, H.; Hochheim, K.; Rabe, N. Applying polarimetric radar imagery for mapping the productivity of wheat crops. *Can. J. Remote Sens.* **2004**, *30*, 517–524. [CrossRef]

23. Jiao, X.; McNairn, H.; Shang, J.; Pattey, E.; Liu, J.; Champagne, C. The sensitivity of RADARSAT-2 polarimetric SAR data to corn and soybean leaf area index. *Can. J. Remote Sens.* **2011**, *37*, 69–81. [CrossRef]

24. Satalino, G.; Mattia, F.; Le Toan, T.; Rinaldi, M. Wheat crop mapping by using ASAR AP data. *IEEE Trans. Geosci. Remote Sens.* **2009**, *47*, 527–530. [CrossRef]

25. Chala, A.; Weinert, J.; Wolf, G.A. An Integrated Approach to the Evaluation of the Efficacy of Fungicides Against Fusarium culmorum, the Cause of Head Blight of Wheat. *J. Phytopathol.* **2003**, *151*, 673–678. [CrossRef]

26. Ferrazzoli, P.; Paloscia, S.; Pampaloni, P.; Schiavon, G.; Sigismondi, S.; Solimini, D. The potential of multifrequency polarimetric SAR in assessing agricultural and arboreous biomass. *IEEE Trans. Geosci. Remote Sens.* **1997**, *35*, 5–17. [CrossRef]
27. Paloscia, S.; Pettinato, S.; Santi, E.; Notarnicola, C.; Pasolli, L.; Reppucci, A. Soil moisture mapping using Sentinel-1 images: Algorithm and preliminary validation. *Remote Sens. Environ.* **2013**, *134*, 234–248. [CrossRef]

Article

AMSR2 Soil Moisture Downscaling Using Temperature and Vegetation Data

Bin Fang [1,*], **Venkat Lakshmi** [1], **Rajat Bindlish** [2] **and Thomas J. Jackson** [3]

[1] School of Earth Ocean and Environment, University of South Carolina, Columbia, SC 29208, USA; vlakshmi@geol.sc.edu
[2] Hydrological Sciences Branch, NASA Goddard Space Flight Center, Greenbelt, MD 20771, USA; rajat.bindlish@nasa.gov
[3] Hydrology and Remote Sensing Laboratory, Beltsville Agricultural Research Center, United States Department of Agriculture, Beltsville, MD 20705, USA; Tom.Jackson@ars.usda.gov
* Correspondence: bfang@geol.sc.edu; Tel.: +1-803-553-7058

Received: 16 July 2018; Accepted: 20 September 2018; Published: 1 October 2018

Abstract: Soil moisture (SM) applications in terrestrial hydrology require higher spatial resolution soil moisture products than those provided by passive microwave remote sensing instruments (grid resolution of 9 km or larger). In this investigation, an innovative algorithm that uses visible/infrared remote sensing observations to downscale Advanced Microwave Scanning Radiometer 2 (AMSR2) coarse spatial resolution SM products was developed and implemented for use with data provided by the Advanced Microwave Scanning Radiometer 2 (AMSR2). The method is based on using the Normalized Difference Vegetation Index (NDVI) modulated relationships between day/night SM and temperature change at corresponding times. Land surface model output variables from the North America Land Data Assimilation System (NLDAS), remote sensing data from the Moderate-Resolution Imaging Spectroradiometer (MODIS), and Advanced Very High Resolution Radiometer (AVHRR) were used in this methodology. The functional relationships developed using NLDAS data at a grid resolution of 12.5 km were applied to downscale AMSR2 JAXA (Japan Aerospace Exploration Agency) SM product (25 km) using MODIS land surface temperature (LST) and NDVI observations (1 km) to produce the 1 km SM estimates. The downscaled SM estimates were validated by comparing them with ISMN (International Soil Moisture Network) in situ SM in the Black Bear–Red Rock watershed, central Oklahoma between 2015–2017. The overall statistical variables of the downscaled AMSR2 SM validation R^2, slope, RMSE and bias, demonstrate good accuracy. The downscaled SM better characterized the spatial and temporal variability of SM at watershed scales than the original SM product.

Keywords: AMSR2; passive microwave soil moisture; soil moisture downscaling

---

1. Introduction

Satellite technology is a practical approach to monitoring Earth surface hydrological characteristics especially in regions with limited ground measurements [1–8]. Studies have shown that passive microwave satellite sensors can provide reliable estimates of surface SM [9–12]. Current low frequency (L-band to X-band) passive radiometers in space provide SM with intrinsic spatial resolutions of ~25–40 km. Data are often interpolated or resampled to grids with higher resolutions (9 km for the Soil Moisture Active Passive (SMAP) mission). These products are valuable for some applications but cannot satisfy all hydrological, agricultural, and weather applications. In contrast, satellite-based radars have much higher spatial resolutions but have limited spatial coverage and infrequent temporal revisit. In addition, SM retrievals from radars are more complex than from radiometers. The spatial variability and sensitivity of active and passive microwave remote sensing SM at different resolutions has been reported in [13–16].

Passive SM products are available from several sources and satellites that include the AMSR-E (Advanced Microwave Scanning Radiometer for the Earth Observing System) and its successor AMSR2, as well as SMOS (Soil Moisture and Ocean Salinity), Aquarius, and SMAP (Soil Moisture Active Passive). Here we focus on SM retrieved from the AMSR2 instrument. The AMSR2 instrument on board GCOM-W [17–19] was launched on 18 May 2012 by JAXA. It has six different frequency microwave bands ranging from 6.925 to 89 GHz (C, X, Ku, and Ka bands), recording the emitted microwave energy from the Earth surface at an altitude of 700 km every 1–2 days [20]. Several research groups produce AMSR2 SM products using different algorithms [21], including the JAXA soil moisture algorithm [22], Single Channel Algorithm (SCA) [23], and Land Parameter Retrieval Model (LPRM) [24].

SM downscaling algorithms attempt to produce higher spatial resolution SM data by integrating coarse-resolution passive microwave SM (or brightness temperature) with high-resolution data from other sources. Several different types of methods have been proposed and evaluated. One type of method utilized high spatial resolution radar backscatter data to downscale coarse resolution radiometer observations [25–28]. The downscaling algorithm of [29,30] used a functional relationship between brightness temperature and backscatter for each pixel based on a time series of observations. In addition to the type of methods mentioned above, other techniques, including data assimilation and machine learning, have been used to develop SM downscaling algorithms [31–36].

Besides the above two types of methods, a number of SM downscaling approaches using high spatial resolution visible/infrared band remotely sensed products have been explored in previous studies. In these approaches, the relationship between SM and soil evaporation efficiency was used to downscale SMOS SM [37–39] using two SM indices: Evaporative Fraction (EF) and Actual Evaporative Fraction (AEF) [40]. Fang et al. [41] developed a variant of this approach and downscaled AMSR-E SM using NLDAS model outputs. Other methods proposed include a downscaling algorithm derived from MODIS NDVI, LST, and brightness temperature, which has been applied to SMOS SM [42], a downscaling model based on vegetation temperature index [43], synergistic approaches using MODIS NDVI and LST to downscale AMSR-E and SMOS SM [44,45], and, using the thermal inertia relationship between daily average SM, temperature variation and NDVI to downscale passive microwave SM from AMSR-E and SMOS in Oklahoma [46–48].

The downscaling algorithm applied in the current investigation has been described in [46] and is based on the assumption/stipulation that SM is related to land surface temperature, vegetation cover, and evapotranspiration (ET) and only requires data that are routinely available. After establishing the necessary relationships between these variables through modeling, downscaling algorithms can be developed to obtain SM at higher spatial resolutions. In this study, we implemented the SM downscaling algorithm from [46] to AMSR2 radiometer-based SM products retrieved from JAXA Version 3 SM [49,50] over the Continental United States (CONUS) region and examined the performance of this algorithm over regions with different SM conditions. This SM downscaling algorithm is based on the relationships between SM, temperature change derived from MODIS Aqua/Terra data, and vegetation index. The vegetation modulated relationships between temperature variation and the SM of AMSR2 ascending/descending overpasses were modeled based on thermal inertia. This method follows prior research from several well-established groups, including the authors of [51,52].

The AMSR2 and other passive microwave satellite instruments can estimate surface SM using the physical relationship between soil dielectric constant and water content [49,50]. As opposed to this, the NLDAS SM is the model output derived from land surface models by different data sources including rain gauge observations, remotely sensed shortwave radiation, and surface meteorology reanalyses. The spatial and temporal data availability of the ground measurements as well as the inconsistency between different data sources may contribute to the uncertainties of NLDAS variables [53–56]. Therefore, it is necessary to develop downscaling algorithms to enhance the spatial resolution of passive microwave SM products instead of using NLDAS SM outputs for various applications.

2. Data and Methodology

2.1. Data Sets

The NLDAS provides near real-time land surface model data at a 1/8° (12.5 km) resolution (http://ldas.gsfc.nasa.gov/nldas/) from available remote sensing observations and land surface model outputs [53,54] covering North America since 1979. In this study, the NLDAS Phase-2 model variables for building the downscaling model include surface skin temperature (K) and SM in the 0–10 cm soil layer, which are coincident to the overpass times of AMSR2. We selected the NLDAS variables derived from the NOAA (National Oceanic and Atmospheric Administration) Noah Land Surface Model Level 4, which was developed for the NCEP (National Centers for Environmental Prediction) meso-scale Eta model [54]. Previous studies have evaluated the NLDAS output data accuracy and uncertainties [55,56].

The NASA LTDR (Long Term Data Record) project produces and publishes global land surface climate data record from the 1980s to the present. The LTDR Version 5 NDVI data comprise a merged collection from AVHRR instruments on NOAA satellites N07~N19 (AVH13C1, 1981–present). In this paper, the 0.05° (5 km) spatial resolution LTDR NDVI daily data produced on the Climate Modeling Grid (CMG) was downloaded from the LTDR website (https://ltdr.nascom.nasa.gov/) and upscaled to be consistent with the NLDAS 12.5 km pixel size 1 using a bilinear interpolation method.

The MODIS sensors are mounted on the NASA satellites Terra/Aqua, which were launched in 1999 and 2002, respectively, providing Earth surface measurements that include temperature and vegetation indices on a daily and global basis [57,58]. The sensors have a total of 36 visible/infrared bands ranging from 0.4 to 14.4 μm at different spatial resolutions 250 m, 500 m and 1 km. The 1 km daily Aqua and Terra MODIS LST (MYD11A1, MOD11A1) and 1 km biweekly Aqua MODIS NDVI (MYD13A2) were downloaded from the online data pool Land Processes Distributed Active Archive Center (LP DAAC) at https://lpdaac.usgs.gov/ as inputs to the downscaling model. The Quality Control layers (QC) were used for removing the cloud contaminated and low quality Day/Night LST estimate pixels. The MODIS products were mosaicked and reprojected using MODIS Reprojection Tool for calculating 1 km LST Day/Night difference and NDVI at the CONUS region.

The AMSR2 instrument is onboard the Global Change Observation Mission–Water 1 (GCOM-W1) mission that was launched by JAXA in 2012. It provides near real-time passive microwave observations for monitoring land, ocean, and atmosphere processes and interactions at global scale from about 700 km altitude above the Earth surface with equatorial overpass times of 1:30 a.m./p.m. [17]. The AMSR2 surface SM products are retrieved using C- and X-band brightness temperature (Tb). As noted above, we used the JAXA Level-3 product which was retrieved using JAXA algorithm and gridded at 25 km spatial resolution. The AMSR2 data were downloaded from JAXA's G-COM website http://suzaku.eorc.jaxa.jp/. The SM used in this study covers various land types including cold and arid areas and excluded vegetated areas over 2 kg/m^2 vegetation water content [50].

The validation of downscaled SM products can be challenging. The most robust evaluation would utilize reliable in situ data over the entire downscaled passive SM product. In this investigation, we utilized unique SM observation data sets available from ISMN for validating the downscaled SM results. The ISMN is a repository collecting in situ SM data at different layers from 0.01 to 1 m which are contributed by 19 international SM networks including more than 500 stations for evaluating and improving SM measurements from satellite and land surface models since 1952 [59–63]. In this study, the in situ SM measurements at a 0–0.05 m soil layer were downloaded from https://ismn.geo.tuwien.ac.at/, from 5 SM monitoring network stations, which have complete daily

SM measurements between 2015 and 2017 located in two watersheds: the Black Bear–Red Rock watershed and the Lower Cimarron Watershed, as shown in Figure 1.

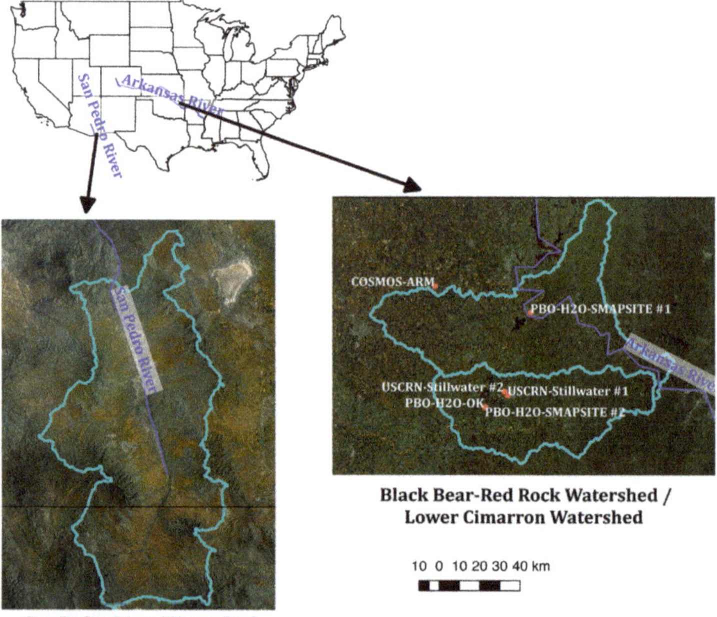

Figure 1. The CONUS region and the sites for mapping and validating the SM downscaling algorithm: Black Bear–Red Watershed/Lower Cimarron Watershed, Oklahoma, and San Pedro River Watershed, Arizona. The ISMN SM monitoring ground stations are denoted in red.

2.2. Soil Moisture Downscaling Algorithm

The JAXA SM algorithm uses X-band observations due to the presence of RFI in C-band observations. The algorithm [22] considers the effects of vegetation cover on SM retrieval and introduces the vegetation fractional parameter into the algorithm. Therefore, the SM downscaling algorithm uses a vegetation-based lookup table to relate microwave polarization index to SM estimates.

Thermal inertia theory as utilized here refers to the time-dependent response of an object to the variation of temperature. The volumetric heat capacity of soil increases when the soil layer becomes wetter, and the greater water content should correspond to a smaller temperature variation. In this study, we assumed that the SM at a particular time (e.g., morning) is inversely proportional to the temperature change 12 h prior to, correspondingly, either the AMSR2 morning or afternoon overpass time. Additionally, it is expected that the SM–temperature change relationship will be modulated by the presence of vegetation. The triangular shape of the SM and temperature relationship changes into a polygonal shape when the vegetation is under water stress [63–69], which indicates that the influence of vegetation on the SM–temperature change relationship is another important factor that must be considered.

Based upon previous studies, the daily average SM is negatively related to the daily temperature difference under various vegetation conditions [46,47]. This is also the research background that models the triangular relationship to downscale the passive microwave SM by using NLDAS derived variables [46–48]. NLDAS provides operational long-term land surface variable outputs at hourly temporal frequency and is able to characterize the soil hydrological properties represented by the

$\theta - \Delta T_s$ relationship. Therefore, for a single month, the relationship between SM and temperature difference for a given NDVI class can be expressed by the following linear regression model:

$$\theta(i,j) = a_0 + a_1 \Delta T_s(i,j) \tag{1}$$

where $\theta(i,j)$ is the NLDAS gridded SM for the AMSR2 morning/afternoon overpasses, and $\Delta T_s(i,j)$ is the NLDAS gridded 12-h temperature difference closest to and before the AMSR2 overpasses. This relationship was built at the scale of the NLDAS pixel using the NLDAS SM and surface skin temperature from all single months between 1981 and 2016. The daily NDVI derived from AVHRR, MODIS data were aggregated and matched-up to the NLDAS pixels by using the nearest neighbor method. The NDVI were binned into different classes from 0 and 1 with an increment of 0.1 (the classes containing fewer than 8 points were not used), which results in the regression lines for the $\theta - \Delta T_s$ scatterplots. The regression equations at the specific NLDAS spatial resolution were applied on a daily basis to the 1 km MODIS pixels within that NLDAS pixel. The 1 km SM was then calculated from the 1 km MODIS LST difference ΔT_s and corresponding NDVI interval.

The AMSR2 SM observations are retrieved from the microwave radiometer, which senses SM at a few centimeters depth and is different from the optical imaging sensor MODIS observations for calculating SM sensed at a few millimeters [70,71]. In order to solve this discrepancy, the 1 km SM θ calculated from the MODIS LST products should be corrected by removing the difference between AMSR2 SM and MODIS LST derived SM using the following relationship:

$$\theta^c(i,j) = \theta(i,j) + [\Theta - \frac{1}{n}\sum \theta_n] \tag{2}$$

where $\theta^c(i,j)$ stands for a corrected 1 km AMSR2 SM using θ calculated from Equation (1), and n is the number of 1 km SM pixels that fall in each AMSR2 pixel. Θ is the 25 km resolution AMSR2 SM.

The θ^c has the following characteristics: (a) the SM at 1 km can be characterized by the relationship between SM and daily temperature variation, (b) the bias between AMSR2 and MODIS-derived SM can be eliminated at the low resolution, and (c) the $\theta - \Delta T_s$ relationship varies in response to different vegetation conditions.

The SM downscaling algorithm modified the original algorithm from [46] as follows: (1) the new algorithm adopted more NDVI classes from 0 to 1 and with an increment of 0.1 for modeling NDVI corresponding $\theta - \Delta T_s$ relationships at regions of different vegetation cover conditions in CONUS; (2) the new algorithm used the IDW (Inverse Distance Weighted) interpolation method for upscaling LTDR NDVI data to coarser spatial resolution; (3) the new algorithm used an averaging method to calculate 1 km SM grids that are overlapped by multiple 12.5 km NLDAS grids; (4) the new algorithm performed a downscaling algorithm on descending/ascending overpasses AMSR2 SM retrievals by building the two models separately, while the original algorithm only downscaled the daily averaged AMSR-E SM retrievals.

Due to a lack of long-term monitoring of very high spatial resolution of land surface variables, the NLDAS outputs as well as AVHRR and MODIS products were normalized to 12.5 km for downscaling the microwave SM to 1 km. The spatial heterogeneity at 1 km within each 12.5 km pixel during the model implementation was also ignored.

3. Evaluation of the Downscaling Algorithm

For examining performance of the downscaling algorithm at different regions and seasons, the R^2 and RMSE (root mean squared error) between θ and corresponding ΔT_s of descending overpass times using the NLDAS data between 1981 and 2016 by each month between April and September are mapped in Figures 2 and 3. The other months from October to March were excluded for building the downscaling model, as the NLDAS SM had poor correlations with in situ observations in the northern CONUS region during cold months and was suggested not to be used, due to the biases caused by frozen soil water content [72]. In these maps, the R^2 and RMSE are averaged from all the NDVI classes for each NLDAS grid. It is found that the western part of CONUS generally has higher R^2 (basically > 0.2 in warm months) than the eastern CONUS. Some possible factors may have an impact on the accuracy of NLDAS variables SM and temperature estimations, such as vegetation type, vegetation structure, and soil texture. This can be referred to the related research on validating NLDAS SM using ground observations over the CONUS region, which found better validation statistical variables in the western region at shallow depth soil layer [72,73]. Low R^2 values are also noted in the Rocky Mountain Region, where is usually covered by snow until mid-summer. The R^2 shows an increasing trend from April to July for both the western/eastern regions and it reaches the peak in July. The southwestern states have overall strong R^2 (0.6–0.8) associated with sparse vegetation coverage and low amounts of precipitation through the entire year. The R^2 slowly decreases from July to September. The western part has greater changes of R^2 than the eastern part through all six months. The above facts may also be determined by the data quality of NLDAS SM estimates: the research [72] found the seasonal variability of validation statistical variables for the NLDAS SM, of which the hottest months (July and August) tend to be more accurate than the other months. These could be explained by the influence of frozen water content in the soil layer or the slow response of the increase in SM in wet seasons (spring and fall).

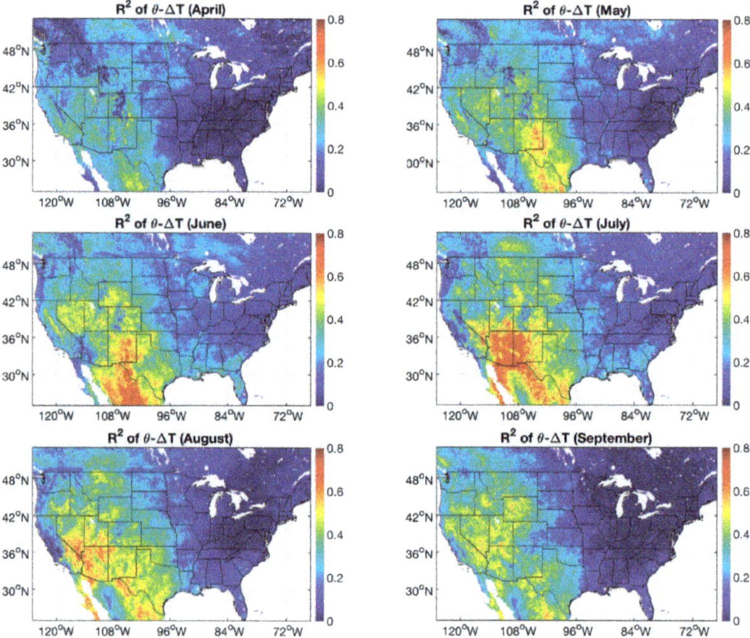

Figure 2. R^2 maps in the CONUS region of AMSR2 descending overpasses from April to September at NLDAS grid spatial resolution 12.5 km. The R^2 for each grid is computed from averaged R^2 values between NLDAS SM and surface skin temperature corresponding to AMSR2 descending overpass times of all NDVI 0–1 classes.

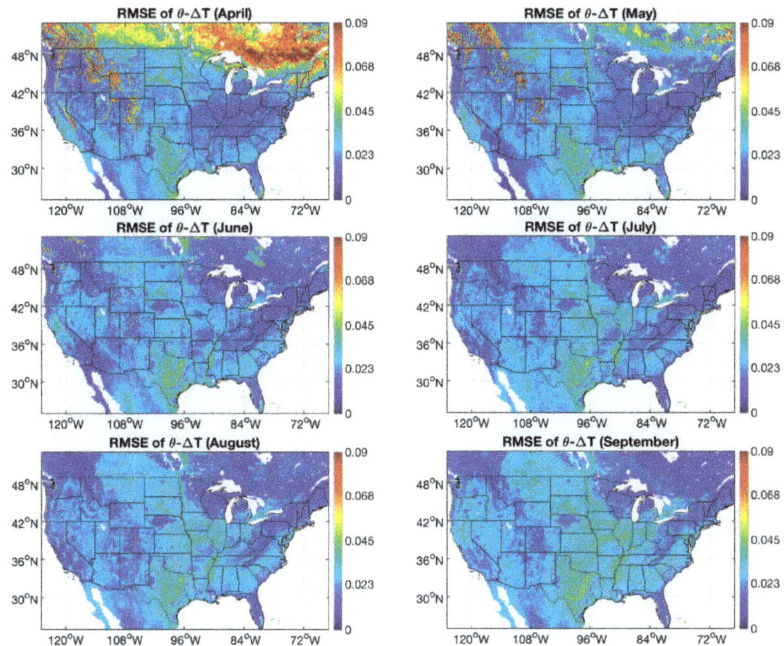

Figure 3. RMSE of SM estimation from the downscaling algorithm (m^3/m^3) maps in the CONUS region of AMSR2 descending overpasses from April to September at NLDAS grid spatial resolution 12.5 km. The RMSE for each grid is computed from averaged RMSE values between NLDAS SM and surface skin temperature corresponding to AMSR2 descending overpass times of all NDVI 0–1 classes.

The RMSE maps for SM prediction do not show clear regional distribution patterns, the only exception being a high RMSE in the north of U.S. and Rocky Mountain region in April–May, where could be covered by snow. Similarly, we do not observe clear seasonal trends of RMSE's variation from cool to warm months. The RMSE values are mostly <0.03 m^3/m^3 throughout the entire CONUS region, except the aforementioned northern and mountainous regions with high RMSEs in cool months.

Figure 4 and Table 1 show four NLDAS grids of modeled $\theta - \Delta T_s$ relationships of different NDVI classes in July between 1981 and 2016. These represent regions of different vegetation coverage and SM conditions. The locations include (1) Walnut Gulch, Arizona, (2) Stillwater, Oklahoma, (3) Reynolds Creek, Idaho, and (4) Ames, Iowa, all of which have permanent long-term SM monitoring stations. Overall, it can be summarized that θ and ΔT_s are negatively correlated, and the fit lines for all NDVI classes have similar slope values. In general, $\theta - \Delta T_s$ correlations have higher R^2 for descending overpasses than ascending overpasses. This fact is also reflected on $\theta - \Delta T_s$ scatterplots, where the descending overpass scatterplots are more concentrated. The R^2 of descending $\theta - \Delta T_s$ relationships range from 0.493 to 0.784 in Walnut Gulch—clearly higher than the other three locations. This is probably attributed to the Walnut Gulch having the overall narrowest NDVI range (0–0.4), so the $\theta - \Delta T_s$ relationship is less complicated by vegetation. In contrast, Ames has the worst overall R^2 (0.092–0.275) and has the widest NDVI range between 0–0.8. On another hand, R^2 values do not show clear decreasing trends as NDVI class increases for all four locations for either descending/ascending overpasses. The outliers are noted for all NDVI classes from the scatterplots in Figure 4, indicating the impact of vegetation on $\theta - \Delta T_s$ relationships. However, such impact does not strengthen when NDVI increases either. In addition, according to the sample numbers shown in Table 1, the NDVI classes with few points may contain more uncertainties with respect to the regression analysis, and the R^2 may be relatively higher or lower than the other classes, such as NDVI class 0.3–0.4 for Walnut Gulch, 0–0.1 for Stillwater, and 0.4–0.5 for Reynolds Creek.

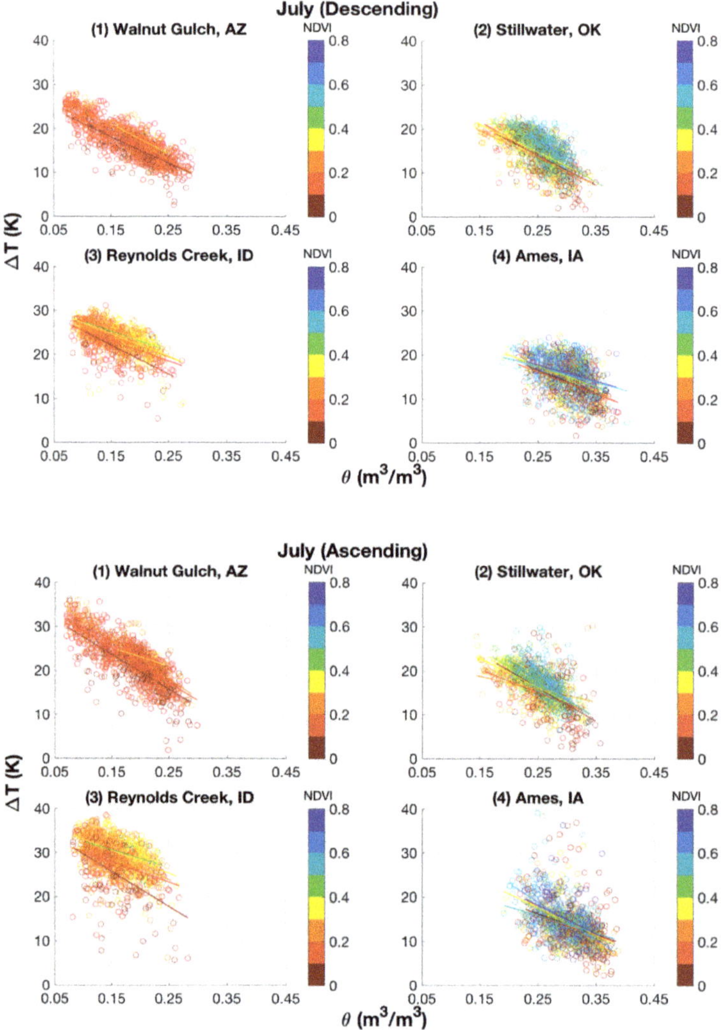

Figure 4. Scatterplots showing the relationships between NLDAS SM and surface skin temperature corresponding to different NDVI classes of descending/ascending overpasses from four selected NLDAS grids, which represents the regions of different vegetation coverage and SM condition: (**1**) Walnut Gulch, Arizona; (**2**) Stillwater, Oklahoma; (**3**) Reynolds Creek, Idaho; (**4**) Ames, Iowa.

Table 1. R^2 and sample numbers between NLDAS surface skin temperature and SM of descending/ascending overpasses in July from four selected NLDAS grids representing different vegetation and SM conditions in the CONUS region, including Walnut Gulch, Arizona; Stillwater, Oklahoma; Reynolds Creek, Idaho; Ames, Iowa.

NDVI	Walnut Gulch	Stillwater	Reynolds Creek	Ames
Descending Overpasses				
0–0.1	0.605	0.341	0.349	0.227
0.1–0.2	0.691	0.445	0.32	0.195
0.2–0.3	0.784	0.448	0.404	0.154
0.3–0.4	0.493	0.376	0.367	0.172
0.4–0.5	-	0.493	0.436	0.142
0.5–0.6	-	0.374	-	0.093
0.6–0.7	-	0.338	-	0.275
Ascending Overpasses				
0–0.1	0.575	0.271	0.232	0.07
0.1–0.2	0.665	0.224	0.189	0.09
0.2–0.3	0.779	0.075	0.323	0.342
0.3–0.4	0.295	0.483	0.247	0.148
0.4–0.5	-	0.493	0.111	0.237
0.5–0.6	-	0.378	-	0.16
0.6–0.7	-	0.428	-	0.357
Sample Numbers				
0–0.1	269	80	120	118
0.1–0.2	594	87	221	101
0.2–0.3	218	100	422	94
0.3–0.4	13	166	291	106
0.4–0.5	-	255	43	131
0.5–0.6	-	348	-	160
0.6–0.7	-	45	-	227

4. Downscaled AMSR2 Soil Moisture Results

4.1. Interpretation of Soil Moisture Maps

In Figure 5, the 1 km downscaled AMSR2 SM are mapped to compare with the original 25 km AMSR2 SM at the CONUS region scale as well as the watershed scale. The 1 km AMSR2 SM from June of morning overpasses (descending overpasses) have very similar spatial distribution patterns as the 25 km SM. However, they show spatial variabilities at a finer spatial scale, especially in the regions with heterogeneous SM conditions, such as the mid-west region in the Mississippi river basin.

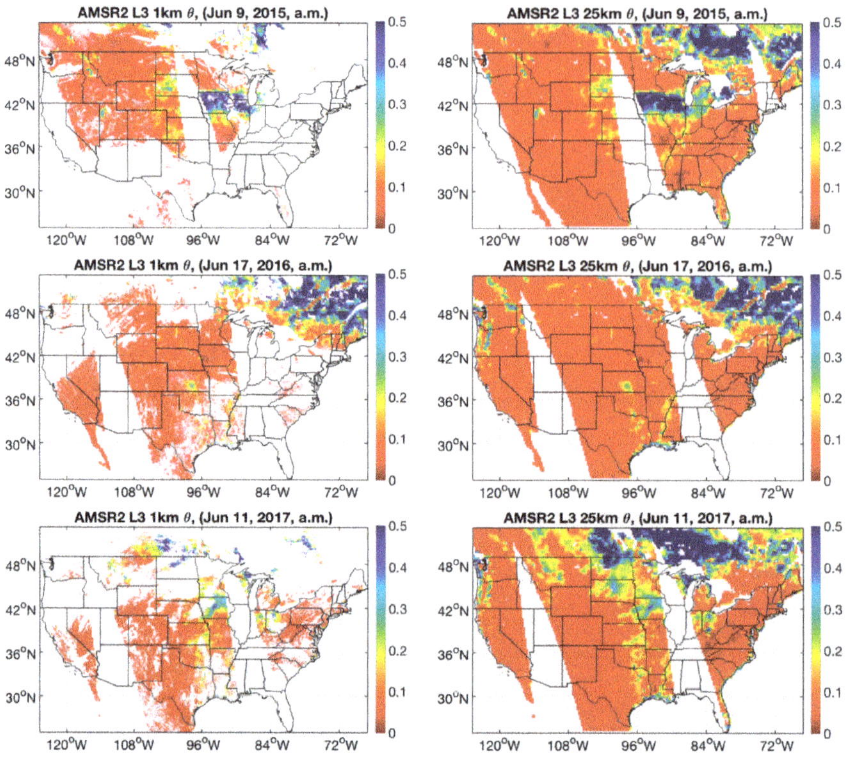

Figure 5. 1 km downscaled and 25 km AMSR2 L3 SM of descending overpasses (morning) in the CONUS region from three days in June 2015–2017.

When one focuses on the watershed scale, as Figure 6 shown the 1 km/25 km AMSR2 SM of morning overpasses in the Black Bear–Red Rock watershed, central Oklahoma from 4–5 September 2014, we can observe the different SM spatial distribution patterns in the west/east side of Arkansas River (The river path is shown in Figure 1) in 1 km SM maps, which are not observed in 25 km maps. The SM dry-down trends are also better captured in 1 km SM maps. The western bank of the Arkansas River demonstrates greater heterogeneity than the eastern bank. This is probably due to the land cover type: the eastern part of the watershed is dominated by grassland, while the agricultural land composes the western part. In addition, wetter regions in the center of the watershed can be noted, which are not captured in the 25 km SM maps. These regions are close to the Sooner Lake and a big bend of Arkansas River. Similarly, another good example in Western U.S. of Figure 7 shows contrast between 1 and 25 km SM of ascending overpasses in the San Pedro River watershed, Arizona, from 3–10 May 2016. The 25 km SM maps show nearly no spatial or temporal variation throughout the entire watershed during this period. In contrast to this, the downscaled 1 km SM clearly demonstrates several wetter regions along the west and east sides of the watershed boundaries, which are parts of the Coronado national forest. Besides this, we can also observe a wetter region across the middle of the watershed which corresponds to the river and alluvial plain. This feature agrees to the topography shown in Figure 1. Another issue about the performance of the downscaling algorithm is worth noticing: the downscaled 1 km SM maps show sharp edges in both Figures 6 and 7.

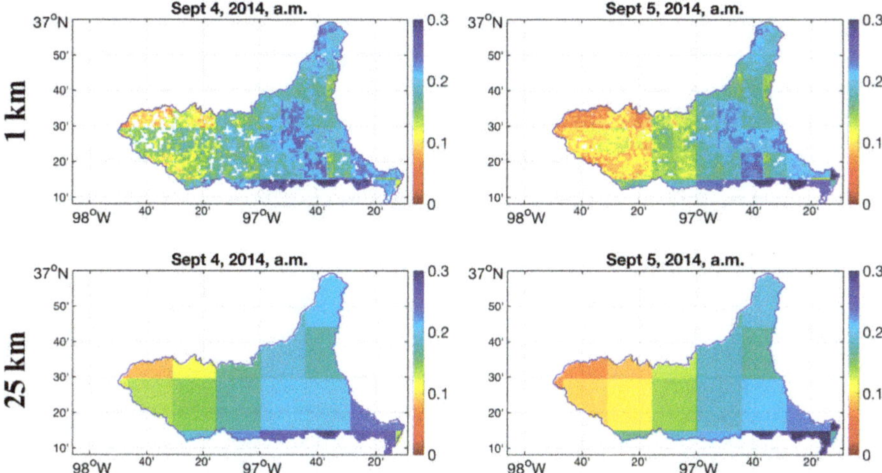

Figure 6. 1 km downscaled AMSR2 L3 SM compared with 25 km original SM estimates of descending overpasses (morning) showing spatial and temporal SM variabilities at the Black Bear–Red Rock watershed, central Oklahoma from 4–5 September 2014.

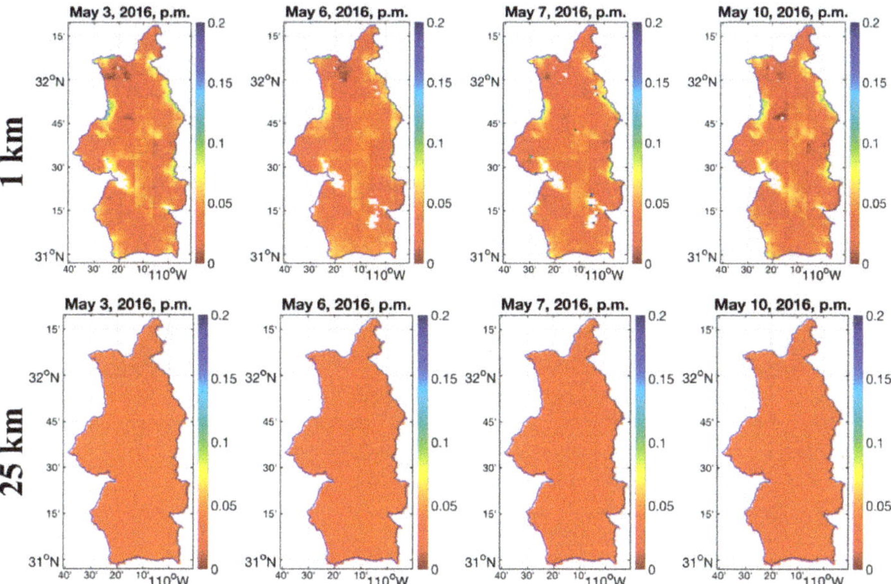

Figure 7. 1 km downscaled AMSR2 L3 SM compared with 25 km original SM estimates of ascending overpasses (afternoon) showing spatial and temporal variabilities at San Pedro river watershed, southeastern Arizona from 4 days of early May 2016.

Particularly, in Figure 6, the sharp-edge issue is more clearly observed once great difference between adjacent 25 km grids. And, the sharp edges exactly correspond to the spatial pattern of 25 km SM maps. In Figure 7, the shape edges are also found at the center of the watershed of the 1 km SM maps, while there is little spatial variability in the 25 km SM maps. The sharp-edge issue can be concluded for two reasons: Firstly, the downscaling algorithm corrects the 25 km resolution AMSR2 SM by using the 1 km SM estimates derived from MODIS LST within the 25 km grid boundaries. However, there is no smoothing step performed on adjacent 25 km grids. The second reason is the discontinuity of the spatial pattern in the 1 km estimated SM grids, which is mainly caused by adjacent 12.5 km NLDAS grids. The downscaling algorithm was built at the 12.5 km NLDAS scale, and adjacent NLDAS grids may have quite different $\theta - \Delta T_s$ relationships.

4.2. Validation

The 1 km downscaled and 25 km AMSR2 SM were validated using ISMN in situ SM measurements in the Black Bear-Red Rock watershed, Oklahoma, between 2015 and 2017, as Figures 8 and 9 shown. In this validation, the ISMN in situ measurements within each 1 km/25 km AMSR2 SM grid were averaged for validating. Please note that there are fewer data points for the 1 km than for the 25 km SM validations for all the stations, as the 1 km downscaled SM results usually have less coverage than 25 km due to the cloud contamination and satellite swath gap issues. From Figure 8, it can be seen that the slopes of 1 km SM validation results of descending overpasses are closer to 1 than the 25 km validation plots. The validation metrics improve from the original 25 km to the 1 km SM for descending overpasses, as shown in Table 2. The R^2 of 1 km SM validations ranges from 0.284 to 0.608, which is much better than the overall range of 0.094–0.449 for 25 km SM. Improvement is also noted in the unbiased RMSE (ubRMSE), which ranges from 0.036 to 0.079 m^3/m^3 for the 1 km SM in comparison to the range of 0.036–0.052 m^3/m^3 for the 25 km SM. Considering the data accuracy of the JAXA algorithm retrieved AMSR2 SM, which has the overall ubRMSE of 0.059 m^3/m^3 validated by core validation sites [64], the performance of the downscaling algorithm in CONUS is fairly good and the overall bias is also improved. For the ascending overpasses, it can be seen that the R^2 and ubRMSE are improved from 25 to 1 km, but degradation for bias is also noted. In addition to this, an overall underestimation trend is observed for the downscaled 1 km SM for both descending/ascending overpasses. From Figure 8 and the bias values of Table 2, it appears that the 1 km SM for the ascending overpasses are underestimated by a larger amount than for the descending overpasses. Additionally, for the 25 km SM validation, descending overpasses generally have better R^2 and ubRMSE than ascending overpasses, but worse bias. By studying Figure 8, the reasons could be that the 25 km SM validation results for ascending are better correlated and the scatter points are more concentrated, but also show stronger dry-biased tendency. By studying the validation results between stations, we found that the COSMOS stations: COSMOS-ARM (Atmospheric Radiation Measurement) and COSMOS-SMAP-OK have improved overall R^2 but worse overall ubRMSE compared with other stations. On the contrary, the stations of USCRN-Stillwater and PBO-H2O-OK have lower overall ubRMSE and the points on the scatterplots are closer to the 1-1 line, but worse R^2. Referring to the previous research on validating in situ SM measurements, the COSMOS SM measurements are recorded by cosmic-ray probes from the wireless sensor network (SoilNET) and the overall accuracy is within 0.04 m^3/m^3 at heterogeneous farmland. Research also found that the vegetation water content, biomass, air pressure, and other factors may have an impact on the accuracy of COSMOS SM [74–76]. On the other hand, USCRN is with an average random error of ~0.012 m^3/m^3 by triple collocation validation method in the CONUS region, and the largest error was noted at shallower depths [77–79]. The USCRN provides more reliable SM measurements than COSMOS, so the downscaled results validated by USCRN show better agreements and smaller overall ubRMSE.

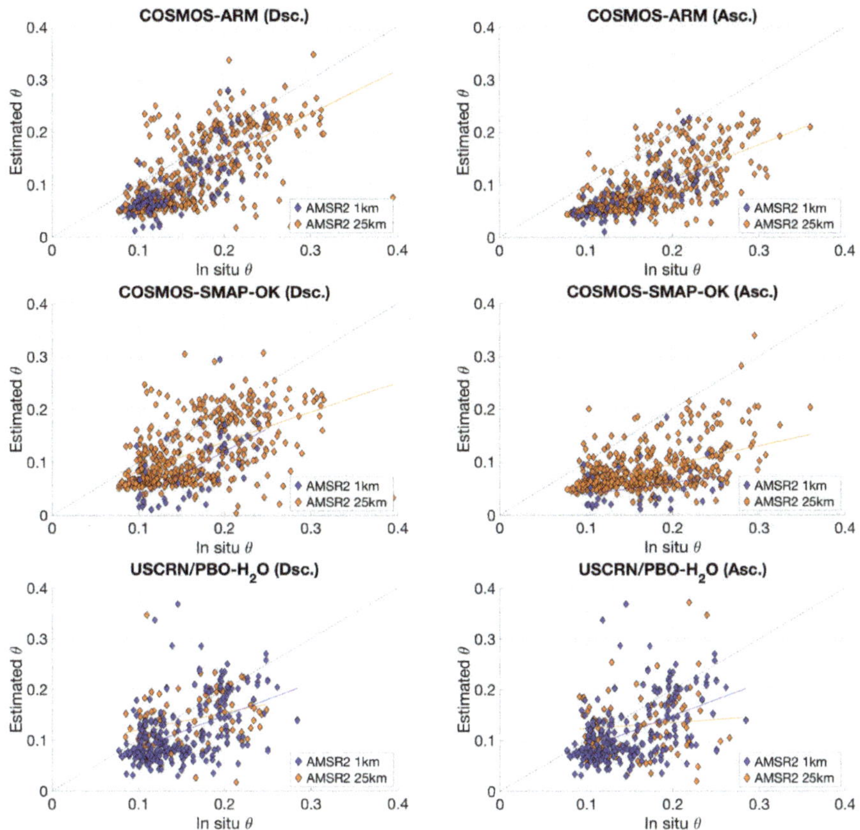

Figure 8. Scatterplots of 1 km/25 km AMSR2 L3 SM of descending (Dsc.)/ascending (Asc.) overpasses between 2015 and 2017, validated using ISMN in situ data from the SM stations near Stillwater, Oklahoma, which are located in the lower Cimarron watershed.

Figure 9 shows time series plots of daily averaged 1 km/25 km AMSR2 SM estimates along with ISMN in situ SM measurements, as well as the corresponding GPM daily precipitation at the COSMOS-SMAP-Oklahoma station between 2015 and 2017. It can be observed that both the 1 km/25 km AMSR SM estimates for the descending/ascending overpasses have similar dry-down and wetting trends when compared to the in situ measurements. In addition to this, the underestimation tendencies are observed for both 1 km and 25 km AMSR2 SM estimates, which correspond to the Figure 8 scatterplots. All three SM data sets show less discrepancy for the descending overpasses than ascending overpasses. Possible reasons for this behavior could be: (1) a remote sensing SM estimate represents the average SM value within the grid domain, while in situ measurements represent point locations, and (2) AMSR2 SM estimates of descending overpasses have better overall accuracy in the validation sites, due to the uniform surface temperature and soil profiles during night time [5,20]. In addition, the precipitation may have an impact on the accuracy of the estimated 1 km/25 km AMSR2 SM. The impact includes contamination on microwave signals, as well as the slow response of soil moisture retrievals to precipitation events [80]. In Figure 9, there are large discrepancies between the AMSR2 SM estimates and the in situ SM for ascending overpasses during rainy seasons, such as mid-April to mid-June of 2015 and mid-April to late-May of 2016.

Table 2. Statistical variables of 1 km downscaled and 25 km AMSR2 SM of descending/ascending overpasses between 2015 and 2017, validated using ISMN in situ measurements which are located in the Black Bear–Red Rock watershed, Oklahoma.

Station	1 km AMSR2 (Dsc.)				1 km AMSR2 (Asc.)			
	R^2	ubRMSE	Bias	Number	R^2	ubRMSE	Bias	Number
COSMOS-ARM	0.608	0.048	0.048	92	0.754	0.056	0.056	45
COSMOS-SMAP-OK	0.484	0.079	0.078	53	0.643	0.062	0.062	16
USCRN-Stillwater #1	0.331	0.036	0.033	143	0.377	0.047	0.046	92
USCRN-Stillwater #2	0.284	0.041	0.037	123	0.239	0.048	0.046	74
PBO-H$_2$O-OK #2	0.382	0.037	0.037	32	0.271	0.039	0.033	24
	25 km AMSR2 (Dsc.)				25 km AMSR2 (Asc.)			
COSMOS-ARM	0.449	0.042	0.074	428	0.490	0.077	0.041	430
COSMOS-SMAP-OK	0.287	0.052	0.083	427	0.229	0.091	0.045	413
USCRN/PBO-H2O	0.094	0.036	0.031	97	0.008	0.053	0.018	91

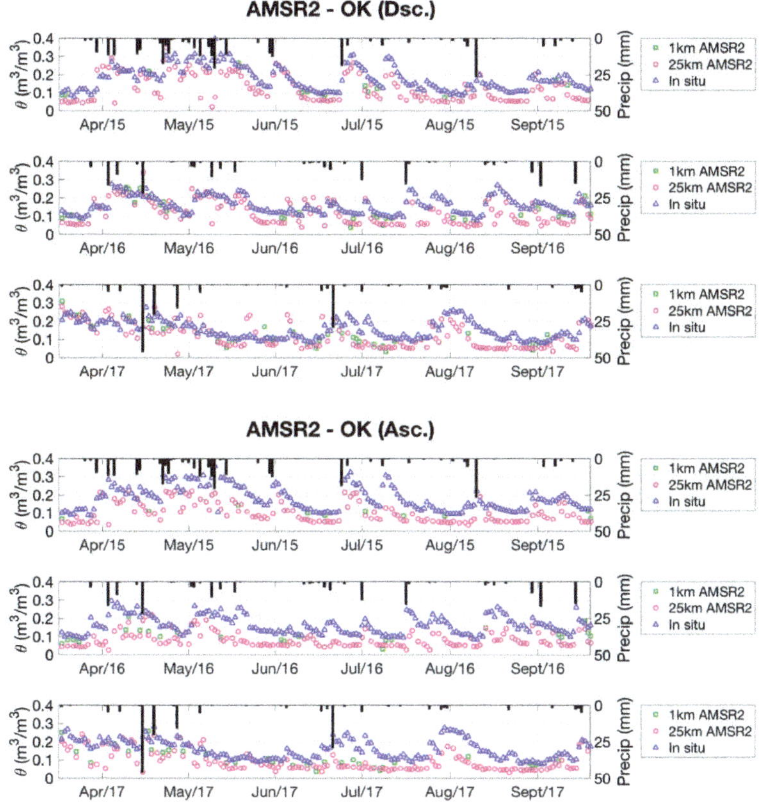

Figure 9. Time-series of daily averaged 1 km/25 km AMSR2 SM estimates of descending/ascending overpasses between 2015 and 2017 along with in situ measurements of corresponding times from the COSMOS station in Oklahoma for SMAP validation. The black bars are daily precipitation derived from a 10 km resolution GPM-IMERG L3 precipitation data.

5. Summary

A SM downscaling algorithm for use with passive microwave satellite estimates was modified, applied as well as validated using in situ ground measurements from an in situ SM database ISMN. The algorithm is based on vegetation modulated relationships between the surface layer SM for descending/ascending overpasses and the corresponding surface temperature difference. The modified algorithm was developed based on only three variables land surface temperature, SM, and NDVI. The use of other related land surface variables, such as precipitation and soil properties, was avoided in the current approach. The algorithm was evaluated in the CONUS region for the months between April and September, and was found generally performed better over the western US than eastern US. The algorithm also demonstrated seasonal variability of performance and the hottest months, July and August, have stronger $\theta - \Delta T_s$ correlations than the other months. The modeled relationships were applied to 1 km MODIS Aqua LST data to downscale coarse resolution 25 km AMSR2 microwave radiometer SM retrieved using the JAXA algorithm. The AMSR2 SM data for the CONUS region between 2013 and 2017 were implemented for producing the 1 km downscaled SM products. SM spatial variability as well as dry-down and wetting trends for selected watersheds in Arizona and Oklahoma were found to be better characterized by the downscaled SM than the coarse SM product. The downscaled SM product was validated using ISMN in situ SM observations in Oklahoma and the assessment results show better overall consistency for the downscaled SM product as compared to the coarser resolution 25 km AMSE2 SM estimates (R^2 increased by 0.141 for descending overpass and increased by 0.215 for ascending overpass). For the downscaled AMSR2 SM, the overall R^2 shows a good correlation with the in situ SM. The precipitation events had an impact on the performance of SM downscaling algorithm due to the reduction in the sensing depth of the microwave signals [11]. The underestimation is noted for both the 1 km downscaled and 25 km AMSR2 SM estimates when compared with in situ data.

In interpreting the validation reported here, there are several issues concerning remotely sensed SM and in situ observations that should be considered [81,82]. First, the passive microwave remotely sensed data originate from an ellipsoidal region at a scale of tens of kilometers, as opposed to the ground observations that record the SM at the point scale. Second, there are potential mismatches in the sensing depths among all the SM data sets. The brightness temperatures sensed by passive microwave sensors (AMSR2) and MODIS, SM outputs from NLDAS, and the in situ SM measured by ground stations are all at different depths. A previous study found that satellite-based SM is noisier than land surface model data [83]. Furthermore, the passive microwave sensors can penetrate a few centimeters, while the MODIS penetrates only a few millimeters, which are compared with the NLDAS SM outputs at a 0–10 cm depth and the ground measurements at a 5 cm depth.

Author Contributions: V.L. coordinated this study and designed the original algorithms for the downscaling of the soil moisture for AMSR-E. B.F. modified the algorithm for AMSR2 and carried out the implementation. R.B. advised the team on the proper implementation of the algorithm and on the validation. T.J.J. was responsible for the validation and the data quality control.

Funding: The work was funded by the NASA Terrestrial Hydrology Program (Program Manager: Jared Entin) and the NASA's Western Water Applications Office (WWAO) Program (Program Manager: Bradley Doorn).

Conflicts of Interest: The authors declare no conflict of interest.

References

1. Njoku, E.G.; Entekhabi, D. Passive microwave remote sensing of soil moisture. *J. Hydrol.* **1996**, *184*, 101–129. [CrossRef]
2. Jackson, T.J.; Schmugge, T.J. Passive microwave remote sensing system for soil moisture: Some supporting research. *IEEE Trans. Geosci. Remote Sens.* **1989**, *27*, 225–235. [CrossRef]
3. Schmugge, T.; Jackson, T.J. Mapping surface soil moisture with microwave radiometers. *Meteorol. Atmos. Phys.* **1994**, *54*, 213–223. [CrossRef]
4. Jackson, T.J.; Le Vine, D.M.; Hsu, A.Y.; Oldak, A.; Starks, P.J.; Swift, C.T.; Isham, J.D.; Haken, M. Soil moisture mapping at regional scales using microwave radiometry: The Southern Great Plains Hydrology Experiment. *IEEE Trans. Geosci. Remote Sens.* **1999**, *37*, 2136–2151. [CrossRef]
5. Jackson, T.J.; Cosh, M.H.; Bindlish, R.; Starks, P.J.; Bosch, D.D.; Seyfried, M.; Goodrich, D.C.; Moran, M.S.; Du, J. Validation of advanced microwave scanning radiometer soil moisture products. *IEEE Trans. Geosci. Remote Sens.* **2010**, *48*, 4256–4272. [CrossRef]
6. Lakshmi, V.; Susskind, J. Comparison of TOVS—Derived land surface variables with ground observations. *J. Geophys. Res. Atmos.* **2000**, *105*, 2179–2190. [CrossRef]
7. Lakshmi, V. The role of satellite remote sensing in the prediction of ungauged basins. *Hydrol. Proc.* **2004**, *18*, 1029–1034. [CrossRef]
8. Lakshmi, V. Remote sensing of soil moisture. *ISRN Soil Sci.* **2013**, *2013*. [CrossRef]
9. Njoku, E.G.; Jackson, T.J.; Lakshmi, V.; Chan, T.K.; Nghiem, S.V. Soil moisture retrieval from AMSR–E. *IEEE Trans. Geosci. Remote Sens.* **2003**, *41*, 215–229. [CrossRef]
10. Chan, S.K.; Bindlish, R.; O'Neill, P.E.; Njoku, E.; Jackson, T.; Colliander, A.; Chen, F.; Burgin, M.; Dunbar, S.; Piepmeier, J.; et al. Assessment of the SMAP passive soil moisture product. *IEEE Trans. Geosci. Remote Sens.* **2016**, *54*, 4994–5007. [CrossRef]
11. Jackson, T.J.; Bindlish, R.; Cosh, M.H.; Zhao, T.; Starks, P.J.; Bosch, D.D.; Seyfried, M.; Moran, M.S.; Goodrich, D.C.; Kerr, Y.H.; et al. Validation of Soil Moisture and Ocean Salinity (SMOS) soil moisture over watershed networks in the US. *IEEE Trans. Geosci. Remote Sens.* **2012**, *50*, 1530–1543. [CrossRef]

12. Bindlish, R.; Jackson, T.; Cosh, M.; Zhao, T.; O'neill, P. Global soil moisture from the Aquarius/SAC–D satellite: Description and initial assessment. *IEEE Geosci. Remote Sens. Lett.* **2015**, *12*, 923–927. [CrossRef]

13. Bindlish, R.; Barros, A.P. Subpixel variability of remotely sensed soil moisture: An inter–comparison study of SAR and ESTAR. *IEEE Trans. Geosci. Remote Sens.* **2002**, *40*, 326–337. [CrossRef]

14. Mladenova, I.; Lakshmi, V.; Walker, J.P.; Panciera, R.; Wagner, W.; Doubkova, M. Validation of the ASAR global monitoring mode soil moisture product using the NAFE'05 data set. *IEEE Trans. Geosci. Remote Sens.* **2010**, *48*, 2498–2508. [CrossRef]

15. Minet, J.; Bogaert, P.; Vanclooster, M.; Lambot, S. Validation of ground penetrating radar full–waveform inversion for field scale soil moisture mapping. *J. Hydrol.* **2012**, *424*, 112–123. [CrossRef]

16. Sridhar, V.; Jaksa, W.T.A.; Fang, B.; Lakshmi, V.; Hubbard, K.G.; Jin, X. Evaluating bias–corrected AMSR–E soil moisture using in situ observations and model estimates. *Vadose Zone J.* **2013**, *12*. [CrossRef]

17. Imaoka, K.; Kachi, M.; Fujii, H.; Murakami, H.; Hori, M.; Ono, A.; Igarashi, T.; Nakagawa, K.; Oki, T.; Honda, Y.; et al. Global Change Observation Mission (GCOM) for monitoring carbon, water cycles, and climate change. *Proc. IEEE* **2010**, *98*, 717–734. [CrossRef]

18. Oki, T.; Imaoka, K.; Imaoka, M. AMSR instruments on GCOM–W1/2: Concepts and applications. In Proceedings of the 2010 IEEE International Geoscience and Remote Sensing Symposium (IGARSS), Honolulu, HI, USA, 25–30 July 2010.

19. Maeda, T.; Taniguchi, Y.; Imaoka, K. GCOM-W1 AMSR2 level 1R product: Dataset of brightness temperature modified using the antenna pattern matching technique. *IEEE Trans. Geosci. Remote Sens.* **2016**, *2*, 770–782. [CrossRef]

20. Cho, E.; Moon, H.; Choi, M. First assessment of the Advanced Microwave Scanning Radiometer 2 (AMSR2) soil moisture contents in Northeast Asia. *J. Meteorol. Soc. Jpn. Ser. II* **2015**, *93*, 117–129. [CrossRef]

21. Bindlish, R.; Cosh, M.H.; Jackson, T.J.; Koike, T.; Fujii, H.; Chan, S.K.; Asanuma, J.; Berg, A.; Bosch, D.D.; Caldwell, T.; et al. GCOM–W AMSR2 soil moisture product validation using core validation sites. *IEEE J. Sel. Top. Appl. Earth Obs. Remote Sens.* **2018**, *11*, 209–219. [CrossRef]

22. Koike, T. *Soil Moisture Algorithm Descriptions of GCOM–W1 AMSR2 (Rev. A)*; Earth Observation Research Center, Japan Aerospace Exploration Agency: Ibaraki, Japan, 2013.

23. Jackson, T.J., III. Measuring surface soil moisture using passive microwave remote sensing. *Hydrol. Proc.* **1993**, *7*, 139–152. [CrossRef]

24. Owe, M.; de Jeu, R.A.; Walker, J. A methodology for surface soil moisture and vegetation optical depth retrieval using the microwave polarization difference index. *IEEE Trans. Geosci. Remote Sens.* **2001**, *39*, 1643–1654. [CrossRef]

25. Lakshmi, V.; Czajkowski, K.; Dubayah, R.; Susskind, J. Land surface air temperature mapping using TOVS and AVHRR. *Int. J. Remote Sens.* **2001**, *22*, 643–662. [CrossRef]

26. Bolten, J.D.; Lakshmi, V.; Njoku, E.G. Soil moisture retrieval using the passive/active L–and S–band radar/radiometer. *IEEE Trans. Geosci. Remote Sens.* **2003**, *41*, 2792–2801. [CrossRef]

27. Narayan, U.; Lakshmi, V.; Njoku, E.G. Retrieval of soil moisture from passive and active L/S band sensor (PALS) observations during the Soil Moisture Experiment in 2002 (SMEX02). *Remote Sens. Environ.* **2004**, *92*, 483–496. [CrossRef]

28. Piles, M.; Entekhabi, D.; Camps, A. A change detection algorithm for retrieving high–resolution soil moisture from SMAP radar and radiometer observations. *IEEE Trans. Geosci. Remote Sens.* **2009**, *47*, 4125–4131. [CrossRef]

29. Narayan, U.; Lakshmi, V.; Jackson, T.J. High–resolution change estimation of soil moisture using L–band radiometer and radar observations made during the SMEX02 experiments. *IEEE Trans. Geosci. Remote Sens.* **2006**, *44*, 1545–1554. [CrossRef]

30. Narayan, U.; Lakshmi, V. Characterizing subpixel variability of low resolution radiometer derived soil moisture using high resolution radar data. *Water Resour. Res.* **2008**, *44*. [CrossRef]

31. Peng, J.; Loew, A.; Merlin, O.; Verhoest, N.E. A review of spatial downscaling of satellite remotely sensed soil moisture. *Rev. Geophys.* **2017**, *55*, 341–366. [CrossRef]

32. Parada, L.M.; Liang, X. Optimal multiscale Kalman filter for assimilation of near-surface soil moisture into land surface models. *J. Geophys. Res. Atmos.* **2004**, *109*. [CrossRef]

33. Pan, M.; Wood, E.F.; McLaughlin, D.B.; Entekhabi, D.; Luo, L. A multiscale ensemble filtering system for hydrologic data assimilation. Part I: Implementation and synthetic experiment. *J. Hydrometeorol.* **2009**, *10*, 794–806. [CrossRef]

34. Sahoo, A.K.; De Lannoy, G.J.; Reichle, R.H.; Houser, P.R. Assimilation and downscaling of satellite observed soil moisture over the Little River Experimental Watershed in Georgia, USA. *Adv. Water Resour.* **2013**, *52*, 19–33. [CrossRef]

35. Mallick, K.; Bhattacharya, B.K.; Patel, N.K. Estimating volumetric surface moisture content for cropped soils using a soil wetness index based on surface temperature and NDVI. *Agric. For. Meteorol.* **2009**, *149*, 1327–1342. [CrossRef]

36. Ranney, K.J.; Niemann, J.D.; Lehman, B.M.; Green, T.R.; Jones, A.S. A method to downscale soil moisture to fine resolutions using topographic, vegetation, and soil data. *Adv. Water Resour.* **2015**, *76*, 81–96. [CrossRef]

37. Merlin, O.; Al Bitar, A.; Walker, J.P.; Kerr, Y. An improved algorithm for disaggregating microwave–derived soil moisture based on red, near–infrared and thermal–infrared data. *Remote Sens. Environ.* **2010**, *114*, 2305–2316. [CrossRef]

38. Merlin, O.; Escorihuela, M.J.; Mayoral, M.A.; Hagolle, O.; Al Bitar, A.; Kerr, Y. Self–calibrated evaporation–based disaggregation of SMOS soil moisture: An evaluation study at 3km and 100m resolution in Catalunya, Spain. *Remote Sens. Environ.* **2013**, *130*, 25–38. [CrossRef]

39. Merlin, O.; Malbéteau, Y.; Notfi, Y.; Bacon, S.; Khabba, S.E.R.S.; Jarlan, L. Performance metrics for soil moisture downscaling methods: Application to DISPATCH data in central Morocco. *Remote Sens.* **2015**, *7*, 3783–3807. [CrossRef]

40. Merlin, O.; Chehbouni, A.; Walker, J.P.; Panciera, R.; Kerr, Y.H. A simple method to disaggregate passive microwave–based soil moisture. *IEEE Trans. Geosci. Remote Sens.* **2008**, *46*, 786–796. [CrossRef]

41. Fang, B.; Lakshmi, V. AMSR–E Soil Moisture Disaggregation Using MODIS and NLDAS Data. *Remote Sens. Terr. Water Cycle* **2014**, *206*, 277.

42. Piles, M.; Camps, A.; Vall–Llossera, M.; Corbella, I.; Panciera, R.; Rudiger, C.; Kerr, Y.H.; Walker, J. Downscaling SMOS–derived soil moisture using MODIS visible/infrared data. *IEEE Trans. Geosci. Remote Sens.* **2011**, *49*, 3156–3166. [CrossRef]

43. Peng, J.; Loew, A.; Zhang, S.; Wang, J.; Niesel, J. Spatial downscaling of satellite soil moisture data using a vegetation temperature condition index. *IEEE Trans. Geosci. Remote Sens.* **2016**, *54*, 558–566. [CrossRef]

44. Choi, M.; Hur, Y. A microwave–optical/infrared disaggregation for improving spatial representation of soil moisture using AMSR–E and MODIS products. *Remote Sens. Environ.* **2012**, *124*, 259–269. [CrossRef]

45. Sánchez–Ruiz, S.; Piles, M.; Sánchez, N.; Martínez–Fernández, J.; Vall–llossera, M.; Camps, A. Combining SMOS with visible and near/shortwave/thermal infrared satellite data for high resolution soil moisture estimates. *J. Hydrol.* **2014**, *516*, 273–283. [CrossRef]

46. Fang, B.; Lakshmi, V.; Bindlish, R.; Jackson, T.J.; Cosh, M.; Basara, J. Passive microwave soil moisture downscaling using vegetation index and skin surface temperature. *Vadose Zone J.* **2013**, *12*, 273–283. [CrossRef]

47. Fang, B.; Lakshmi, V. Soil Moisture at Watershed Scale: Remote Sensing Techniques. *J. Hydrol.* **2014**, *516*, 258–272. [CrossRef]

48. Fang, B.; Lakshmi, V.; Bindlish, R.; Jackson, T.J. Downscaling of SMAP soil moisture using land surface temperature and vegetation data. *Vadose Zone J.* **2018**. accepted. [CrossRef]

49. Fujii, H.; Koike, T.; Imaoka, K. Improvement of the AMSR–E algorithm for soil moisture estimation by introducing a fractional vegetation coverage dataset derived from MODIS data. *J. Remote Sens. Soc. Jpn.* **2009**, *29*, 282–292.

50. Imaoka, K.; Kachi, M.; Kasahara, M.; Ito, N.; Nakagawa, K.; Oki, T. Instrument performance and calibration of AMSR–E and AMSR2. *Int. Arch. Photogram. Remote Sens. Spat. Inf. Sci.* **2010**, *8*, 13–18.

51. Minacapilli, M.; Iovino, M.; Blanda, F. High resolution remote estimation of soil surface water content by a thermal inertia approach. *J. Hydrol.* **2009**, *379*, 229–238. [CrossRef]

52. Zhao, W.; Li, A. A Downscaling Method for Improving the Spatial Resolution of AMSR–E Derived Soil Moisture Product Based on MSG–SEVIRI Data. *Remote Sens.* **2013**, *5*, 6790–6811. [CrossRef]

53. Cosgrove, B.A.; Lohmann, D.; Mitchell, K.E.; Houser, P.R.; Wood, E.F.; Schaake, J.C.; Robock, A.; Marshall, C.; Sheffield, J.; Duan, Q.; et al. Real-time and retrospective forcing in the North American Land Data Assimilation System (NLDAS) project. *J. Geophys. Res. Atmos.* **2013**, *108*. [CrossRef]

54. Mitchell, K.E.; Lohmann, D.; Houser, P.R.; Wood, E.F.; Schaake, J.C.; Robock, A.; Cosgrove, B.A.; Sheffield, J.; Duan, Q.; Luo, L.; et al. The multi-institution North American Land Data Assimilation System (NLDAS):

Utilizing multiple GCIP products and partners in a continental distributed hydrological modeling system. *J. Geophys. Res. Atmos.* **2004**, *109*. [CrossRef]

55. Robock, A.; Luo, L.; Wood, E.F.; Wen, F.; Mitchell, K.E.; Houser, P.R.; Schaake, J.C.; Lohmann, D.; Cosgrove, B.; Sheffield, J.; et al. Evaluation of the North American Land Data Assimilation System over the southern Great Plains during the warm season. *J. Geophys. Res. Atmos.* **2003**, *108*. [CrossRef]

56. Schaake, J.C.; Duan, Q.; Koren, V.; Mitchell, K.E.; Houser, P.R.; Wood, E.F.; Robock, A.; Lettenmaier, D.P.; Lohmann, D.; Cosgrove, B.; et al. An intercomparison of soil moisture fields in the North American Land Data Assimilation System (NLDAS). *J. Geophys. Res. Atmos.* **2004**, *109*. [CrossRef]

57. Wan, Z.; Li, Z.L. A physics–based algorithm for retrieving land–surface emissivity and temperature from EOS/MODIS data. *IEEE Trans. Geosci. Remote Sens.* **1997**, *35*, 980–996.

58. Tucker, C.J. Red and photographic infrared linear combinations for monitoring vegetation. *Remote Sens. Environ.* **1979**, *8*, 127–150. [CrossRef]

59. Dorigo, W.A.; Xaver, A.; Vreugdenhil, M.; Gruber, A.; Hegyiova, A.; Sanchis–Dufau, A.D.; Zamojski, D.; Cordes, C.; Wagner, W.; Drusch, M. Global automated quality control of in situ soil moisture data from the International Soil Moisture Network. *Vadose Zone J.* **2013**, *12*. [CrossRef]

60. Gruber, A.; Dorigo, W.A.; Zwieback, S.; Xaver, A.; Wagner, W. Characterizing coarse–scale representativeness of in situ soil moisture measurements from the International Soil Moisture Network. *Vadose Zone J.* **2013**, *12*. [CrossRef]

61. Larson, K.M.; Small, E.E.; Gutmann, E.D.; Bilich, A.L.; Braun, J.J.; Zavorotny, V.U. Use of GPS receivers as a soil moisture network for water cycle studies. *Geophys. Res. Lett.* **2008**, *35*. [CrossRef]

62. Herring, T.A.; Melbourne, T.I.; Murray, M.H.; Floyd, M.A.; Szeliga, W.M.; King, R.W.; Phillips, D.A.; Puskas, C.M.; Santillan, M.; Wang, L. Plate Boundary Observatory and related networks: GPS data analysis methods and geodetic products. *Rev. Geophys.* **2016**, *54*, 759–808. [CrossRef]

63. Schmugge, T.; Gloersen, P.; Wilheit, T.; Geiger, F. Remote sensing of soil moisture with microwave radiometers. *J. Geophys. Res.* **1974**, *79*, 317–323. [CrossRef]

64. Choudhury, B.J.; Schmugge, T.J.; Chang, A.; Newton, R.W. Effect of surface roughness on the microwave emission from soils. *J. Geophys. Res. Ocean.* **1979**, *84*, 5699–5706. [CrossRef]

65. Carlson, T.N.; Gillies, R.R.; Schmugge, T.J. An interpretation of methodologies for indirect measurement of soil water content. *Agric. For. Meteorol.* **1995**, *77*, 191–205. [CrossRef]

66. Gillies, R.R.; Kustas, W.P.; Humes, K.S. A verification of the'triangle'method for obtaining surface soil water content and energy fluxes from remote measurements of the Normalized Difference Vegetation Index (NDVI) and surface e. *Int. J. Remote Sens.* **1997**, *18*, 3145–3166. [CrossRef]

67. Hong, S.; Lakshmi, V.; Small, E.E.; Chen, F.; Tewari, M.; Manning, K.W. Effects of vegetation and soil moisture on the simulated land surface processes from the coupled WRF/Noah model. *J. Geophys. Res. Atmos.* **2009**, *114*. [CrossRef]

68. Lakshmi, V.; Hong, S.; Small, E.E.; Chen, F. The influence of the land surface on hydrometeorology and ecology: New advances from modeling and satellite remote sensing. *Hydrol. Res.* **2011**, *42*, 95–112. [CrossRef]

69. Pablos, M.; Piles, M.; Sánchez, N.; Vall–llossera, M.; Martínez–Fernández, J.; Camps, A. Impact of day/night time land surface temperature in soil moisture disaggregation algorithms. *Eur. J. Remote Sens.* **2016**, *49*. [CrossRef]

70. Owe, M.; Van de Griend, A.A. Comparison of soil moisture penetration depths for several bare soils at two microwave frequencies and implications for remote sensing. *Water Resour. Res.* **1998**, *34*, 2319–2327. [CrossRef]

71. Justice, C.O.; Vermote, E.; Townshend, J.R.; Defries, R.; Roy, D.P.; Hall, D.K.; Salomonson, V.V.; Privette, J.L.; Riggs, G.; Strahler, A. The Moderate Resolution Imaging Spectroradiometer (MODIS): Land remote sensing for global change research. *IEEE Trans. Geosci. Remote Sens.* **1998**, *36*, 1228–1249. [CrossRef]

72. Xia, Y.; Ek, M.B.; Wu, Y.; Ford, T.; Quiring, S.M. Comparison of NLDAS–2 simulated and NASMD observed daily soil moisture. Part I: Comparison and analysis. *J. Hydrometeorol.* **2015**, *16*, 1962–1980. [CrossRef]

73. Luo, L.; Robock, A.; Mitchell, K.E.; Houser, P.R.; Wood, E.F.; Schaake, J.C.; Lohmann, D.; Cosgrove, B.; Wen, F.; Sheffield, J.; et al. Validation of the North American land data assimilation system (NLDAS) retrospective forcing over the southern Great Plains. *J. Geophys. Res. Atmos.* **2003**, *108*. [CrossRef]

74. Han, X.; Jin, R.; Li, X.; Wang, S. Soil moisture estimation using cosmic–ray soil moisture sensing at heterogeneous farmland. *IEEE Geosci. Remote Sens. Lett.* **2014**, *11*, 1659–1663.

75. Rosolem, R.; Shuttleworth, W.J.; Zreda, M.; Franz, T.E.; Zeng, X.; Kurc, S.A. The effect of atmospheric water vapor on neutron count in the cosmic–ray soil moisture observing system. *J. Hydrometeorol.* **2013**, *14*, 1659–1671. [CrossRef]

76. Zreda, M.; Shuttleworth, W.J.; Zeng, X.; Zweck, C.; Desilets, D.; Franz, T.; Rosolem, R. COSMOS: The cosmic–ray soil moisture observing system. *Hydrol. Earth Syst. Sci.* **2012**, *16*, 4079–4099. [CrossRef]

77. Coopersmith, E.J.; Cosh, M.H.; Bell, J.E.; Crow, W.T. Multi–profile analysis of soil moisture within the US Climate Reference Network. *Vadose Zone J.* **2016**, *15*. [CrossRef]

78. Bell, J.E.; Palecki, M.A.; Baker, C.B.; Collins, W.G.; Lawrimore, J.H.; Leeper, R.D.; Hall, M.E.; Kochendorfer, J.; Meyers, T.P.; Wilson, T.; et al. US Climate Reference Network soil moisture and temperature observations. *J. Hydrometeorol.* **2013**, *14*, 977–988. [CrossRef]

79. Diamond, H.J.; Karl, T.R.; Palecki, M.A.; Baker, C.B.; Bell, J.E.; Leeper, R.D.; Easterling, D.R.; Lawrimore, J.H.; Meyers, T.P.; Helfert, M.R.; et al. US Climate Reference Network after one decade of operations: Status and assessment. *Bull. Am. Meteorol. Soc.* **2013**, *94*, 485–498. [CrossRef]

80. Jin, K.-W.; Njoku, E.; Chan, S. Impact of Rainfall on the Retrieval of Soil Moisture using AMSR–E data. In Proceedings of the 2006 IEEE International Symposium on Geoscience and Remote Sensing, Denver, CO, USA, 31 July–4 August 2006.

81. Zhang, X.; Zhang, T.; Zhou, P.; Shao, Y.; Gao, S. Validation analysis of SMAP and AMSR2 soil moisture products over the United States using ground–based measurements. *Remote Sens.* **2017**, *9*, 104. [CrossRef]

82. Cosh, M.H.; Jackson, T.J.; Bindlish, R.; Prueger, J.H. Watershed scale temporal and spatial stability of soil moisture and its role in validating satellite estimates. *Remote Sens. Environ.* **2004**, *92*, 427–435. [CrossRef]

83. Fang, L.; Hain, C.R.; Zhan, X.; Anderson, M.C. An inter–comparison of soil moisture data products from satellite remote sensing and a land surface model. *Int. J. Appl. Earth Obs. Geoinf.* **2016**, *48*, 37–50. [CrossRef]

 *remote sensing*

Article

Analysis of the Radar Vegetation Index and Potential Improvements

Christoph Szigarski [1,*], Thomas Jagdhuber [1], Martin Baur [2], Christian Thiel [3], Marie Parrens [4,5], Jean-Pierre Wigneron [6], Maria Piles [7] and Dara Entekhabi [8]

[1] German Aerospace Center, Microwaves and Radar Institute, P.O. Box 1116, 82234 Wessling, Germany; thomas.jagdhuber@dlr.de

[2] Bayreuth Center for Ecology and Environmental Research, University of Bayreuth, Universitätsstraße 30, 95447 Bayreuth, Germany; martin.baur@uni-bayreuth.de

[3] German Aerospace Center, Institute of Data Science, Mälzerstraße 3, 07745 Jena, Germany; christian.thiel@dlr.de

[4] Cesbio, 18 avenue Edouard Belin, 31401 Toulouse, France; marie.parrens@cesbio.cnes.fr

[5] E.I Purpan, 75 voie du TOEC, BP57611, 31076 Toulouse CEDEX 3, France

[6] INRA, Centre de Bordeaux Aquitaine, 71 avenue E. Bourlaux, 33882 Villenave d'Ornon CEDEX, France; jean-pierre.wigneron@inra.fr

[7] Image Processing Lab, Parc Cientific, University of Valencia, 46980 Paterna Valencia, Spain; maria.piles@uv.es

[8] Department of Civil and Environmental Engineering, MIT, Vassar Street 15, Cambridge, MA 02139, USA; darae@mit.edu

* Correspondence: chszigarski@aol.de; Tel.: +49-176-47194985

Received: 7 October 2018; Accepted: 7 November 2018; Published: 9 November 2018

Abstract: The Radar Vegetation Index (RVI) is a well-established microwave metric of vegetation cover. The index utilizes measured linear scattering intensities from co- and cross-polarization and is normalized to ideally range from 0 to 1, increasing with vegetation cover. At long wavelengths (L-band) microwave scattering does not only contain information coming from vegetation scattering, but also from soil scattering (moisture & roughness) and therefore the standard formulation of RVI needs to be revised. Using global level SMAP L-band radar data, we illustrate that RVI runs up to 1.2, due to the pre-factor in the standard formulation not being adjusted to the scattering mechanisms at these low frequencies. Improvements on the RVI are subsequently proposed to obtain a normalized value range, to remove soil scattering influences as well as to mask out regions with dominant soil scattering at L-band (sparse or no vegetation cover). Two purely vegetation-based RVIs (called RVII and RVIII), are obtained by subtracting a forward modeled, attenuated soil scattering contribution from the measured backscattering intensities. Active and passive microwave information is used jointly to obtain the scattering contribution of the soil, using a physics-based multi-sensor approach; simulations from a particle model for polarimetric vegetation backscattering are utilized to calculate vegetation-based RVI-values without any soil scattering contribution. Results show that, due to the pre-factor in the standard formulation of RVI the index runs up to 1.2, atypical for an index normally ranging between zero and one. Correlation analysis between the improved radar vegetation indices (standard RVI and the indices with potential improvements RVII and RVIII) are used to evaluate the degree of independence of the indices from surface roughness and soil moisture contributions. The improved indices RVII and RVIII show reduced dependence on soil roughness and soil moisture. All RVI-indices examined indicate a coupled correlation to vegetation water content (plant moisture) as well as leaf area index (plant structure) and no single dependency, as often assumed. These results might improve the use of polarimetric radar signatures for mapping global vegetation.

Keywords: microwaves; radiometer; radar; vegetation index; soil scattering; roughness; soil moisture; SMAP; SMOS

1. Introduction

Mapping the density of woody vegetation cover at large scales is required in applications such as estimating terrestrial carbon stocks. Vegetation dynamics are also linked to global climate processes [1,2]. The forest ecosystems exchange water, energy, carbon and other biogeochemical species with the atmosphere and especially play a critical role in atmospheric carbon dioxide uptake [2–4]. Continuous mapping of these ecosystems delivers crucial information to quantify and understand change in the environment from local to global scales [1].

Over the past decades, remote sensing of vegetation cover has mainly been based on measurements in visible, near infrared and shortwave bands [1,5,6]. Optical and infrared indices have limitations associated with opacity of the atmosphere due to clouds, water vapor and aerosols. Active (send & receive) and passive (only receive) microwave remote sensing operates in a wavelength range between 1 mm and 1 m. Unlike optical and infrared, microwave remote sensing is independent from solar illumination and penetrates areas covered by clouds, haze and rain (study frequency: L-band at about 1.3 GHz [7]) [6,8]. They thus allow gap-free data streams for vegetation mapping.

Moreover, while optical methods mostly sense surface and top of canopy conditions, microwaves can penetrate into the vegetation canopy especially at longer wavelengths (e.g., L-band at about 21 cm wavelength). In this regard, observations in the microwave spectrum provide information related to plant moisture content and structure and offer a distinct view on plant conditions [5]. A parameter frequently used in vegetation mapping is the vegetation optical depth (VOD) [9], which can be retrieved for instance from AMSR or SMAP radiometer data. In order to use radar for mapping vegetation cover, a robust index is desirable. However, at longer wavelengths (e.g., L-band), microwave indices are potentially influenced by soil moisture and surface roughness contributions, also in presence of dense vegetation [10]. When applying microwave indices as in [11], these soil contaminations have not been taken into account, resulting in misleading interpretations.

One of the more established active microwave indices for mapping vegetation cover is the Radar Vegetation Index (RVI) [12]. A sensitivity analysis of RVI at L-band was presented in [13], and it has been utilized in studies such as [14–16]. In particular, prior investigations on RVI have compared it to optical-based measures like the Normalized Difference Vegetation Index (NDVI), indicating greenness of vegetation and Leaf Area Index (LAI), used to detect canopy layers from vegetation [14]. Also, the RVI index has been shown to be related to the Vegetation Water Content (VWC) [14–16] that is directly linked to the wetness of vegetation. Optical and infrared indices are however mostly sensitive to scattering and absorption on leaf surfaces. The influence of wetness (e.g., VWC) and structure (e.g., LAI) of a canopy on RVI is not yet understood sufficiently. In addition, the extent to which the RVI index is solely connected to vegetation at long wavelengths (e.g., L-band) is not fully known.

This study investigates the characteristics of the RVI and assesses the potential for its improvement. We investigate the degree to which RVI isolates the scattering by vegetation and is independent of soil surface reflectivity, soil moisture and roughness. In the following sections we analyze these characteristics of the RVI index and introduce two improved formulations using a multi-sensor (active-passive microwave) approach.

2. Test Sites and Experimental Data

2.1. Test Sites

Investigations regarding the RVI-value range are initially carried out at global level. Beside the investigation on global data, an in-depth study will be conducted over a few regions, in order to span the entire range of land cover types without biasing the results towards a few dominant land cover types. To define a set of suitable test sites, the following criteria are considered:

- Coverage of different climate and vegetation zones, based on the International Geosphere-Biosphere Programme (IGBP) classification from [17],
- Diversely vegetated areas to cover high and low RVI mean,
- Varying degrees of complexity in vegetation structure,
- Inclusion of areas with distinct vegetation cover,
- Availability of polarimetric radar data and auxiliary datasets.

The following sites are selected for the statistical analyses: Eastern U.S., Yucatan Peninsula, Sweden, East Europe, Eastern South Africa, Kalimantan and Southeast China (Figure 1). The dataset is compiled based on the above-mentioned criteria and includes vegetation canopies from: tropical climate with evergreen broadleaf forests, mangroves, nipa palms (Kalimantan, Central parts of Yucatan); warm temperate areas that grow deciduous broadleaf forests, mixed forests and cultivated patterns (Eastern Europe, Southeast of China, Eastern U.S.); warm temperate and fully humid lands covered by woody savannas and grassland (Eastern South Africa); continental and fully humid territories characterized by extensive deciduous broadleaf forest (Central Sweden); polar regions that contain robust needleleaf species and coverage by single pine and spruce trees.

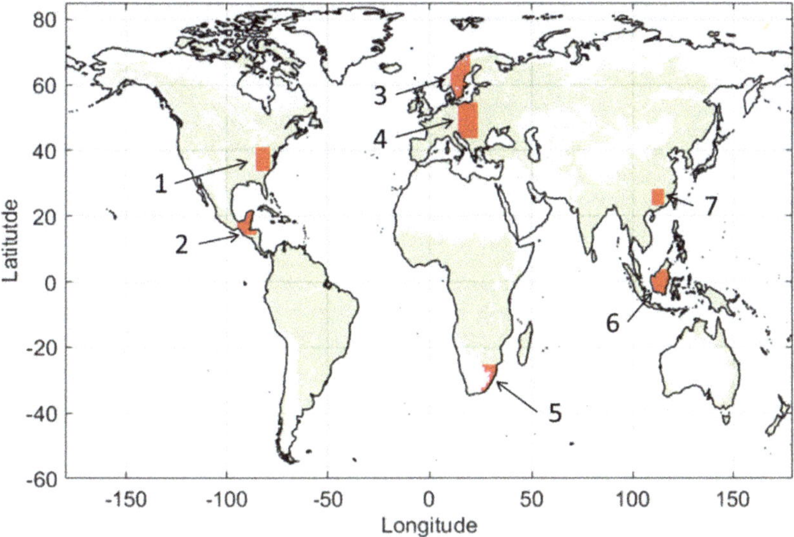

Figure 1. Locations of the in-depth test sites are marked as red polygons. The following sites are selected: Eastern U.S. (1), Yucatan Peninsula (2), Sweden (3), East Europe (4), Eastern South Africa (5), Kalimantan (6) and Southeast China (7). The light green color indicates the areas of dominant vegetation scattering (result of masking areas with dominant soil scattering (cf. Section 3)).

2.2. Experimental Data

The L-band, polarimetric, space-borne radar data acquired by the NASA SMAP mission (Soil Moisture Active Passive [7]) is the central dataset used in this study. The SMAP active-passive dataset was recorded from 13 April 2015 to 7 July 2015. This period begins with the post-launch start of science data acquisition by SMAP and ends on the date of the SMAP radar failure [7]. The used data is the global level 3 product (SPL3SMAP) [18] that contains the linear backscattering intensities for the polarizations HH, VV and aggregation of cross-polarizations (HV or VH) from the L-band radar on a 9 km grid, as well as soil moisture and vegetation optical depth from SMAP L-band radiometer retrievals on a 36 km grid for a global scale [7]. For the investigations of the RVI, time-averaged values (April to July 2015) were considered because of the short period of record.

In addition, soil roughness information (vertical component) from Parrens et al. [19] are used in the forward modelling of the surface scattering component. This dataset is computed from SMOS (Soil Moisture and Ocean Salinity [20]) L-band passive microwave observations with a spatial resolution of ~43 km at global scale for the year 2011 [21]. A gap-filling procedure using mean roughness values per IGBP land cover class is used to provide spatially continuous information, as suggested in [22]. This was necessary due to the data gaps in the original product of [19] caused by strong vegetation cover in regions where no soil roughness could be estimated.

Furthermore, the MODIS-based vegetation water content (VWC) [23] included in the SMAP product (SPL3SMAP) [18] is used for the correlation analysis in the results section. The MODIS-based leaf area index (LAI) [24] will be used in the same section. The LAI data were available for all study areas with 500 m spatial resolution gridded on 0.5° in latitude and longitude [25].

3. Modelling and Retrieval of Standard and Improved RVI

The radar vegetation index (RVI) has been used in a number of previous studies (e.g., [13,14]) mostly for predicting the growth level of crop vegetation over time [14]. The RVI, as firstly introduced by Kim & van Zyl, is a measure of volume scattering (from randomly oriented dipoles), a scattering mechanism usually caused by complex structural elements of vegetation (e.g., combination of leaves, branches & trunks). The index is defined as [26]:

$$\mathrm{RVI} = \frac{8\,\sigma_{HV}}{\sigma_{HH} + \sigma_{VV} + 2\sigma_{HV}}, \tag{1}$$

where σ_{HH}, σ_{HV} and σ_{VV} represent the measured linear backscattering intensities [-]. The 8 in the numerator is referred to the pre-factor in this study. The RVI is meant to be a normalized index that ideally varies between zero and one.

For L-band SAR applications at global scale it is likely that the measured quantities contain not only scattering from vegetation cover, but also from the soil underneath. This is especially critical for sparsely vegetated areas where there is soil backscattering in the co- and cross-polarizations. In some cases, the RVI may over- or under-estimate vegetation volume scattering because it is based on a ratio. Soil scattering contributions need to be removed from the measured intensities in (1). As stated in [14], the RVI should typically vary between zero and one. Values near zero should correspond to bare surfaces. Values near one should indicate a dense vegetation canopy. A previous study showed that the index tends to overestimate biomass in dry regions [13]. To gain insight into the behavior of the index, a polarimetric vegetation scattering forward model needs to be developed.

Scattering from a vegetation canopy can be modelled using a cloud of randomly oriented thin cylinders and discs with lossy dielectric properties [12]. A polarimetric spheroidal particle model is used here [27] to ensure flexibility for application to vegetation canopies with different structures and geometries. This model is driven by the particle anisotropy (Ap) and orientation distribution width (ψ) of the spheroidal particles that form the scattering canopy media. The particle anisotropy represents the main plant constituents (e.g., stem, branches, leaves) forming the uniformly shaped volume. The orientation distribution width embodies the degree of orientation and organization of the constituents within the volume. Further details of the model are given in [27,28]. The linear backscattering coefficients σ_{PP} [-], as a function of Ap [-] and ψ [rad], are given as follows [27]:

$$\sigma_{HH} = \frac{1}{1 + A_P^2} \cdot \frac{1}{8} \left(3A_P^2 + 2A_P + 3 + 4\left(A_P^2 - 1\right) Sinc(2\psi) + (A_P - 1)^2 Sinc(4\psi)\right), \tag{2}$$

$$\sigma_{VV} = \frac{1}{1 + A_P^2} \cdot \frac{1}{8} \left(3A_P^2 + 2A_P + 3 - 4\left(A_P^2 - 1\right) Sinc(2\psi) + (A_P - 1)^2 Sinc(4\psi)\right), \tag{3}$$

$$\sigma_{HV} = \frac{1}{1 + A_P^2} \cdot \frac{1}{8} (A_P - 1)^2 (1 - Sinc(4\psi)). \tag{4}$$

The backscattering intensities are calculated using three sets of Ap-values, varying between zero and 10^6 (0–1; 1–100; 100–10^6). The first set (0–1) capture particle shapes from vertical dipoles to spheres, the second (1–100) and third range (100–10^6) simulate particles from spheres (values around 1) to horizontal dipoles (values that tend to infinity). The evaluation of the standard RVI index using the forward modelled linear backscattering intensities leads to an upper limit of 1.2 for the index. Analysis of the entire Ap- and ψ-range with the canopy model can result in an optimized pre-factor for the cross-polarization component within the RVI formulation (numerator in (1)). The optimization of this pre-factor can shift the resulting values of RVI to be confined in zero to one range. The standard pre-factor (A = 8) in (1) assumes randomly oriented dipoles that do not normalize the full range of potentially occurring σ_{HV}–intensities due to variety in vegetation shape and structure. The white line in Figure 2 marks the maximum of modelled σ_{HV}-values (0.125) from vegetation that can be normalized by the standard pre-factor of 8. The global maximum when modelled with the particle model was located at a higher value of 0.152. Therefore, the optimum pre-factor for an improved RVI is found to be 6.57.

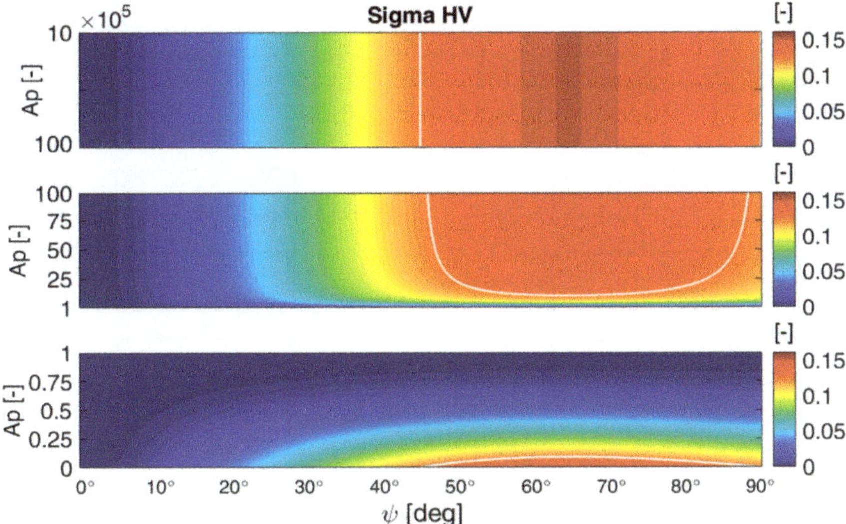

Figure 2. Density plot of modelled σ_{HV}-values [-] from (4) as a function of vegetation parameters ψ [rad] (orientation distribution width = degree of orientation in canopy) and Ap (particle anisotropy = plant shape) [-]. The white line indicates the maximum of modelled σ_{HV}-values (0.125) that can be normalized by the standard pre-factor of 8 in (1). The three panels correspond to different ranges of Ap (0–1; 1–100; 100–10^6).

In addition to modelling the structural component of the RVI, we investigate the permittivity component of the forward vegetation scattering model. This modelling approach is still based on the variation in ψ, but starts on permittivity input level. To simulate the dielectric constant ε_v of the vegetation, the semi-physical method of Ulaby & El-Rayes is used [29]. The resulting complex dielectric constant ε_v is input to the modelling of the particle shape, leading to variations in the parameter Ap. This is done by solving the particle shape functions L_1, L_2 for a given major ($x_1 = 0.01$ m) and minor axes ($x_2 = 0.05$ m) [30,31]:

$$L_1 = \begin{cases} \text{prolate} : \frac{1-e^2}{e^2}\left(-1 + \frac{1}{2e}ln\frac{1+e}{1-e}\right) & x_1 > x_2 & e^2 = 1 - \frac{x_2^2}{x_1^2} \\ \text{oblate} : \frac{1-f^2}{f^2}\left(1 - \frac{1}{f}arctan(f)\right) & x_1 < x_2 & f^2 = \frac{x_2^2}{x_1^2} - 1 \end{cases} \tag{5}$$

$$L_2 = \frac{1}{2}\left(1 - L_1\right). \tag{6}$$

Both functions differentiate between a prolate and an oblate scenario to capture the entire shape variation of the spheroidal particles. The particle anisotropy A_p can be calculated using a ratio of the polarizabilities ρ_{ee1} and ρ_{ee2} calculated from shape functions L_1 and L_2 [28]:

$$A_p = \frac{\rho_{ee1}}{\rho_{ee2}} = \frac{L_2 + 1/(\varepsilon_r - 1)}{L_1 + 1/(\varepsilon_r - 1)} \quad \begin{cases} A_p < 1 & \text{oblate spheroids} \\ A_p = 1 & \text{spheres} \\ A_p > 1 & \text{prolate spheroids} \end{cases} \tag{7}$$

This enables the determination of the shape A_p in dependence of a permittivity input. Consequently, the RVI is simulated using the A_p-ψ model in (1) with the retrieved particle anisotropy A_p as well as the modelled orientation distribution width ψ (full extend: 0° to 90°). The conceptual workflow of the model-based RVI, starting at permittivity input level is shown in Figure 3.

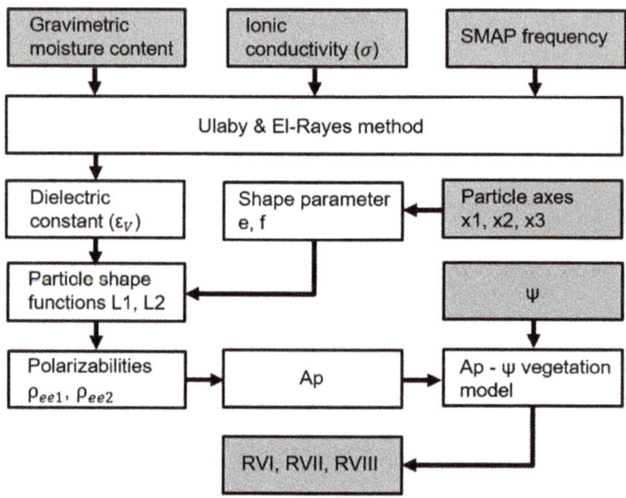

Figure 3. Conceptual workflow of the model-based RVI, starting at permittivity input level. Input and output parameters are marked in grey boxes.

The modelled results for the RVI are shown in Figure 4 (prolate scenario). For low ψ–values (20–30°) there are no variations of the RVI with increasing vegetation permittivity. With an increasingly complex vegetation structure (40–90°), there are the widest variations in RVI along vegetation permittivity (real & imaginary part). The variations of RVI for the distribution interval from 30 to 60° are similar and between 15–50 in real part and between 5–15 in imaginary part of the permittivity. It also clearly shows that only for very low permittivity (real & imaginary part), there are no variations of the RVI over this complexity range. The oblate scenario exhibited similar results (not shown).

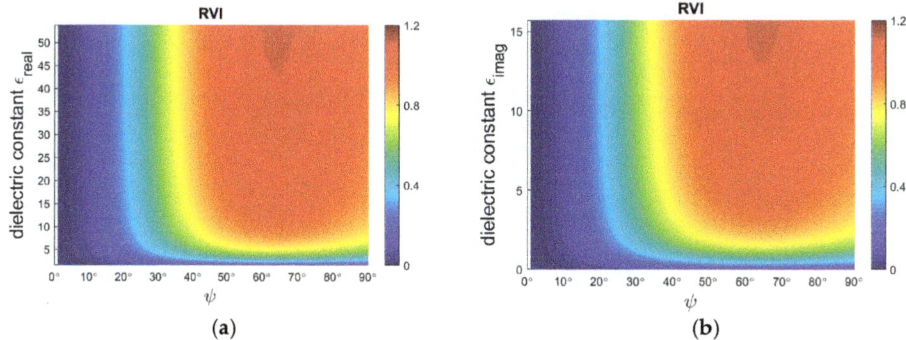

Figure 4. Density plots of modeled RVI [-] as a function of the vegetation structure parameter ψ [deg] and the complex dielectric constant ε_v [-] of vegetation (after Ulaby & El-Rayes [29]), split in real (**a**) and imaginary (**b**) part for the prolate spheroid scenario.

With the availability of the forward model of the σ_{HV}–component, there is now potential for improvements in the pre-factor of RVI (normalization of index). Further improvements are introduced by including an active-passive microwave-based correction for a soil scattering component within the cross polarization σ_{HV} in (1), originating from soil roughness effects [27]. This improvement is realized using a multi-sensor approach based on a forward modelling of the soil component. Hence, an improved definition of (1) is found given to be:

$$RVII = \frac{6.57\left(\sigma_{HV}{}^{m} - \sigma_{HV}{}^{s}\gamma^{2}\right)}{\left(\sigma_{HH}{}^{m}\right) + \left(\sigma_{VV}{}^{m}\right) + 2\left(\sigma_{HV}{}^{m}\right)}, \tag{8}$$

where the parameter γ represents the one-way attenuation of the soil signal while passing through the vegetation media. The one-way attenuation is derived from SMAP radiometer-based retrievals of vegetation optical depth (VOD), using the MT-DCA algorithm from [32]. The estimates are assumed polarization independent for L-band at global scales [32].

In (8) $\sigma_{pq}{}^{m}$ is the measured linear backscattering intensities and $\sigma_{pq}{}^{s}$ is the soil scattering intensity, originating from forward modelling. Here, the extended Bragg model is used [33]. This polarimetric bare soil scattering model for longer wavelength (L-band) requires input of the roughness and relative permittivity of the soil. The latter is obtained from passive microwave soil moisture retrievals (e.g., SMAP baseline algorithm), using the dielectric mixing model of Mironov & Fomin for the conversion to permittivity [22,34]. For the soil roughness input, a SMOS-based dataset is utilized, as introduced in Chapter 2.2. The conceptual workflow of the improved RVI (RVII and RVIII) retrieval is shown in Figure 5.

An alternative approach of incorporating corrections for soil contributions in (5) is that all measured input intensities are corrected for their respective soil scattering component. In this way, soil contribution terms σ^{s} need to be known for all co- and cross polarizations. The maximum soil correction of the measured intensities, is expressed as follows:

$$RVIII = \frac{6.57\left(\sigma_{HV}{}^{m} - \sigma_{HV}{}^{s}\gamma^{2}\right)}{\left(\sigma_{HH}{}^{m} - \sigma_{HH}{}^{s}\gamma^{2}\right) + \left(\sigma_{VV}{}^{m} - \sigma_{VV}{}^{s}\gamma^{2}\right) + 2\left(\sigma_{HV}{}^{m} - \sigma_{HV}{}^{s}\gamma^{2}\right)}. \tag{9}$$

However, the quality and spatial resolution of the RVII and RVIII can only be as good as its remotely-sensed (SMAP & SMOS) input data for $\sigma_{pq}{}^{s}$ and γ. The improved indices incorporates ((8) and (9)) combine information from active and passive microwave sensor data.

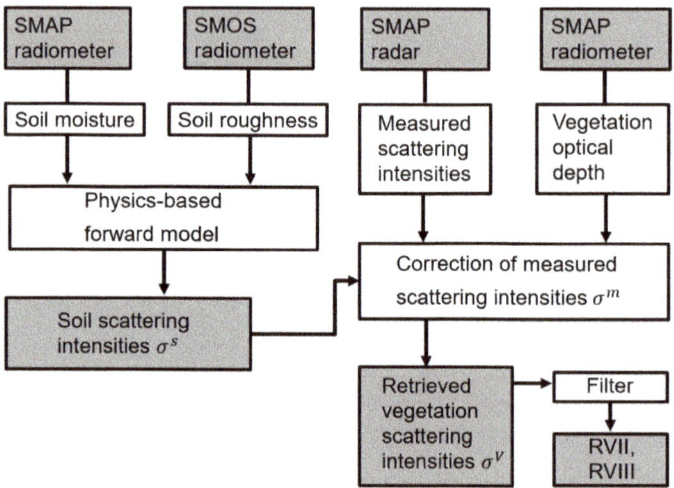

Figure 5. Conceptual workflow of retrieving the improved radar vegetation index (RVII) by a multi-sensor (active-passive microwave) approach. The input sensors and the output parameters are marked in grey boxes.

In addition to the correction of the soil scattering component, the original RVI and the improved indices (RVII and RVIII) need to be applied to the regions where they are valid (non-bare and non-ice and snow-covered surfaces). The definitions of the excluded regions are described in more detail in Section 4. The IGBP classification of land cover is used and the exclusion regions are shown in Figure 6. Bare soil and sparse vegetation regions (deserts, sparse shrublands and steppes) are not within the range of applicability of radar vegetation indices. If the removal of σ_{pq}^s from σ_{pq}^m led to negative values, an indicator of the dominance of soil scattering over vegetation scattering, the area is masked. This check for negative-valued intensities after removal of soil scattering is done for all polarizations.

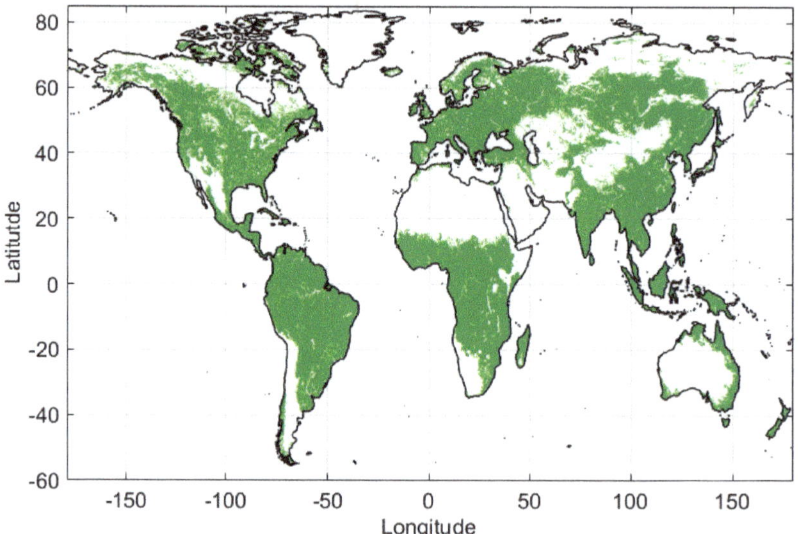

Figure 6. Retrieved mask of valid (dominant vegetation scattering) RVI-values (green) and excluded regions where soil scattering is dominant (white) at L-band.

In the next section correlations between standard and improved RVI with soil- and vegetation-related parameters are conducted in order to assess the degree to which RVI is an indicator of vegetation structure and permittivity and independent of surface soil reflectivity. The model analyses in this section showed that there is permittivity as well as structure dependence in radar vegetation indices. Therefore, the following parameters are selected for a comparison: volumetric soil moisture and vegetation water content (VWC) for permittivity dependency; vertical surface roughness (*ks*) and leaf area index (LAI) for structure related correlation.

4. Results

4.1. Global Results

The global overview of the standard RVI, calculated with (1) and using SMAP radar data is shown in Figure 7a. The Radar Vegetation Index defined in (1) shows a global pattern corresponding to known global distribution of land cover. It is based on the short period (April to July 2015) where the SMAP radar was operating. But cloud cover and illumination do not result in any data losses because it is based on L-band radar measurements. The panel in Figure 7b shows that the index can exceed the value of one. Especially in the tropical latitudes (e.g., Amazon basin) the maximum values reach up to 1.2. The global maximum of the histogram in Figure 7b is around 0.8 and there is also a local maximum close to one. There may be an overall overestimation of RVI towards higher values (>0.8 [-]), which was already reported empirically in [14] and affirmed and analyzed with forward modelling of a polarimetric vegetation scattering model in the previous section. The mask in Figure 6 is not applied to RVI because in this study we used the RVI-values in the excluded region to diagnose soil surface scattering and reflectivity contributions.

The pre-factor of the numerator in (1) can result in improved dynamic range and interpretability of RVI. Figure 8 illustrates the result of the improved RVI (RVII) computed with (8). Figure 9 shows the results of the fully corrected RVI (RVIII) calculated with (9). The incorporation of the forward model-derived pre-factor and the active-passive microwave-based removal of the attenuated soil contribution leads to an indexing between zero and one.

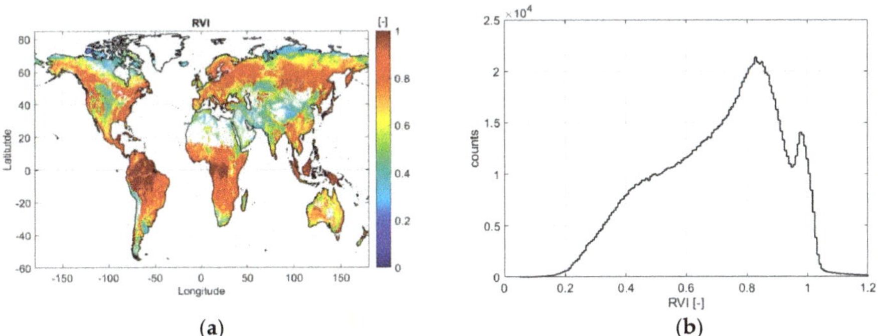

(**a**)   (**b**)

Figure 7. Result of the averaged (April–July 2015) standard RVI [-] calculated with (1): (**a**) global map; (**b**) histogram of values ranging beyond one for high amounts of biomass.

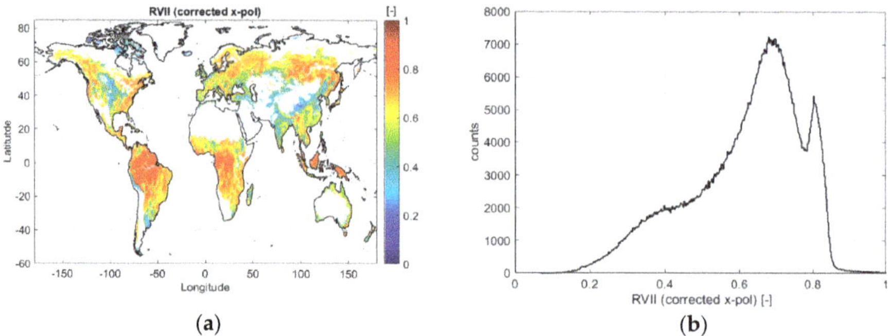

(a)   **(b)**

Figure 8. Result of the averaged (April–July 2015) RVII [-] calculated with (8): (**a**) global map; (**b**) histogram of values ranging only between zero and one.

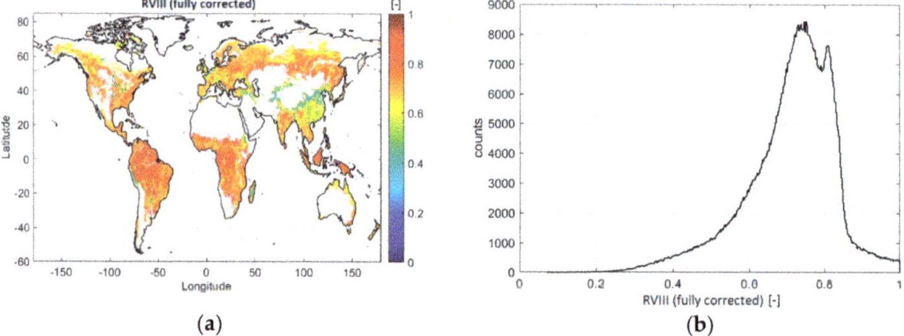

(a)   **(b)**

Figure 9. Result of the averaged (April–July 2015) RVIII [-] calculated with (9): (**a**) global map; (**b**) histogram of values ranging only between zero and one.

There is a general shift towards lower values of the radar vegetation index as shown in the histograms in Figures 8b and 9b. The global maximum in the histogram of the RVII is located at 0.7 and a local one can be found at 0.8. The dry and bare surfaces of Central and Western Australia show RVI-values up to 0.7 (see Figure 7a). However, these values are unlikely to be representative of the vegetation cover in the region. This is due to soil roughness-induced high values of $\sigma_{HV}{}^m$ instead of any cross-polarization contribution from vegetation. Both RVII and RVIII remove the soil scattering contribution and, in cases where the resulting backscatter is negative, the region is considered to have a too strong surface soil contribution and is masked out. Radar-based vegetation indices at L-band are not suited (valid) for these regions.

4.2. Correlation of Radar Vegetation Indices with Soil- and Vegetation-Related Parameters

Correlations of soil related parameters are shown in Figures 10 and 11. Figure 10 demonstrates the relation between the RVI, RVII, RVIII and the volumetric soil moisture derived from SMAP radiometer measurements [18]. The data have soil moisture values that are limited to a range between 0 and 0.6 [cm^3/cm^3] [35]. Ideally there should be no correlation between soil moisture and the indices. The dependence on soil moisture is already small with RVI (R^2 = 0.08). The new indices reduce the dependence further as expected from their formulation (RVII: R^2 = 0.06; RVIII: R^2 = 0.02).

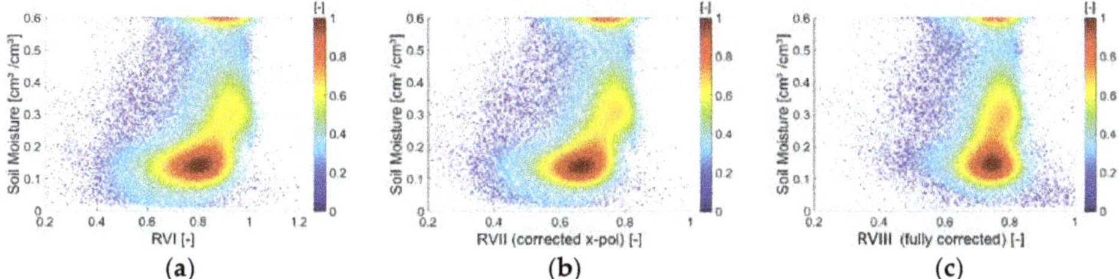

Figure 10. Correlation for the combined study areas between: (**a**) RVI [-]; (**b**) x-pol corrected RVII [-]; (**c**) fully corrected RVIII [-] and volumetric surface soil moisture [cm^3/cm^3] derived from SMAP [18,35].

Figure 11 shows the relation of soil roughness height standard deviation non-dimensionalized with wave-length (*ks*) to RVI, RVII and RVIII. The standard index (RVI) exhibits a dependency to the surface roughness parameter ($R^2 = 0.37$) as shown in Figure 11a. The relation improves (by reduction in correlation) slightly but definitely not adequately for the cross-polarization-corrected RVII ($R^2 = 0.36$) as shown in Figure 11b. The fully corrected index (RVIII) shows much reduced correlation to the soil roughness ($R^2 = 0.01$) as shown in Figure 11c. This affirms that a correction of the soil roughness contribution is necessary, as a pure vegetation-based radar vegetation index should be independent of surface roughness influences.

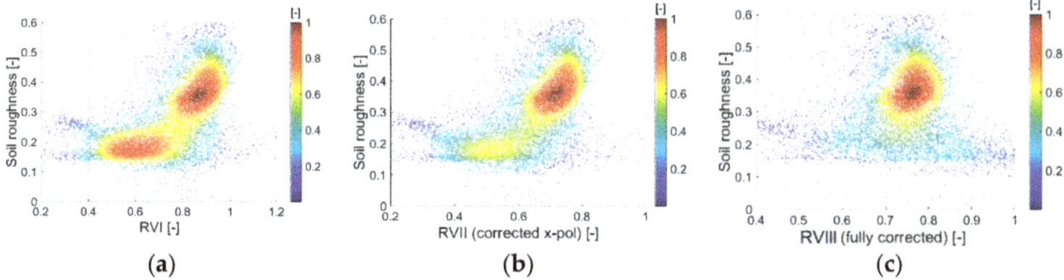

Figure 11. Correlation for the combined study areas between: (**a**) RVI [-]; (**b**) x-pol corrected RVII [-]; (**c**) fully corrected RVIII [-] and SMOS-based soil surface roughness (*ks*) [-] (derived from Parrens et al. from [19]).

The relation with vegetation-related parameters is shown in Figures 12 and 13. Figure 12 shows the correlation of the RVI, RVII and RVIII with the volumetric water content of the vegetation (VWC). Unlike the previous analyses with surface soil moisture and roughness (where the ideal was low correlation), the analysis with vegetation properties ideally show high correlation.

The scatterplots indicate a distinct link to the water content of vegetation in all three cases (RVI, RVII and RVIII). The distributions reveal a positive (linear) relationship for the standard RVI ($R^2 = 0.53$) as shown in Figure 12a. The correlation is most clearly evident for RVI between 0.5 and 1 [-]. The correlations do not improve appreciably for RVII and RVIII. All three indices are equally linked to wetness and structure of vegetation.

Results for the correlation with the plant structure and geometry-related leaf area index (LAI) are shown in Figure 13. However, the structural part of the forward modelling indicates a dependency of RVI (RVII, RVIII) to the leaf structure. The resulting scatterplots point towards a weak but positive linear relation. The statistics however show reduced correlation in the case of Figure 13c (RVI: $R^2 = 0.56$; RVII: $R^2 = 0.53$; RVIII: $R^2 = 0.35$). Hence, RVIII indicates lower sensitivity under high biomass after elimination of soil contribution than RVI and RVII. Whether LAI should be dominantly reflected in a low-frequency microwave vegetation index, from L-band waves penetrating into canopy and below the leaf layer, is open to debate. Therefore, it is likely possible that the optical LAI and the microwave radar vegetation indices provide information about rather different volumes of the vegetation canopy. The former is more indicative of canopy leaves and their abundance [36], while the latter is also including woody elements (trunk and branch) [37].

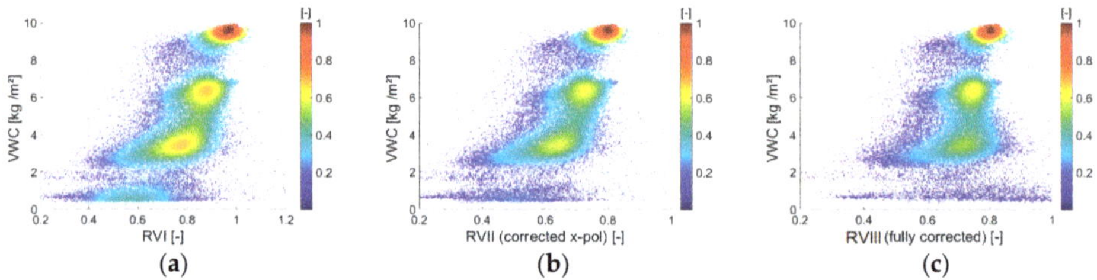

Figure 12. Correlation for the combined study areas between: (**a**) RVI [-]; (**b**) x-pol corrected RVII [-]; (**c**) fully corrected RVIII [-] and MODIS-based volumetric vegetation water content (VWC) [kg/m^2] [18].

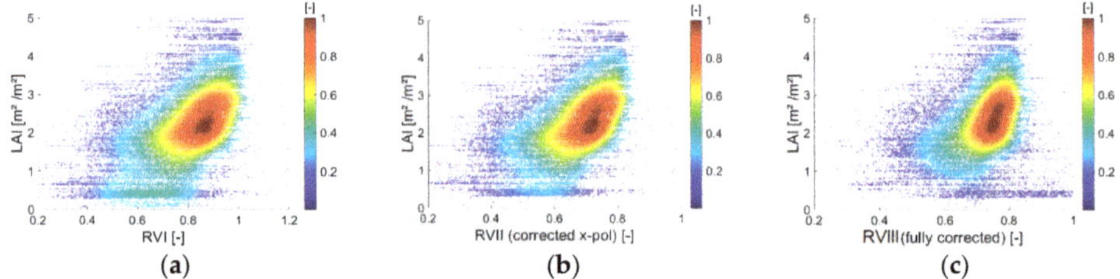

Figure 13. Correlation for the combined study areas between: (**a**) RVI [-]; (**b**) x-pol corrected RVII [-]; (**c**) fully corrected RVIII [-] and Leaf Area Index (LAI) [m^2/m^2] from Myneni et al. [24].

5. Summary and Conclusions

The aim of this study is to better understand the behavior of the microwave radar vegetation indices (RVI) at low microwave frequencies. We use a multi-sensor (active-passive microwave) approach to isolate vegetation and surface soil contributions to the indices. This is achieved using forward models that predict the linear backscattering intensity from the vegetation and from the soil underneath, as well as the attenuation within the vegetation. The backscatter from vegetation is calculated using a polarimetric single layer model of spheroids also known as Ap-ψ model. The soil contribution originates from a polarimetric soil scattering model for longer wavelength, the extended Bragg model, which includes a passive microwave-based static map of surface roughness [19]. The vegetation attenuation estimates are also derived from passive microwaves (through the vegetation optical depth parameter) [32]. The structural and dielectric components contributing to volume scattering are estimated with a polarimetric backscattering (Ap-ψ) model of vegetation. Results show that the standard pre-factor in the definition of the standard RVI does not normalize the index and does not limit it to be between zero and one, which makes its interpretation difficult and applicability. A new pre-factor is found that properly normalizes RVI and it is 6.57 instead of 8 based on the forward model analyses. Two improved versions of the RVI (called RVII and RVIII) are proposed to provide a vegetation index that reduces the soil scattering contribution present in the standard index at L-band. The degree to which the radar vegetation indices isolate the volume scattering from vegetation from surface contributions is analyzed by performing correlation analyses of the indices with surface soil moisture and surface roughness estimates as well as vegetation properties.

In known sparsely or non-vegetated areas, the RVI can be as high as 0.7 [-] due to soil roughness-induced cross-polarization ($\sigma_{HV}{}^m$) instead of cross-polarization from vegetation. The improved indices combine active and passive microwave data to confine their dynamic range between zero and one. The new indices remove the surface contributions (scattering from rough surfaces and surface reflectivity due to soil moisture). However, where the surface contribution is large compared to any existing vegetation, i.e., deserts, open shrublands, steppes, etc. then the residual backscatter (after removal) may become negative. This indicates a lower limit of applying microwave backscatter indices due to significant penetration of L-band waves into sparse and open canopies [38]. In this respect the microwave vegetation indices represent a different fraction of the vegetation volume than optical- and infrared-based ones. The latter indices also work in sparsely vegetated regions, but their sensitivity is significantly diminished in closed and denser canopies.

Author Contributions: C.S. carried out the data analysis and led the manuscript writing. T.J. developed the methodological framework, discussed the results with C.S. and supervised the research in all development stages. M.B. contributed to the data processing and especially on analyzing the RVI by forward modelling. C.T., J.-P.W., M.P. (Maria Piles) and D.E. supervised the research efforts. M.P. (Marie Parrens) delivered crucial soil roughness data and knowledge for the forward modelling of soil scattering. All authors compiled and approved the final manuscript.

Funding: This research received no external funding.

Acknowledgments: The authors gratefully acknowledge Stefan Kern from the Integrated Climate Data Center (ICDC) for providing the MODIS data and Kaighin McColl for discussions on RVI modeling and retrieval. The authors thank MIT for supporting this research with the MIT-Germany Seed Fund "Global Water Cycle and Environmental Monitoring using Active and Passive Satellite-based Microwave Instruments". M.P. (Maria Piles) is supported by a Ramón y Cajal contract (MINECO). The authors want to offer here that the RVII, RVIII and the vegetation-dominance mask are available on request.

Conflicts of Interest: The authors declare no conflict of interest.

References

1. Xie, Y.; Sha, Z.; Yu, M. Remote Sensing Imagery in Vegetation Mapping: A Review. *J. Plant Ecol.* **2008**, *1*, 9–23. [CrossRef]

2. Xiao, X.; Zhang, Q.; Braswell, S.; Urbanski, S.; Boles, S.; Wolfsy, B.; Moore, B.; Ojima, D. Modelling gross primary production of temperate deciduous broadleaf forest using satellite and climate data. *Remote Sens. Environ.* **2004**, *91*, 256–270. [CrossRef]

3. IPCC. Climate Change 2013: The Physical Science Basis. In *Contribution of Working Group I to the Fifth Assessment Report of the Intergovernmental Panel on Climate Change*; Stocker, T.F., Qin, D., Plattner, G.K., Tignor, M., Allen, S.K., Boschung, J., Nauels, A., Xia, Y., Bex, V., Midgley, P.M., Eds.; Cambridge University Press: Cambridge, MA, USA, 2013.

4. Frank, D.; Reichstein, M.; Bahn, M.; Thonicke, K.; Frank, D.; Mahecha, M.D.; Smith, P.; Van der Velde, M.; Vicca, S.; Babst, F.; et al. Effects of Climate Extremes on the Terrestrial Carbon Cycle: Concepts, Processes and Potential Future Impacts. *Glob. Chang. Biol.* **2015**, *21*, 2861–2880. [CrossRef] [PubMed]

5. Vereecken, H.; Weihermueller, L.; Jonard, F.; Montzka, C. Characterization of Crop Canopies and Water Stress Related Phenomena using Microwave Remote Sensing Methods: A Review. *Vadose Zone J.* **2012**, *11*. [CrossRef]

6. Steele-Dunne, S.C.; McNairn, H.; Monsivais-Hubertero, A.; Judge, J.; Liu, P.-W.; Papathanassiou, K. Radar Remote Sensing of Agricultural Canopies: A Review. *Sel. Top. Appl. Earth Obs. Remote Sens.* **2017**, *10*, 2249–2273. [CrossRef]

7. Entekhabi, D.; Das, N.; Yueh, S.; De Lannoy, G.; O'Neill, P.; Dunbar, R.S.; Kellogg, K.H.; Edelstein, W.N.; Allen, A.; Entin, J.K.; et al. *SMAP Handbook—Soil Moisture Active Passive—Mapping Soil Moisture and Freeze/Thaw from Space*; Jet Propulsion Laboratory: Pasadena, CA, USA, 2014.

8. Ulaby, F.T.; Long, F.G.; Blackwell, W.; Elachi, C.; Fung, A.; Ruf, C.; Sarabandi, K.; Zebker, H.; Van Zyl, J. *Microwave Radar and Radiometric Remote Sensing*; Artech House: Boston, MA, USA, 2015; ISBN 9780472119356.

9. Du, J.; Kimbal, J.S.; Jones, L.A.; Kim, Y.; Glassy, J.; Watts, J.D. A Global Satellite Environmental Data Record derived from AMSR-E and AMSR2 Microwave Earth Observations. *Earth Syst. Sci. Data* **2017**, *9*, 791–808. [CrossRef]

10. Tao, J.; Shi, J.; Jackson, J.; Du, J.; Bindlish, R.; Zhang, L. Monitoring Vegetation Water Content using Microwave Vegetation Indices. In Proceedings of the 2008 IEEE International Symposium on Geoscience and Remote Sensing, Boston, MA, USA, 7–11 July 2008; pp. 197–200.

11. Srivastava, P.; O'Neill, P.; Cosh, M.; Lang, R.; Jospeh, A. Evaluation of Radar Vegetation Indices for Vegetation Water Content Estimation using Data from a Ground-Based SMAP Simulator. In Proceedings of the 2015 IEEE International Geoscience and Remote Sensing Symposium (IGARSS), Milan, Italy, 26–31 July 2015; pp. 1296–1299.

12. Motofumi, A.; van Zyl, J.; Kim, Y. A General Characterization for Polarimetric Scattering from Vegetation Canopies. *IEEE Trans. Geosci. Remote Sens.* **2010**, *48*, 3349–3357. [CrossRef]

13. McColl, K.A.; Entekhabi, D.; Piles, M. Uncertainty Analysis of Soil Moisture and Vegetation Indices using Aquarius Scatterometer Observations. *IEEE Trans. Geosci. Remote Sens.* **2013**, *52*, 4259–4272. [CrossRef]

14. Kim, Y.; Jackson, T.; Bindlish, R.; Lee, H.; Hong, S. Radar Vegetation Index for Estimating the Vegetation Water Content of Rice and Soybean. *IEEE Geosci. Remote Sens. Lett.* **2012**, *9*, 564–568. [CrossRef]

15. Kim, Y.; Jackson, T.; Bindlish, R.; Hong, S.; Jung, G.; Lee, K. Retrieval of Wheat Growth Parameters with Radar Vegetation Indices. *IEEE Geosci. Remote Sens. Lett.* **2014**, *11*, 4259–4272. [CrossRef]

16. Huang, Y.; Walker, J.P.; Gao, Y.; Wu, X.; Monerris, A. Estimation of Vegetation Water Content from the Radar Vegetation Index at L-band. *IEEE Trans. Geosci. Remote Sens.* **2015**, *54*, 981–989. [CrossRef]

17. Belward, E.A. The IGBP-DIS global 1 km Land Cover Dataset "DISCover": Proposal and Implementation Plans. In *Report of the Land Cover Working Group of the IGBP-DIS. IGBP-DIS Working Paper*; IGBP-DIS Office: Stockholm, Sweden, 1996.

18. Entekhabi, D.; Das, N.; Njoku, E.G.; Johnson, J.T.; Shi, J. *SMAP L3 Radar/Radiometer Global Daily 9 km EASE-Grid Soil Moisture*; Version 3; NASA National Snow and ICE Data Center Distribution Active Archive Center: Boulder, CO, USA, 2016. [CrossRef]

19. Parrens, M.; Wigneron, J.-P.; Richaume, P.; Mialon, A.; Al Bitar, A.; Fernandez-Moran, R.; Al-Yaari, A.; Kerr, Y.H. Global-Scale Surface Roughness Effects at L-band as Estimated from SMOS observations. *Remote Sens. Environ.* **2016**, *181*, 122–136. [CrossRef]

20. Kerr, Y.H.; Waldteufel, P.; Wigneron, J.-P.; Martinuzzi, J.-M.; Font, J.; Berger, M. Soil Moisture Retrieval from Space: The Soil Moisture and Ocean Salinity (SMOS) Mission. *IEEE Trans. Geosci. Remote Sens.* **2001**, *39*, 727–744. [CrossRef]

21. Kerr, Y.H.; Al-Yaari, A.; Rodriguez-Fernandez, N.; Parrens, M.; Molero, B.; Leroux, D.; Bircher, S.; Mahmoodi, A.; Mialon, A.; Richaume, P.; et al. Overview of SMOS Performance in Terms of Global Soil Moisture Monitoring after Six Years in Operation. *Remote Sens. Environ.* **2016**, *180*, 40–63. [CrossRef]

22. Wigneron, J.-P.; Jackson, T.J.; O'Neill, P.; De Lannoy, G.; De Rosnay, P.; Walker, J.P.; Ferrazzoli, P.; Mironov, V.; Bircher, S.; Grant, J.P.; et al. Modelling the Passive Microwave Signature from Land Surfaces: A Review of Recent Results and Application to the L-band SMOS & SMAP Soil Moisture Retrieval Algorithms. *Remote Sens. Environ.* **2017**, *192*, 238–262. [CrossRef]

23. Chan, S.; Bindlish, R.; Hunt, R.; Jackson, T.; Kimball, J.S. *SMAP Ancillary Data Report Vegetation Water Content*; Version 1; Jet Propulsion Laboratory: Pasadena, CA, USA, 2013.

24. Myneni, R.; Knyazikhin, Y.; Park, T. *MOD15A2H MODIS/Terra Leaf Area Index/FPAR 8-Day L4 Global 500 m SIN Grid V006*; Version 6; NASA EOSDIS Land Processes DAAC: Boulder, CO, USA, 2015.

25. Myneni, R.; Knyazikhin, Y.; Glassy, J.; Votova, P.; Shabanov, N. *FPAR, LAI (ESDT:MOD15A2) 8-Day Composite NASA MODIS Land Algorithm*; Terra MODIS Land Team: Greenbelt, MD, USA, 2003.

26. Kim, Y.; van Zyl, J. On the Relationship between Polarimetric Parameters. In Proceedings of the 2000 IEEE International Symposium on Geoscience and Remote Sensing, Honolulu, HI, USA, 24–28 July 2000; pp. 1298–1300.

27. Jagdhuber, T. *Soil Parameter Retrieval under Vegetation Cover Using SAR Polarimetry*; University of Potsdam: Potsdam, Germany, 2012.

28. Cloude, S.R. *Polarisation: Applications in Remote Sensing*; Oxford University Press: Oxford, UK, 2010; ISBN 9780199569731.

29. Ulaby, F.T.; El-Rayes, M.A. Microwave dielectric spectrum of vegetation-Part II: Dual-dispersion model. *IEEE Trans. Geosci. Remote Sens.* **1987**, *GE-25*, 550–557. [CrossRef]

30. Van de Hulst, H.C. *Light Scattering by Small Particles*; Dover Publications: Dover, UK, 1981; ISBN 9780486642284.

31. Bohren, C.F.; Huffman, D.R. *Absorption and Scattering of Light by Small Particles*; Wiley: Weinheim, Germany, 2004; ISBN 9780471293408.

32. Konings, A.G.; Piles, M.; Rötzer, K.; McColl, K.; Chan, S.K.; Entekhabi, D. Vegetation Optical Depth and Scattering Albedo Retrieval using Time Series of Dual-Polarized L-band Radiometer Observations. *Remote Sens. Environ.* **2016**, *172*, 178–189. [CrossRef]

33. Hajnsek, I.; Pottier, E.; Cloude, S.R. Inversion of Surface Parameters from Polarimetric SAR. *IEEE Trans. Geosci. Remote Sens.* **2003**, *41*, 727–744. [CrossRef]

34. Mironov, V.L.; Fomin, S.V. Temperature and Mineralogy Dependable Model for Microwave Dielectric Spectra of Moist Soils. *PIERS Online* **2009**, *5*, 411–415. [CrossRef]

35. Das, N.N.; Scott Dunbar, R. *SMAP Level 3 Active/Passive Soil Moisture Product Specification Document*; Jet Propulsion Laboratory: Pasadena, CA, USA, 2015.

36. Zheng, G.; Moskal, L.M. Retrieving Leaf Area Index (LAI) using Remote Sensing: Theories, Methods and Sensors. *Sensors* **2009**, *9*, 2719–2745. [CrossRef] [PubMed]

37. Le Toan, T.; Beaudoin, A.; Riom, J.; Guyon, D. Relating Forest Biomass to SAR Data. *IEEE Trans. Geosci. Remote Sens.* **1992**, *30*, 403–411. [CrossRef]

38. Baur, M.; Jagdhuber, T.; Link, M.; Piles, M. Multi-Frequency Estimation of Canopy Penetration Depths from SMAP/AMSR2 Radiometer and IceSAT Lidar Data. In Proceedings of the IEEE International Symposium Geoscience and Remote Sensing, Valencia, Spain, 23–27 July 2018.

remote sensing

Review

Radiometric Microwave Indices for Remote Sensing of Land Surfaces

Simonetta Paloscia *, Paolo Pampaloni and Emanuele Santi

Institute of Applied Physics "Nello Carrara" (IFAC-CNR), via Madonna del Piano, 10, 50019 Firenze, Italy; p.pampaloni@ifac.cnr.it (P.P.); e.santi@ifac.cnr.it (E.S.)
* Correspondence: s.paloscia@ifac.cnr.it; Tel.: +39-0555-226-494; Fax: +39-0555-226-434

Received: 10 October 2018; Accepted: 14 November 2018; Published: 22 November 2018

Abstract: This work presents an overview of the potential of microwave indices obtained from multi-frequency/polarization radiometry in detecting the characteristics of land surfaces, in particular soil covered by vegetation or snow and agricultural bare soils. Experimental results obtained with ground-based radiometers on different types of natural surfaces by the Microwave Remote Sensing Group of IFAC-CNR starting from '80s, are summarized and interpreted by means of theoretical models. It has been pointed out that, with respect to single frequency/polarization observations, microwave indices revealed a higher sensitivity to some significant parameters, which characterize the hydrological cycle, namely: soil moisture, vegetation biomass and snow depth or snow water equivalent. Electromagnetic models have then been used for simulating brightness temperature and microwave indices from land surfaces. As per vegetation covered soils, the well-known tau-omega (τ-ω) model based on the radiative transfer theory has been used, whereas terrestrial snow cover has been simulated using a multi-layer dense-medium radiative transfer model (DMRT). On the basis of these results, operational inversion algorithms for the retrieval of those hydrological quantities have been successfully implemented using multi-channel data from the microwave radiometric sensors operating from satellite.

Keywords: microwave radiometry; microwave indices; soil moisture content; vegetation biomass; snow cover characteristics

---

1. Introduction

Microwave radiometry has been used since the first space Earth's observations to investigate some important surface phenomena over the oceans and land at global scale. The early experiments demonstrated that parameters such as ice concentration, wind speed and precipitations over the ocean, as well as some physical characteristics of soil, snow and vegetation can be retrieved at different levels of accuracy and reliability with more or less sophisticated instruments and algorithms developed in several times since the '80s. (e.g., [1]).

Further studies have shown that, as expected, combining data collected at different frequencies and polarizations in appropriate indices made it possible to significantly improve the accuracy of the measured quantities, with respect to the one achievable with single frequency/polarization observations. In particular, some Microwave Indices have been successfully related to the main geophysical parameters associated to land hydrological cycle, such as soil moisture (SMC), Plant Water Content (PWC), and Snow Depth (SD) or Snow Water Equivalent (SWE). These indices have therefore been used for implementing operational retrieval algorithms based on data from different channels of satellite radiometric sensors (e.g., SMMR, SSM/I, AMSR-E, AMSR2).

Presently, most of the operational algorithms for monitoring land surfaces are based on visible and infrared indices, such as the Normalized Difference Vegetation Index (NDVI) [2] and Enhanced

Vegetation Index (EVI) [3], which is sensitive to vegetation "greenness" and consequently related to its biomass, or the MODIS Snow Cover Fraction. However, observations in optical bands, besides being bound to the light diurnal cycle, are significantly influenced by the presence of clouds and can give information of the observed surface layer only. On the other hand, microwaves are slightly affected by atmospheric perturbations and, depending on the observation frequency and incidence angle, can penetrate in vegetation cover, snow and even the underlying soil. Moreover, the high sensitivity of microwaves to the water content of the observed bodies allows a direct estimate of the SMC, PWC and SWE (e.g., [4,5]).

Investigations on the use of the difference between two linear polarizations for monitoring land surfaces have been carried out since '80s by several groups of scientists analyzing passive microwave data from both ground-based and satellite sensors. In particular, a Polarization Index (PI) was defined as the difference between the two linear polarizations (Tbv − Tbh) normalized to their average value [(Tbv + Tbh)/2] [6,7].

As it is well known, the microwave radiation emitted from a specular surface at an angle different from the zenith is partially polarized. The degree of polarization depends on the soil dielectric constant and can be estimated by means of the Fresnel coefficients. When the soil is characterized by a random rough surface the degree of polarization depends on the roughness parameters as well, and decreases as the roughness increases [8]. Moreover, experimental and theoretical investigations have shown that the radiation from a canopy is much less polarized than that from bare soil. The different polarization characteristics of a smooth bare soil and vegetation suggest the possibility of using a polarization measurement, such as PI, as an indicator of vegetation cover.

First studies focused on estimating the sensitivity of the microwave brightness temperature (Tb) to vegetation biomass were carried out since late '70s on the basis of ground based experiments and model simulations (e.g., [9,10]) .

The reason for using polarization indices to estimate vegetation biomass was that the measurement at single polarization is influenced by the geometry of plants, providing different results according to the crop type. On the other hand, polarization indices were found to be mostly related to plant water content (PWC) without being significantly influenced by plant structure and surface temperature.

Ref [11], in 1990, identified different combinations of the Special Sensor Microwave Instrument (SSM/I) brightness temperature channels by statistically analyzing satellite data on a global scale, thus allowing the classification of several land classes, such as dense vegetation, rangeland and agricultural soils, deserts, snow, precipitation, and soil surface moisture.

In addition to those based on the polarization difference, other approaches for retrieving PWC from multi-frequency satellite data have been examined combining data at two or more frequencies (e.g., [12]). More recently, [13] noted that the brightness temperatures measured at a given polarization with two adjacent AMSR-E frequency channels can be described by a linear function, which includes two coefficients, both independent of the underlying soil/surface signals and dependent only on vegetation properties. One is positively correlated to NDVI and is affected by the vegetation properties and the surface physical temperature, the other is negatively correlated to NDVI and is only affected by the vegetation properties.

A field of investigations where microwave indices are really useful for implementing retrieval algorithms is the one of snow cover. First investigations on the capability of satellite microwave sensors for snow monitoring took place in early '80s by using Nimbus-7 SMMR data over Finland (e.g., [14,15]). Many operative algorithms for the retrieval of the main parameters of snow cover have been implemented since then and are mostly based on multiple combinations of polarizations and frequencies. [16] and [17] developed an operational algorithm for the retrieval of snow depth from SSM/I and AMSR-E data basically using the difference in brightness temperature between Ku and Ka bands in horizontal or vertical polarizations. The Ka band channel is sensitive to scattering by the snowpack while the Ku band channel is relatively unaffected by the snow and is responsive to the surface under the snow [18,19].

Refs [16,20] provided operational algorithms based on microwave indices for their spatial agencies (NASA and JAXA, respectively) focused on the distribution of snow products. In general, the retrieval algorithms are supported by direct theoretical or semi-empirical models, which simulate microwave emission and related indices of land surfaces in different conditions of vegetation, soil moisture, and snow cover. These models are subsequently inverted with greater or lesser success by using different approaches in order to retrieve the main surface parameters.

In this paper, the main results obtained by the Microwave Remote Sensing Group since early 1980s on the retrieval of soil, vegetation and snow parameters using passive microwave data have been reviewed.

2. Experimental Relationships between Microwave Emission and Land Surface Parameters

Data presented in this paper were collected on different times and sites from ground-based and airborne platforms by using microwave radiometers, operating at L, C, Ku, and Ka bands in both vertical and horizontal (V&H) polarizations, over bare, vegetated, and snow-covered soils since early '80s. Examples of installations of microwave radiometers are shown in Figure 1.

Figure 1. Installations of IROE microwave radiometers on several platforms: in a shelter on snow, on hydraulic booms on forest and agricultural fields, and on helicopter and ultra-light aircraft.

The IFAC microwave instruments were total-power, self-calibrating, dual-polarized radiometers with an internal calibrator based on two loads at different temperatures (cold, 250 K $\pm$ 0.2 K, and hot, 370 K $\pm$ 0.2 K). The beamwidth of the corrugated conical horns was 20° at −3 dB and 56° at −20 dB for all frequencies and polarizations. Calibration checks were performed during the field experiments by means of absorbing panels of known emissivity and temperature and an internal noise source. Moreover, frequent observations of clear sky were performed. The measurement accuracy (repeatability) was better than ±1 K, with an integration time of 1 s [21].

During the experiments, in-situ measurements of the parameters of soil (moisture, SMC, and surface roughness, denoted by the Height Standard Deviation, Hstd, and correlation length, Lc), vegetation (plant geometry, vegetation water content, PWC), and snow (Depth, SD, Water Equivalent, SWE, density, Dn, Water Liquid Content, WLC, and grain size, GS) were collected to be compared with microwave data acquired simultaneously.

2.1. Non Vegetated Land Surfaces

Microwave emission from non-vegetated soils is primarily sensitive to soil moisture due to the high contrast between the permittivity of dry matter and water. Besides, soil emission is influenced by surface roughness too, whose importance depends on the relative dimensions of the roughness parameters of the surface profile (i.e., Hstd, and Lc), and the observation wavelength, λ. Hence, the same surface can be "seen" as more or less rough depending on the observation frequency, as stated by the Rayleigh criterion. As predicted by theoretical models and confirmed by experiments, the effect of surface roughness is to increase emissivity and reduce the sensitivity to soil moisture. As an example, measurements carried out with ground based radiometer at L (λ = 21 cm), X (λ = 3.2 cm) and Ka (λ = 0.8 cm) bands on a sandy soil sample with a very smooth surface (Hstd < 1cm) are represented in Figure 2, which shows the normalized temperature (Tn), i.e., the brightness temperature (Tb) normalized to the thermometric surface temperature, as a function of soil moisture (SMC, in %) of the first soil layers. Due to the different penetration depths of the three frequency signals, data at L, X, and Ka bands have been correlated to the first 5.0, 2.5 and 1.0 cm layers, respectively. We can see that, for this very smooth surface, the sensitivity of Tn to SMC is almost the same at L and X bands (slope $\cong$ −0.0085), whereas it is significantly smaller at Ka band (−0.002), with rather low determination coefficient (R^2 = 0.47) [21].

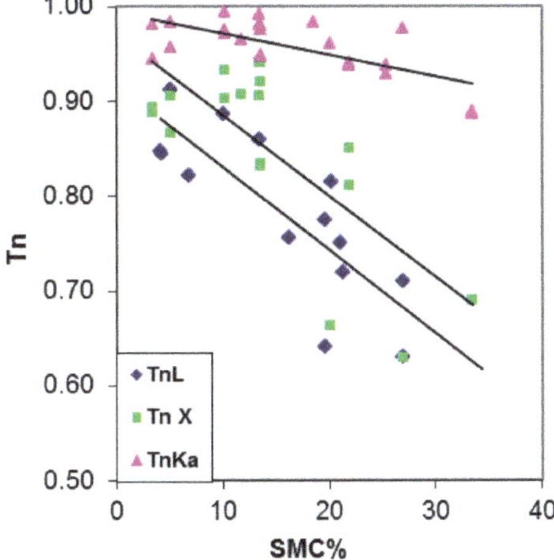

Figure 2. The normalized Temperature (Tn, i.e., the ratio between brightness temperature, and thermal surface temperature, at Ka, X, and L as a function of SMC% of a bare smooth sandy soil.

Also polarization is influenced by the moisture content. The behavior of the Polarization Index (the difference between the vertical, V, and horizontal, H, components of Tb normalized to their average value), PI = (Tbv − Tbh)/(1/2) (Tbv + Tbh) at L and X bands vs. SMC is represented in the diagrams of Figure 3a (X band) and Figure 3b (L band). PI at X band is significantly sensitive to SMC for smooth soils only (Hstd < 0.5cm), with R^2 = 0.87 and slope 0.016, whereas, when Hstd is higher than

0.5 cm, the sensitivity to SMC becomes very low ($R^2 = 0.34$ and slope 0.002). At L band the relationship between PI and SMC is similar for both types of surfaces ($R^2 = 0.46$ and slope 0.004), confirming the scarce influence of surface roughness in this range of Hstd to the emission at this frequency.

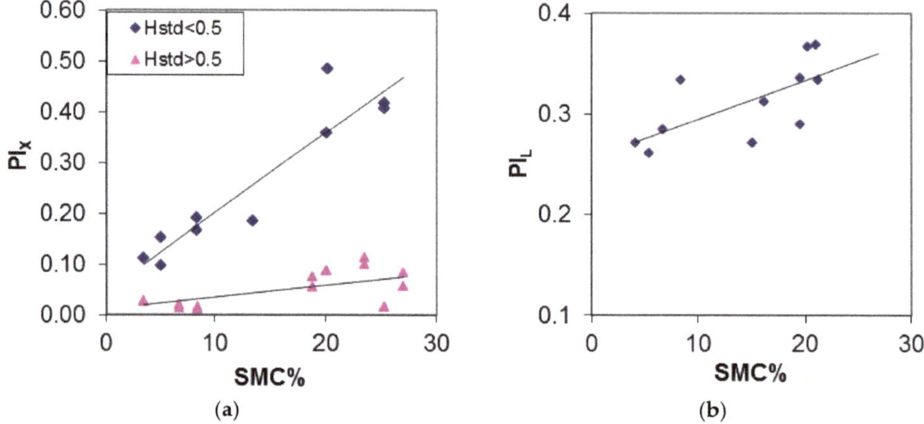

(a)   (b)

Figure 3. (a) PI at X band (PI_X) vs. SMC for 2 surface types; (b) PI at L band (PI_L) vs. SMC.

These results confirm that emission from natural terrains is influenced not only by SMC, but by the surface roughness too, which in general, increases the value of brightness temperature and reduces the sensitivity to SMC [22]. As an example, Figure 4 shows the Tn at L band as a function of SMC for three classes of roughness (Hstd < 0.4 cm, 0.7–1.2 cm, and 1.2–3.0 cm). We can note that even L band emission, in spite of the long wavelength, is influenced by surface roughness, especially when Hstd is higher than 1.2 cm. Although R^2 remains almost the same for the 3 roughness classes (between 0.7 and 0.8), the slope significantly decreases (from −0.009 for smooth soils, to −0.0024 for the rougher surface), confirming that, as said, the same surface appears rougher at the smaller wavelengths.

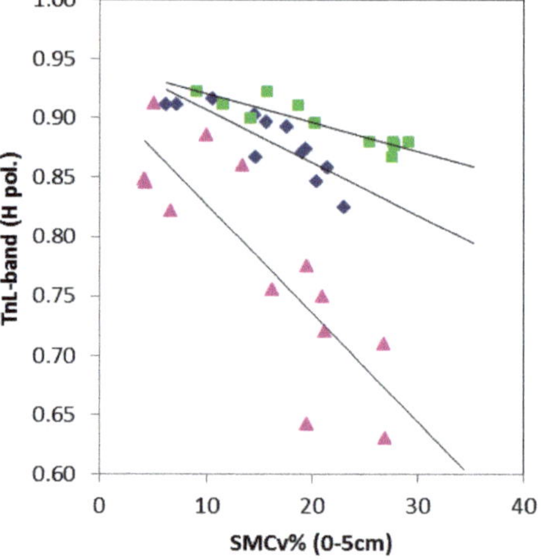

Figure 4. Tn (L band, Hpol, $\Theta = 10°–20°$) as a function of volumetric SMC (0–5 cm) for three different ranges of surface roughness Hstd ($\cdot \leq 0.4$ cm, $\Diamond = 0.7–1.2$ cm, $\Delta = 1.2–3.0$ cm).

Hence, a refinement of the measurements of SMC would require some knowledge of the surface roughness. A simple parametric model, which approximates fairly well the emissivity of a rough surface with Hsdt between 0 and 2.5 cm, in a frequency range between L and Ka bands, was developed by [23] by correcting the reflection coefficient with an exponential factor function of the square root of the wavelength. Other interesting approaches to account for the roughness effect were suggested by [24] and [25].

If dual or multi-frequency measurements are available, the effect of roughness on the measurement of SMC can be more easily evaluated. As an example, the index δTb (i.e., the difference $Tb_{Ka} - Tb_X$), measured on surfaces with similar SMC but different roughness, shows a gradual decrease as the roughness increases, as it is shown in Figure 5.

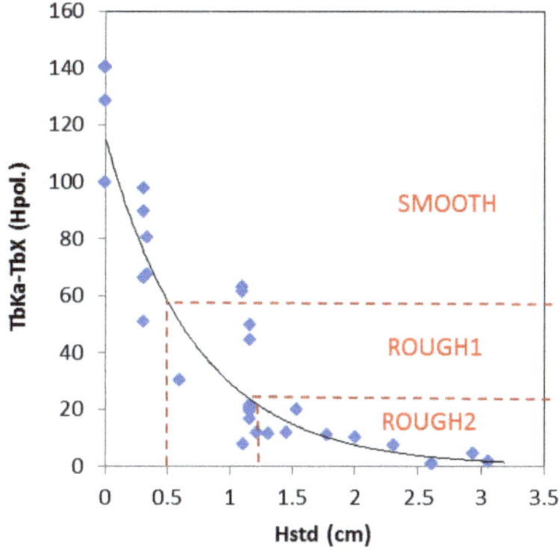

Figure 5. $\delta Tb = Tb_{Ka} - Tb_X$ at $\theta = 40°$ and H pol. as a function of Hstd (After [21]).

In [21], this frequency index was related to the Hstd with an exponential function as: $\delta Tn = 114.7 \exp(-1.36\ Hstd)$, which approximates experimental data with $R^2 = 0.83$. This approach allowed the identification of almost three ranges of roughness from Hstd < 1 cm to 2.5 cm and can provide a correction of the relationship between Tn at L band and SMC by separating measurements on surfaces characterized by different roughness.

Another approach to evaluate the surface roughness is based on the measurements of the PI. Emission from a smooth flat surface at an incidence angle far from zenith is different for the two polarization V&H components as predicted by the Fresnel reflection coefficients. The presence of surface roughness tends to reduce or destroy this polarization difference, so that the measurement of PI can give a direct estimate of the surface Hstd. A direct relationship between PI, at both X and Ka bands, and Hstd is shown in Figure 6. We can see that in the range of Hstd between 0 and 3 cm, typical of most agricultural fields, PI at X band gradually decreases as Hstd increases ($R^2 = 0.65$), although the experimental points are largely spread, whereas it quickly saturates at Ka band, as soon as Hstd becomes slightly > 0 cm ($R^2 = 0.55$). From this diagram, it can be concluded that, the most appropriate frequency to perform this estimate of the surface Hstd in the range of roughness usually encountered in the agricultural fields is close to X band, which can allow the identification of 2–3 levels of roughness.

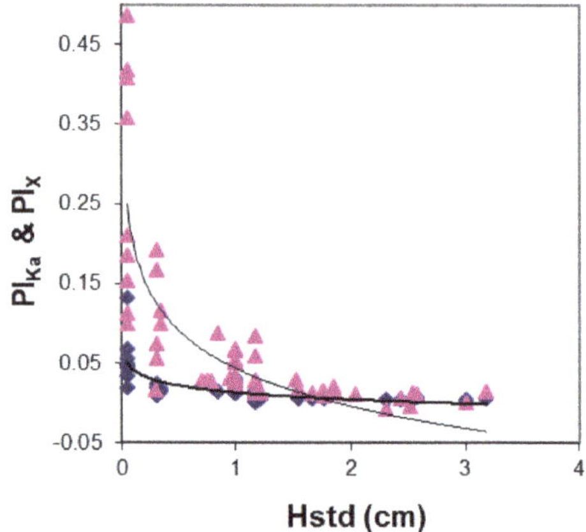

Figure 6. PI, at X (triangles) and Ka (rhombs) bands, vs. Hstd.

In summary, a combination of dual frequency/polarization data at Ka and X bands makes it possible to improve the accuracy of the SMC measurements based on L band data.

2.2. Vegetation

On vegetated surfaces, vegetation can be at the same time a disturbing factor for the estimate of soil moisture and a target for the measurements of vegetation biomass. In remote sensing, the latter is usually expressed by Plant Water Content (PWC, in Kg/m^2), i.e., the total amount of water in plant elements per unit area. It should be noted that instead of our original notation PWC, most authors are now using the term Vegetation Water Content (VWC) (e.g., [10]).

Emission from vegetated surfaces is a combination of soil emission attenuated by the canopy with the emission from plant elements. In general, the contribution of vegetation increases with the observation frequency, f, and depends on the structure and dimensions of plant elements. The most commonly used for modeling microwave emission from soil covered by vegetation is the tau-omega (τ-ω) model, which is a simple formulation of RT transfer theory [26].

Also in this case, multi-frequency, dual polarization measurements can provide significantly more information than single channel observations. Indeed, depending on the type of plants and observation wavelength, Tb can increase or decrease as the biomass increases. This corresponds to different types of electromagnetic interactions. In general, absorption occurs for plant elements that are small with respect to observation wavelength, whereas scattering dominates in the opposite case [27].

On the other hand, the trend of the difference between the two linear polarization components (and then the PI), was found to be independent of the vegetation type and always decreasing as biomass increases [12]. Indeed, the polarized emission from an almost homogeneous and smooth soil is attenuated by the volumetric effect of any vegetation type [6]. Thus, significant information on vegetation biomass can be obtained by using PI, making it possible establishing an inversion approach to retrieve vegetation biomass independently of crop type.

Figure 7 shows experimental values of Tn (in H pol.) (left) and PI at X and Ka bands (right) as a function of the PWC of two crop types: narrow-leaf crops (e.g., wheat and alfalfa), and broad-leaf crops (e.g., corn, sugar-beet and sunflower). In case of narrow-leaf crops, the mechanism of absorption is significant and Tb increases as PWC increases; whereas on broad-leaf crops scattering is dominant and Tb decreases with PWC. In all cases PI decreases as a function of increasing vegetation biomass, with a trend that is gradual at X band and rather steep at Ka band.

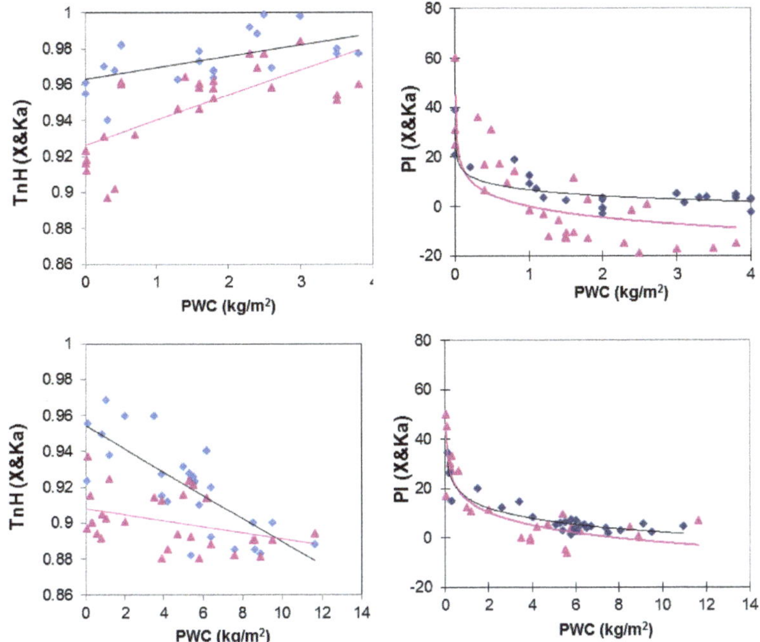

Figure 7. left: Normalized Temperature (Tn, in H pol.) at X (triangles) and Ka (rhombs) bands as a function of PWC, and right: PI at X and Ka bands as a function of PWC, for two different crop types: narrow-leaf crops (e.g., wheat and alfalfa) (**top**), and broad-leaf crops (e.g., corn, sugar beet and sunflower) (**bottom**).

Table 1 shows the regression equations with the determination coefficients (R^2) at the two frequencies for the two crop types.

Table 1. Regression equations and R^2 between Tn and PWC, and PI and PWC, at X and Ka bands.

Crop Type	Tn Regression Lines	R^2	PI Regression Lines	R^2
Narrow-leaf	TnX = 0.014PWC + 0.93	0.5	PIX = −0.5ln(PWC) + 0.077	0.6
Narrow-leaf	TnKa = 0.0063PWC + 0.96	0.3	PIKa = −3.52ln(PWC) + 6.77	0.74
Broad-leaf	TnX = −0.002PWC + 0.91	0.13	PIX = −7.51ln(PWC) + 15.35	0.79
Broad-leaf	TnKa = −0.0065PWC + 0.95	0.57	PIKa = −6.17ln(PWC) + 16.34	0.85

These data obtained with ground-based sensors where positively compared with model simulations based on tau-omega (τ-ω) model in [28].

The results obtained from ground-based or airborne sensors have been confirmed by satellite investigations: a global map of vegetation cover based on the polarization difference at Ka band, obtained from Nimbus 7 data, was first shown in [7]. More recently, maps of PWC retrieved from PI at X band were obtained in the context of an algorithm based on an Artificial Neural Network (ANN) developed for generating simultaneous maps of SMC, PWC, and SD from the Advanced Multifrequency Scanning Radiometer (AMSR-E) [29,30].

2.3. Snow

The sensitivity of microwave emission to snow cover has been evident since the early experimental (e.g., [31,32]) and theoretical (e.g., [33]) studies.

Radiation emitted at lower frequencies of the microwave band (lower than about 6–10 GHz) by soil covered with a shallow layer of dry snow is mostly influenced by soil conditions below the snowpack. At higher frequencies and for thick snow layers, however, the role played by volume scattering increases, and microwave emission becomes sensitive to the presence of snow.

The most interesting parameter for hydrological applications is the snow water equivalent (SWE) equal to the product of snow depth (SD) by its density. As past research has demonstrated, the key-frequency channels for detecting the presence of snow and estimating SWE or SD are Ku and Ka bands. Measurements collected over several winter seasons on a relatively flat area located in Northeast Italy on Mount Cherz, by using ground-based radiometers at Ku and Ka bands, showed a decrease of Tb as the SWE increases up to about 260 mm at Ka band and 300mm at Ku. For SWE values beyond this value, Tb tends to increase again due to emission from the snowpack itself which masks the large scattering from the deep hoar (e.g., [34,35]). This trend, with some variability due to the snow characteristics, was observed in several other studies (e.g., [36–38]). Moreover, the range of SWE in which the minimum of Tb occurs depends on the penetration depths of radiation inside the snowpack.

This reversal of brightness temperature at increasing SWE can cause ambiguity in the retrieval. In our measurements, after the inflection point, Tb shows a sharp increase at both frequencies and then tends to fluctuate with a relatively flat behavior. However, the difference between Tb at the two frequencies also tends to slightly increase after the threshold. Hence, we can speculate that, by using an appropriate combination of observation frequency and polarization the retrieval of SWE can be extended beyond the range 0–300 mm (e.g., [17,39,40]). For example, the Frequency Index (FI = ((Tb_{KuV} − Tb_{KaV}) + (Tb_{KuH} − Tb_{KaH}))/2) is sensitive to SWE and SD due to the fact that, in the case of dry snow, radiation at Ku band penetrates the snowpack with smaller attenuation and more deeply than the emission at the higher frequency (Ka band), which is more influenced by the scattering inside the snowpack [29]. The difference between the brightness at the two frequencies can therefore be linearly related, to some extent, to SD (and/or SWE). Other combinations of frequency channels and polarizations have also been tested to evaluate their sensitivity to SWE and, among these indices, the Spectral Polarization Difference defined as SPD = (Tb_{KuV} − Tb_{KaV}) + (Tb_{KuV} − Tb_{KaH}) was identified as the best correlated quantity to SD and SWE [38]. In summary, both FI and SPD present rather high correlation (in terms of R^2) to SD and SWE, as demonstrated in [41], where the comparison of radiometric data with ground truth has shown the following logarithmic regressions: FI = 9.4 Ln(SWE) − 27.59 (R^2 = 0.71); SPD = 22.76 Ln(SWE) − 58.32 (R^2 = 0.76) in a range of SWE up to 500 mm. This result confirms that the use of dual-frequency/dual-polarization indices allows investigating snow properties, even beyond the inversion limit (Figure 8).

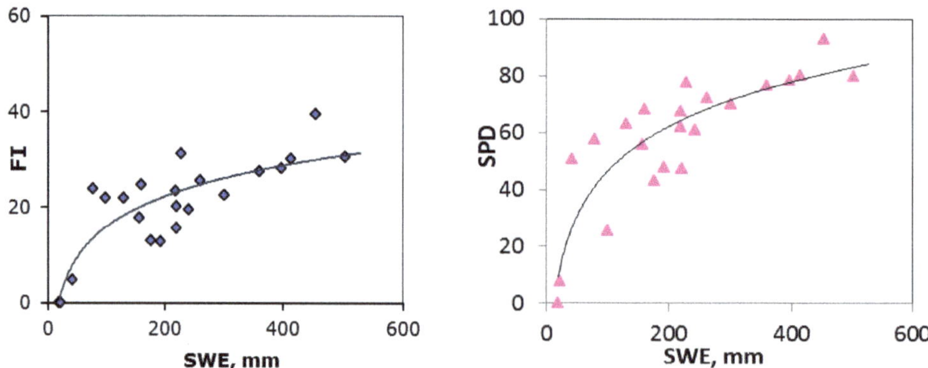

Figure 8. FI and SPD as a function of SWE (mm). The regression lines are: FI = 9.4 Ln(SWE) − 27.6 (R^2 = 0.71); SPD = 22.76 Ln(SWE) − 58.3 (R^2 = 0.76).

3. Model Simulations

3.1. Soil and Vegetation

Several models have been used for simulating brightness temperature and microwave indices from land surfaces. As per vegetation covered soils, the most used model is an approximate solution of the radiative transfer equation for a homogeneous soil overlaid by a medium at uniform temperature characterized either by small scattering (ks $\ll$ ka) or scattering "mainly forward". In this approach, well-known as tau-omega (τ-ω) model [26], the parameters that characterize the absorbing and scattering properties of vegetation are the optical depth (τ) and the "albedo (ω). The radiation component due to vegetation is assumed to be unpolarized, whereas the radiation emitted from the smooth soil, and then by the whole canopy-soil system, is partially polarized.

The key parameter of the (τ-ω) equation related to the vegetation biomass is vegetation optical depth (VOD or τ). This quantity increases as the canopy grows and, at L band, it has usually been related to the VWC/PWC with a linear relationship, for several crop types [42,43]. However, early studies at higher frequencies (X and Ka bands), had shown that experimental data can be fairly well approximated ($R^2 > 0.8$) by the following logarithmic function [28,44], as it is shown in Figure 9:

$$\frac{\tau}{\sqrt{\lambda}} = k \times \ln(1 + PWC) \tag{1}$$

where k is a constant depending on crop type, and λ is the wavelength of the emitted radiation. Equation (1) is represented in Figure 9 compared with experimental data for some crop types (alfalfa, corn, sugar beet and sunflower). The lines refer to the model obtained using two values of k (0.16 for alfalfa) and 0.4 for corn and sugar-beet) for better simulating the different crop types.

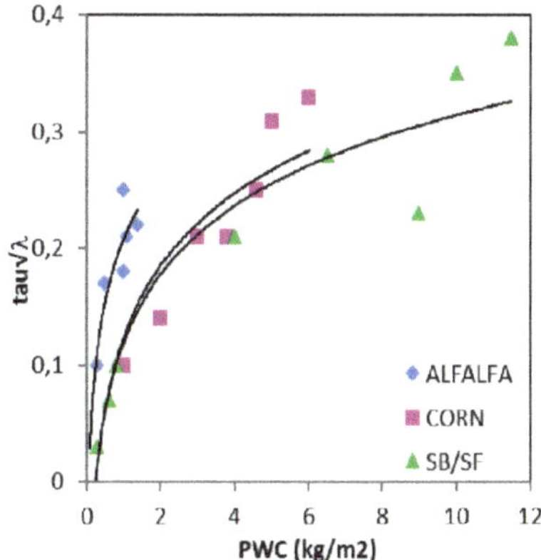

Figure 9. Relationship between $\tau\sqrt{\lambda}$ and PWC for some crop types: alfalfa, corn, sugar-beet (SB) and sunflower (SF).

The conflict between the linear or logarithmic trend of τ versus PWC was clarified in [10] by expanding Equation (1) into power series (Equation (2)) and showing that this corresponds to the power expansion of the extinction coefficient, γ, of a collection of discrete scatterers computed with the radiative transfer theory. Hence, the linear relation between optical depth and PWC, frequently used at L band, agrees with the first term of this series and can be considered valid for low values of vegetation water content and long wavelengths:

$$\tau = k \times PWC - \frac{1}{2}k \times PWC^2 + \ldots = k \times PWC \times \left(1 - \frac{1}{2}PWC + \ldots\right) = \tau_0\left(1 - \frac{1}{2}k \times \tau_0 + \ldots\right) \tag{2}$$

where τ_0 is the optical depth at low values of PWC.

In spite of the reduced range of the experimented PWC values (up to 2.5 Kg/m^2), the progressive shift from linear to logarithmic relationship as the frequency increases is demonstrated in a study by [10,45] (Figure 10).

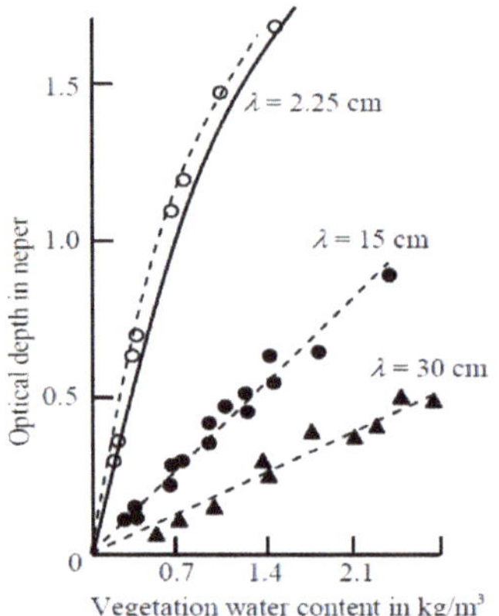

Figure 10. Dependence of optical depth on PWC. Solid curve is calculated by the model of Equation (1) (After [10]).

Figure 11 shows model simulations of PI (X band) as function of PWC compared with experimental data of two crop types: narrow-leaf (alfalfa and wheat) and broad-leaf (corn, sugar-beet and sunflower) crops. Here simulations are obtained by means of the τ-ω solution of the RT model, relating τ to PWC as in Equation (1) and using two values of k (0.16 and 0.40) for the two crop types. In the model, the scattering albedo, ω, the surface temperature, Ts, and the soil moisture, SMC, are kept constant and equal to 0.01, 290K, and 15%, respectively [6].

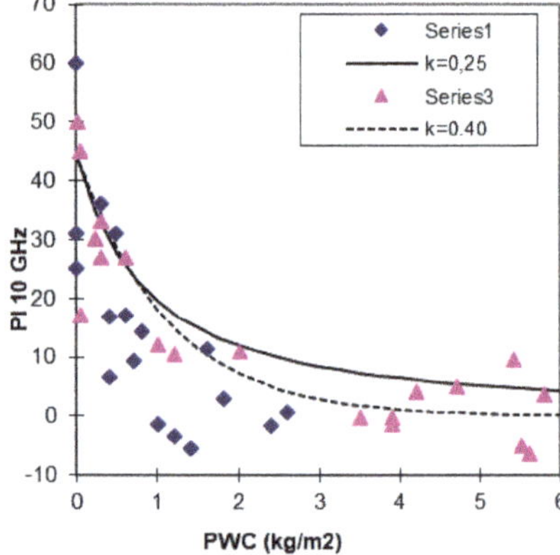

Figure 11. Simulations of PI (at X band) as a function of PWC compared with experimental data of two crop types: narrow-leaf (alfalfa and sugar beet, rhombs) and broad-leaf (corn and sunflower, triangles).

3.2. Snow

Microwave emission from for several types of terrestrial snow cover has been simulated using a multi-layer dense-medium radiative transfer model (DMRT) [46] implemented under the quasi crystalline approximation (ML-QCA). In particular, the model evaluated the sensitivity of the two FI and SPD indices on SWE comparing simulations to radiometric dual frequency/polarization measurements collected over three winter seasons between 2007 and 2011 in the Eastern Italian Alps [40].

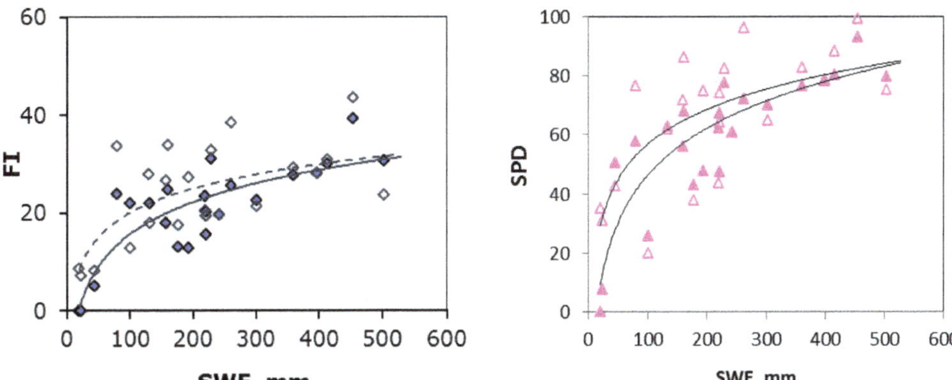

Figure 12. FI and SPD vs. SWE. The empty dots represent the model simulations carried out with the DMRT model, and the full dots the experimental data. Lines represent the regression equations for both simulated (dashed lines) and experimental (continuous lines) data.

In these simulations, inputs to the model were taken from ground data considering the changes in the snow structure (grain size, density, stickiness of snow particles, number and depth of layers) as SWE increased. The soil contribution was accounted for by using the Advanced Integral Equation Model (AIEM) [47] with a permittivity corresponding to frozen or moderately wet soil depending on the measured temperature.

The simulations confirmed the sensitivity of these indices to SWE up to a value of 500 mm water equivalent (Figure 12). The regression equations obtained for both experimental and simulated data are the following:

FI = 9.4ln(SWE) − 27.6 (R^2 = 0.71) (measured).
FI= 7.1ln(SWE) − 12.63 (R^2 = 0.44) (modeled).
SPD = 22.76ln(SWE) − 58.32 (R^2 = 0.76) (measured).
SPD = 16.89ln(SWE) −20.88 (R^2 = 0.47) (modeled).

4. Observation from Satellite

Results obtained from satellite data (SSM/I, AMSR-E, AMSR2, SMAP) confirmed those obtained from ground-based and airborne sensors [48], by exploiting the potential of microwave indices in a global scale estimation of geophysical parameters, provided appropriate retrieval procedures are used. As an example, by including PI at X and Ku band in a SMC retrieval algorithm based on Artificial Neural Networks (ANN), a correlation coefficient R^2 > 0.7 between retrieved and target SMC was obtained, while the correlation achievable on the same dataset by using only Tb at C band would have been lower, i.e., $R \simeq 0.5$ [49]. Other studies demonstrated the possibility of estimating SD in the Scandinavian peninsula, using PI and FI derived from AMSR-E, with RMSE = 9.13 cm and $R^2 \simeq 0.8$ [29].

Another algorithm based on the joint use of PI at C, X and Ku band data derived from AMSR2 was able to produce global maps of vegetation biomass with a RMSE < 1 kg/m^2 [30]. The validation of the latter algorithm, carried out on the entire Australian continent, demonstrated that the microwave data from AMSR2 can be legitimately used to produce vegetation maps on a global scale by separating several levels of biomass on low and medium dense vegetation (up to 8 Kg/m^2), without any need of further information from other sensors and guaranteeing an all-weather monitoring.

5. Summary and Conclusions

Passive microwave remote sensing has been proved to be an important technique for monitoring land surfaces, and, in particular, three important parameters: soil moisture (SMC), vegetation biomass (PWC), and snow water equivalent (SWE). Unfortunately, all these parameters, together with some others (e.g., soil surface roughness/texture, vegetation/snow type), simultaneously affect microwave emission, so that the retrieval of the requested parameters is a typically ill-posed problem.

The complexity of the algorithms to be developed for retrieving spatial variations in land surface parameters depends on the auxiliary information available and on the direct models selected for the inversion procedures. Indeed, the more information (auxiliary or model derived) available, the more accurate but more complex algorithm can be developed. The analytical inversion of EM models is a complicated procedure, and generally, unfeasible without setting several boundary conditions. Different approaches have therefore been studied to provide information on all the factors that affect emission and reduce the effects of the undesired parameters, using ancillary or a-priori data. In this framework, the synergy between observations at different frequencies, polarizations, and incidence angles, significantly helps in improving the reliability of the inversion methods. A typical example is the polarization index (PI) at X band, which was confirmed to be the most suitable parameter for estimating vegetation biomass, and, furthermore, to be able to significantly increase the accuracy of the estimate of soil moisture based on L and C band data.

Concerning the retrieval of snow parameters, the Frequency Index (FI) and the Spectral Polarization Difference (SPD) demonstrated to be able to overcome the ambiguity introduced by the non-linear relationship between Tb and SWE, making it possible estimating SWE up to 500 mm.

Author Contributions: P.P. and S.P. conceived and performed the experiments, and analyzed the data through EM models; E.S. analyzed the satellite data and developed the retrieval algorithms.

Conflicts of Interest: The authors declare no conflicts of interest.

References

1. Hollinger, J.; Lo, R.; Poe, G.; Savage, R.; Pierce, J. *Special Sensor Microwave/Imager User's Guide*; Naval Res. Lab: Washington, DC, USA, 1987.
2. Tucker, C.J. Red and Photographic Infrared Linear Combinations for Monitoring Vegetation. *Remote Sens. Environ.* **1979**, *8*, 127–150. [CrossRef]
3. Huete, A.; Didan, K.; Miura, T.; Rodriguez, E.P.; Gao, X.; Ferreira, L.G. Overview of the radiometric and biophysical performance of the MODIS vegetation indices. *Remote Sens. Environ.* **2002**, *83*, 195–213. [CrossRef]
4. Ulaby, F.T.; Moore, R.K.; Fung, A.K. *Microwave Remote Sensing: Active and Passive*; Addison-Wesley: Reading, MA, USA, 1982; Volume II.
5. Ulaby, F.T.; Moore, R.K.; Fung, A.K. *Microwave Remote Sensing: Active and Passive. Vol III*; Artech House: Nordwood, MA, USA, 1986.
6. Paloscia, S.; Pampaloni, P. Microwave Polarization Index for Monitoring Vegetation Growth. *IEEE Trans. Geosci. Remote Sens.* **1988**, *26*, 617–621. [CrossRef]
7. Choudhury, B.J. Monitoring global land surface using Nimbus-7 37 GHz. Theory and examples. *Int. J. Remote Sens.* **1989**, *10*, 1579–1605. [CrossRef]
8. Wang, J.; Choudhury, B. Remote sensing of soil moisture content over bare fields at 1.4 GHz frequency. *J. Geophys. Res.* **1981**, *86*, 5277–5282. [CrossRef]
9. Kirdiashev, K.P.; Chukhlantsev, A.A.; Shutko, A.M. Microwave radiation of the Earth's surface in the presence of vegetation cover. *Radio Eng. Electron. Phys. Engl. Transl.* **1979**, *24*, 256–264.
10. Chukhlantsev, A.A. *Microwave Radiometry of Vegetation Canopies*; Springer: Dordrecht, The Netherlands, 2006.

11. Neale, C.M.U.; Farland, M.J.M.; Chang, K. Land surface type classification using microwave brightness temperatures from the SMM/I. *IEEE Trans. Geosci. Remote Sens.* **1990**, *28*, 829–838. [CrossRef]
12. Paloscia, S.; Pampaloni, P. Microwave Vegetation Indexes for detecting biomass and water conditions of agricultural crops. *Remote Sens. Environ.* **1992**, *40*, 15–26. [CrossRef]
13. Shi, J.; Jackson, T.; Tao, J.; Du, J.; Bindlish, R.; Lu, L.; Chen, K.S. Microwave vegetation indices for short vegetation covers from satellite passive microwave sensor AMSR-E. *Remote Sens. Environ.* **2008**, *112*, 4285–4300. [CrossRef]
14. Hallikainen, M.T.; Jolma, P.A. Comparison of algorithms for retrieval of snow water equivalent from Nimbus-7 SMMR data in Finland. *IEEE Trans. Geosci. Remote Sens.* **1992**, *30*, 124–131. [CrossRef]
15. Hallikainen, M.T. Retrieval of snow water equivalent from Nimbus-7 SMMR data: Effect of land-cover categories and weather conditions. *IEEE J. Ocean. Eng.* **1984**, *9*, 372–376. [CrossRef]
16. Kelly, R.; Chang, A.T.C. Development of a passive microwave global snow depth retrieval algorithm for Special Sensor Microwave Imager (SSM/I) and Advanced Microwave Scanning Radiometer-EOS (AMSR-E) data. *Radio Sci.* **2003**, *38*, 8076. [CrossRef]
17. Chang, T.C.; Foster, J.L.; Hall, D.K.; Rango, A.; Hartline, B.K. Snow water equivalent estimation by microwave radiometry. *Cold Reg. Sci. Technol.* **1982**, *5*, 259–267. [CrossRef]
18. Foster, J.L.; Chang, A.T.C.; Hall, D.K. Comparison of snow mass estimates from a prototype passive microwave snow algorithm, a revised algorithm and a snow depth climatology. *Remote Sens. Environ.* **1997**, *62*, 132–142. [CrossRef]
19. Aschbacher, J. Microwave Signatures from Land surface Radiometry. In Proceedings of the 10th Annual International Symposium on Geoscience and Remote Sensing IGARSS 1990, College Park, MD, USA, 20–24 May 1990; pp. 1149–1152.
20. Tsutsui, H.; Koike, T. Development of Snow Retrieval Algorithm Using AMSR-E for the BJ Ground-Based Station on Seasonally Frozen Ground at Low Altitude on the Tibetan Plateau. *J. Meteorol. Soc. Jpn.* **2012**, *90*, 99–112. [CrossRef]
21. Paloscia, S.; Pampaloni, P.; Chiarantini, L.; Coppo, P.; Gagliani, S.; Luzi, G. Multifrequency passive microwave remote sensing of soil moisture and roughness. *Int. J. Remote Sens.* **1993**, *14*, 467–483. [CrossRef]
22. Choudhury, B.J.; Schmugge, T.J.; Chang, A.; Newton, R.W. Effect of Surface Roughness on the Microwave Emission from Soils. *J. Geophys. Res.* **1979**, *84*, 5699–5706. [CrossRef]
23. Coppo, P.; Luzi, G.; Paloscia, S.; Pampaloni, P. Effect of soil roughness on microwave emission: Comparison between experimental data and model. In Proceedings of the Remote Sensing: Global Monitoring for Earth Management, IGARSS '91, Espoo, Finland, 3–6 June 1991; pp. 1167–1170. [CrossRef]
24. Wegmuller, U.; Maetzler, C. Rough bare soil reflectivity model. *IEEE Trans. Geosci. Remote Sens.* **1999**, *37*, 1391–1395. [CrossRef]
25. Chen, M.; Weng, F. Modeling Land Surface Roughness Effect on Soil Microwave Emission in Community Surface Emissivity Model. *IEEE Trans. Geosci. Remote Sens.* **2016**, *54*, 1716–1726. [CrossRef]
26. Mo, T.; Choudhury, B.J.; Schmugge, T.J.; Wang, J.R.; Jackson, T.J. A model for microwave emission from vegetation-covered fields. *J. Geophys. Res.* **1982**, *87*, 11229–11237. [CrossRef]
27. Ferrazzoli, P.; Paloscia, S.; Pampaloni, P.; Schiavon, G.; Solimini, D.; Coppo, P. Sensitivity of Microwave Measurements to Vegetation Biomass and Soil Moisture Content: A Case Study. *IEEE Trans. Geosci. Remote Sens.* **1992**, *30*, 750–756. [CrossRef]
28. Paloscia, S. Microwave emission from vegetation. In *Passive Microwave Remote Sensing of Land-Atmosphere Interactions*; Choudhury, B., Njoku, E., Kerr, Y., Pampaloni, P., Eds.; VSP Press: Zeist, The Netherlands, 1995; pp. 358–374.
29. Santi, E.; Pettinato, S.; Paloscia, S.; Pampaloni, P.; Macelloni, G.; Brogioni, M. An algorithm for generating soil moisture and snow depth maps from microwave spaceborne radiometers: HydroAlgo. *Hydrol. Earth Syst. Sci.* **2012**, *16*, 3659–3676. [CrossRef]
30. Santi, E.; Paloscia, S.; Pampaloni, P.; Pettinato, S.; Tomoyuki, N.; Seki, M.; Sekiya, K.; Maeda, T. Vegetation Water Content Retrieval by Means of Multifrequency Microwave Acquisitions from AMSR2. *IEEE J. Sel. Top. Appl. Earth Obs. Remote Sens.* **2017**, *10*, 3861–3873. [CrossRef]
31. Kunzi, K.F.; Fisher, A.D.; Staelin, D.H.; Waters, J.W. Snow and ice surfaces measured by the Nimbus 5 microwave spectrometer. *J. Geophys Res.* **1976**, *81*, 4965–4980. [CrossRef]

32. Hofer, R.; Mätzler, C. Investigation of snow parameters by radiometry in the 3- to 60-mm wavelength region. *J. Geophys. Res.* **1980**, *85*, 453–460. [CrossRef]

33. Jin, Y.Q.; Kong, J.A. Strong fluctuation theory of electromagnetic wave scattering by a layer of random discrete scatterers. *J. Appl. Phys.* **1984**, *55*, 1364–1369. [CrossRef]

34. Dong, J.; Walker, J.P.; Houser, P.R. Factors affecting remotely sensed snow water equivalent uncertainty. *Remote Sens. Environ.* **2005**, *97*, 68–82. [CrossRef]

35. Markus, T.; Powell, D.C.; Wang, J.R. Sensitivity of passive microwave snow depth retrievals to weather effects and snow evolution. *IEEE Trans. Geosci. Remote Sens.* **2006**, *44*, 68–77. [CrossRef]

36. Langlois, A.; Royer, A.; Derksen, C.; Montpetit, B.; Dupont, F.; Goïta, K. Coupling the snow thermodynamic model SNOWPACK with the microwave emission model of layered snowpacks for subarctic and arctic snow water equivalent retrievals. *Water Resour. Res.* **2012**, *48*, 1–14. [CrossRef]

37. Schanda, E.; Matzler, C.; Kunzi, K. Microwave remote sensing of snow cover. *Int. J. Remote Sens.* **1982**, *4*, 149–158. [CrossRef]

38. Takala, M.; Luojus, K.; Pulliainen, J.; Derksen, C.; Lemmetyinen, J.; Kärnä, J.P.; Koskinen, J.; Bojkov, B. Estimating northern hemisphere snow water equivalent for climate research through assimilation of space-borne radiometer data and ground-based measurements. *Remote Sens. Environ.* **2011**, *115*, 3517–3529. [CrossRef]

39. Tait, A.B. Estimation of Snow Water Equivalent Using Passive Microwave Radiation Data. *Remote Sens. Environ.* **1988**, *64*, 286–291. [CrossRef]

40. Santi, E.; Paloscia, S.; Pampaloni, P.; Pettinato, S.; Brogioni, M.; Xiong, C.; Crepaz, A. Analysis of Microwave Emission and Related Indices Over Snow using Experimental Data and a Multilayer Electromagnetic Model. *IEEE Trans. Geosci. Remote Sens.* **2017**, *99*, 1–14. [CrossRef]

41. Rott, H.; Aschbacher, J. On the use of satellite microwave radiometers for large-scale hydrology. In Proceedings of the IAHS 3rd International Assembly on Remote Sensing and Large-Scale Global Processes, Baltimore, MD, USA, 10–19 May 1989.

42. Chukhlantsev, A.A. Microwave Emission from Vegetation Canopies. Ph.D. Thesis, The Moscow Institute of Physics and Technology (State University), Moscow, Russia, 1981. (In Russian)

43. Jackson, T.J.; Schmugge, T.J.; Wang, J.R. Passive microwave sensing of soil moisture under vegetation canopies. *Water Resour. Res.* **1982**, *18*, 1137–1142. [CrossRef]

44. Pampaloni, P.; Paloscia, S. Microwave emission and plant water content: A comparison between field measurement and theory. *IEEE Trans. Geosci. Remote Sens.* **1986**, *24*, 900–905. [CrossRef]

45. Chukhlantsev, A.A.; Golovachev, S.P.; Shutko, A.M. Experimental study of vegetable canopy microwave emission. *Adv. Space Res.* **1989**, *9*, 317–321. [CrossRef]

46. Tsang, L.; Pan, J.; Liang, D.; Li, Z.X.; Cline, D.; Tan, Y.H. Modeling active microwave remote sensing of snow using dense media radiative transfer (DMRT) theory with multiple scattering effects. *IEEE Trans. Geosci. Remote Sens.* **2007**, *45*, 990–1004. [CrossRef]

47. Wu, T.D.; Chen, K.S. A Reappraisal of the Validity of the IEM Model for Backscattering from Rough Surfaces. *IEEE Trans. Geosci. Remote Sens.* **2004**, *42*, 743–753.

48. Macelloni, G.; Paloscia, S.; Pampaloni, P.; Santi, E. Global scale monitoring of soil and vegetation using active and passive sensors. *Int. J. Remote Sens.* **2003**, *24*, 2409–2425. [CrossRef]

49. Santi, E.; Paloscia, S.; Pettinato, S.; Brocca, L.; Ciabatta, L. Robust Assessment of an Operational Algorithm for the Retrieval of Soil Moisture from AMSR-E Data in Central Italy. *IEEE J. Sel. Top. Appl. Earth Obs. Remote Sens.* **2016**, *9*, 2478–2492. [CrossRef]

 *remote sensing*

Article

Soil Moisture in the Biebrza Wetlands Retrieved from Sentinel-1 Imagery

Katarzyna Dabrowska-Zielinska [1,*], Jan Musial [1], Alicja Malinska [1], Maria Budzynska [1], Radoslaw Gurdak [2], Wojciech Kiryla [1], Maciej Bartold [2] and Patryk Grzybowski [1]

1   Institute of Geodesy and Cartography, Jacka Kaczmarskiego 27, 02-679 Warsaw, Poland;
    jan.musial@igik.edu.pl (J.M.); alicja.malinska@igik.edu.pl (A.M.); maria.budzynaka@igik.edu.pl (M.B.);
    wojciech.kiryla@igik.edu.pl (W.K.); patryk.grzybowski@igik.edu.pl (P.G.)
2   Department of Geoinformatics, Cartography and Remote Sensing, Faculty of Geography and Regional
    Studies, University of Warsaw, Krakowskie Przedmiescie 30, 00-927 Warsaw, Poland;
    radoslaw.gurdak@igik.edu.pl (R.G.); maciej.bartold@igik.edu.pl (M.B.)
*   Correspondence: katarzyna.dabrowska-zielinska@igik.edu.pl; Tel.: +48-22-3291974

Received: 19 October 2018; Accepted: 5 December 2018; Published: 7 December 2018

Abstract: The objective of the study was to estimate soil moisture (SM) from Sentinel-1 (S-1) satellite images acquired over wetlands. The study was carried out during the years 2015–2017 in the Biebrza Wetlands, situated in north-eastern Poland. At the Biebrza Wetlands, two Sentinel-1 validation sites were established, covering grassland and marshland biomes, where a network of 18 stations for soil moisture measurement was deployed. The sites were funded by the European Space Agency (ESA), and the collected measurements are available through the International Soil Moisture Network (ISMN). The SAR data of the Sentinel-1 satellite with VH (vertical transmit and horizontal receive) and VV (vertical transmit and vertical receive) polarization were applied to SM retrieval for a broad range of vegetation and soil moisture conditions. The methodology is based on research into the effect of vegetation on backscatter (σ°) changes under different soil moisture and Normalized Difference Vegetation Index (NDVI) values. The NDVI was derived from the optical imagery of a MODIS (Moderate Resolution Imaging Spectroradiometer) sensor onboard the Terra satellite. It was found that the state of the vegetation expressed by NDVI can be described by the indices such as the difference between σ° VH and VV, or the ratio of σ° VV/VH, as calculated from the Sentinel-1 images in the logarithmic domain. The most significant correlation coefficient for soil moisture was found for data that was acquired from the ascending tracks of the Sentinel-1 satellite, characterized by the lowest incidence angle, and SM at a depth of 5 cm. The study demonstrated that the use of the inversion approach, which was applied to the newly developed models using Water Cloud Model (WCM) that includes the derived indices based on S-1, allowed the estimation of SM for wetlands with reasonable accuracy (10 vol. %). The developed soil moisture retrieval algorithms based on S-1 data are suited for wetland ecosystems, where soil moisture values are several times higher than in agricultural areas.

Keywords: Sentinel-1 backscatter; polarization; Terra MODIS; NDVI; soil moisture

---

1. Introduction

The soil moisture (SM) is an essential variable in environmental studies related to wetlands as it controls the biophysical processes that influence water, energy, and carbon exchanges. Hence, there is the need for SM constant monitoring. The SAR satellite imagery is an important source to fulfill this objective regardless of cloud cover and, especially in the areas, in which deployment of in situ SM measurements is not possible or economically unprofitable The possibility of using high temporal and spatial resolution of the Sentinel-1 (S-1) imagery motivated authors to develop the methodology for its retrieval based on backscattering coefficient ($\sigma°$), as calculated from the VH and VV polarizations.

The study was conducted in the Biebrza Wetlands, situated in north-eastern Poland, with a total area of 59,233 ha. The wetlands are unique in Europe for their non-drained floodplains, marshes, and fens, surrounded by a post-glacial landscape [1]. The Biebrza Wetlands were designated as a wetland site of global importance, as part of NATURA 2000, and since 1995 it has been under the protection of the RAMSAR Convention. Changes in soil moisture towards depletion cause peat mineralization, and the release of substantial amounts of carbon into the atmosphere [2,3]. Therefore, monitoring of soil moisture is very important for the management of the wetlands, to prevent peat degradation. The retrieval of soil moisture (SM) estimates by the means of satellite data is of great interest for a wide range of hydrological applications. The demand for operational SM monitoring was raised in numerous studies, and this was emphasized by the Global Climate Observing System (GCOS) by endorsing SM as an Essential Climate Variable (ECV).

Wetlands are often areas of limited access, where field sampling is difficult due to the inaccessible terrain and the seasonally dynamic nature of the area, and therefore satellites can provide information on the types of wetland vegetation and the dynamics of the local water cycle, in which soil moisture is a significant factor. Controlling soil moisture content is essential for the protection of peat-forming plant communities and for slowing down the drying processes against mineralization [4].

There are numerous studies that describe different remote sensing techniques for the assessment of soil moisture; however the SAR data give very good possibility for frequent spatial monitoring because of their independence from the weather conditions. Kornleson and Coulibaly [5] conducted a comprehensive literature review to provide soil moisture retrieval methodology from SAR data. The researchers have proved that microwave backscatter ($\sigma°$) is affected by the moisture and roughness of the canopy-soil layer. It is further affected by satellite sensor configurations such as the incident angle and the electromagnetic wave polarization [6,7]. The strong interactions of the backscatter signal with the soil and vegetation may not be expressed by simple linear functions. Atema and Ulaby [8] proposed a water cloud model (WCM) that characterized vegetation as the cloud, and represented the total backscatter from the canopy as the sum of the contribution of the vegetation $\sigma°_{veg}$, and of the underlying soil $\sigma°_{soil}$. The WCM model was adopted by Dabrowska-Zielinska et al. [9] for agricultural fields. The separation of the soil and vegetation components is not straightforward due to the complex interactions between them, which simultaneously affect SAR backscatter. The signal strongly depends on the type of vegetation, the amount of moisture, and the type of ecosystem [9]. Wetlands are characterized by deep peat layers, and it is not possible to compare agriculture ecosystems to wetlands, which are wet and very different. Thus, the models derived for wetlands have to be treated separately from models that are designated for agriculture soils and agriculture vegetation.

The C-band SAR on board the ERS-1/2 (European Remote Sensing) satellite, and also on board the ENVISAT (ENVIronmental SATellite), and following the Sentinel-1 satellite, has been applied for soil moisture retrieval [5,10]. The researchers used different models to distinguish the influence of vegetation and soil moisture on the microwave signal. Most of the methods that are applied for soil moisture retrieval have been developed for bare soils and agricultural areas [5,11–15], and only a few have been found for natural environments such as wetlands. Mattia et al. [16] and Balenzano et al. [17] present the SMOSAR (Soil MOisture retrieval from multi-temporal SAR data) algorithm for soil moisture retrieval using the multi-temporal SAR data from Sentinel-1 data. Paloscia et al. [18] developed a soil moisture content (SMC) algorithm for Sentinel-1 characteristics, based on an artificial neural network (ANN), which was tested and validated in several test areas in Italy, Australia, and Spain. Also, ANN-based algorithms for the SMC retrieval applying C-band SAR data (ENVISAT/ASAR, Cosmo-SkyMed) have been adapted and presented by Santi et al. [19]. The overview of the retrieval algorithms presented in [19] demonstrated that ANN is a very powerful tool for estimating the soil moisture at both local and global scales. The proposed model simulates the backscatter of the vegetated areas as a function of the soil backscatter, and the vegetation water content as computed from the NDVI. Kasischke et al. [20] conducted an investigation on the response of the ERS C-band SAR backscatter to variations in soil moisture and surface inundation in Alaskan wetlands, and found a positive correlation between the backscatter and soil moisture in sites that were dominated by herbaceous vegetation cover. Multi-temporal C-band SAR data, HH, and VV polarized, available from ERS-2 and ENVISAT satellites were used by Lang et al. [21] for the investigation of inundations and soil moisture determination at wetlands. Gao et al. [22] presented two methods for the retrieval of soil moisture over irrigated crop fields based on Sentinel-1 data recorded in the VV polarization combined with Sentinel-2 optical data. The first method used minimum and maximum values of backscattering coefficient calculated from Sentinel-1 data, whereas the second one was based on the analysis of backscattering differences on two consecutive acquisition days. With both methods, the Sentinel-1 data was combined with NDVI index computed from Sentinel-2 data. They obtained estimated RMS soil moisture errors of approximately 0.087 m^3m^{-3} and 0.059 m^3m^{-3} for the first and second methods, respectively. El Hajj et al. [23] used a neural network technique to develop an operational method for soil moisture estimates in agricultural areas based on the synergistic use of Sentinel-1 and Sentinel-2 data. They found that VV polarization alone as well as both VV and VH provides better accuracy on the soil moisture calculation than VH alone. The method developed by them could be applied for agricultural plots with an NDVI lower than 0.75 and allows for the soil moisture estimates with an accuracy of approximately 5 vol. %. Baghdadi at al. [24] applied the Water Cloud Model for estimating surface soil moisture of crop fields and grasslands from Sentinel-1/2 data. They simulated the soil contribution (moisture content and surface roughness) applying Integral Equation Model and used NDVI values as the vegetation descriptor. They obtained that the soil contribution to the total radar signal is large in VV polarization when soil moisture is between 5 and 35 vol. %, and NDVI between 0 and 0.8. Tomer et al. [25] developed an algorithm to retrieve surface soil moisture based on the Cumulative Density Function Transformation of multi-temporal RADARSAT-2 backscattering coefficient. The algorithm, which was tested in a semi-arid tropical region in South India and validated with the in situ data showed RMSE of soil moisture estimates ranging from 0.02 to 0.06 m^3m^{-3} depending on soil information used and development of vegetation. Dabrowska-Zielinska et al. [26] conducted an investigation on soil moisture monitoring in the Biebrza Wetlands using Sentinel-1 data, and found, that LAI dominates the influence on σ° when soil moisture is low. They developed models for soil moisture assessment under different wetland vegetation habitat types (non-forest communities) applying VH polarization (R^2 = 0.70 to 0.76). There are not many studies for wetlands SM retrieval applying S-1 data, as can be seen from the literature review. Most of the publications refer to agriculture crops or bare soils. The difference and the ratio of the VH and VV backscatter as the proxy of vegetation conditions has been recently studied and published by several researchers. Vreugdenhil et al. [27] examined Sentinel-1 VV and VH backscatter and their ratio VH/VV to monitor crop conditions with special reference to vegetation water content (VWC) of agriculture crop. Greifeneder et al. [28] demonstrated that the ratio of VH/VV calculated from AQUARIUS L-band scatterometer allows a good compensation of vegetation dynamics for the retrieval of soil moisture. Hosseini et al. [29] used RADARSAT-2 to estimate Leaf Area Index (LAI) for corn and soybeans fields. They found high correlation coefficients between ground measured and estimated LAI values, when dual like-cross polarizations were used (either HH–HV or VV–HV). Also, it has been found that RADARSAT-2 (HH-HV) can be used for the retrieval of soil moisture and the total biomass, while RADARSAT-2 (VV-HV) can be used for the retrieval of the biomass of the wheat heads [30].

The aim of this research study was to examine the sensitivity of Sentinel-1 backscatter (σ°) to SM variation under vegetation, as characterized by different biomasses, and to develop the new models for SM retrieval under wetland vegetation cover (non-forest communities), by applying the C-band SAR data VH and VV polarized, which are available from the Sentinel-1 (S-1) satellite. The vegetation biomass was represented by NDVI, which was calculated by applying the Terra MODIS data. The authors present the approach, which applies the SAR indices such as the difference of σ° VH-VV and the ratio VV/VH as vegetation descriptors in SM retrieval using modified version of WCM.

Application of these descriptors, as dual polarization, give better results to separate the influence of vegetation from the soil moisture impact on backscatter. The modification consists in linearization of WCM model applying Least Squares Method.

The authors are motivated to undertake this study due to the lack of operational methods for the monitoring of SM based on Sentinel-1 data in the Central European wetlands areas. The presented study is a new approach to the previous one [26] on SM modelling based on S-1 data. Due to the temporal frequency of the two S-1 satellites' (S-1A and S-1B) acquisitions, it is possible to monitor soil moisture changes every six days with high spatial resolution (10 × 10 m). The results will highlight the contribution of S-1 data to soil moisture assessment, improving hydrological studies carried out in wetlands, which have so far very often been based on in–situ observations.

2. Materials and Methods

2.1. Study Area

The Biebrza Wetlands holds 25,494 ha of peatlands, much biodiversity in the rich plant habitats, as well as highly diversified fauna, especially for birds [1]. This is still one of the wildest areas in Europe, and one of the areas that has been least destroyed, damaged, or changed by human activity. The Biebrza Wetlands belong to the largest of Poland's National Parks—Biebrza National Park (BNP), which was created on September 9, 1993 [31]. It is located in Podlaskie Voivodeship, northeastern Poland, and it is situated along the Biebrza River. The geographical position of the study area is: UL: N54° E22°10′ and LR: N53°10′ E23°30′. The Biebrza Wetland area is flat with an average altitude of about 105 m above sea level (m a.s.l.). To the north, the altitude increases, reaching approximately 120 m a.s.l. The main river is the Biebrza River, which flows out near the eastern border of Poland. The Biebrza River drainage basin area is 7051 km^2, the river length is 155 km, and its mean flow is 35.3 m^3 s^{-1}. The Wetlands are flooded annually in the spring, and besides precipitation, flooding is the main supply of moisture into the peat soil. The weather in the Biebrza River Valley is one of the coolest in Poland—the mean year daily temperature is 6.5 °C. The mean sum of the yearly precipitation ranges between 550–650 mm, and is one of the lowest in Poland. The length of the growing season is less than 200 days, and this is one of the shortest in Poland. Generally, summer is warm but short; winter is cold and long. The coldest month is January, with a mean temperature of −4.2 °C, and with temperatures dropping as low as −50 °C. Snow cover can last up to 140 days. July is the warmest month in the Biebrza Valley, with mean temperatures of 17.5 °C, and with temperatures increasing up to 35.3 °C. The length of the summer ranges between 77–85 days [32].

At the Biebrza Wetlands, two sites for Sentinel-1 (S-1) soil moisture (SM) retrieval were established (grassland and marshland), where a network of soil moisture ground stations was built (Figure 1).

Both sites had a flat topography and homogeneous land cover, which ensured the representativeness of average SM estimates across the sites. The environmental conditions between both sites varied with respect to the SM level, vegetation density, and the type of vegetation community cover. The soil moisture for these two sites differed. For the same years, the SM median for the grassland site was equal to 35 vol. % and it was much higher for the marshlands—close to 60 vol. %. The grassland site (Figure 2) was located on an intensively mowed, drained meadow with semi–organic soil (muck-peat soil). The marshland site (Figure 3) was located within the Biebrza National Park, and covered unmanaged sedges with more moist organic soil (peat soil).

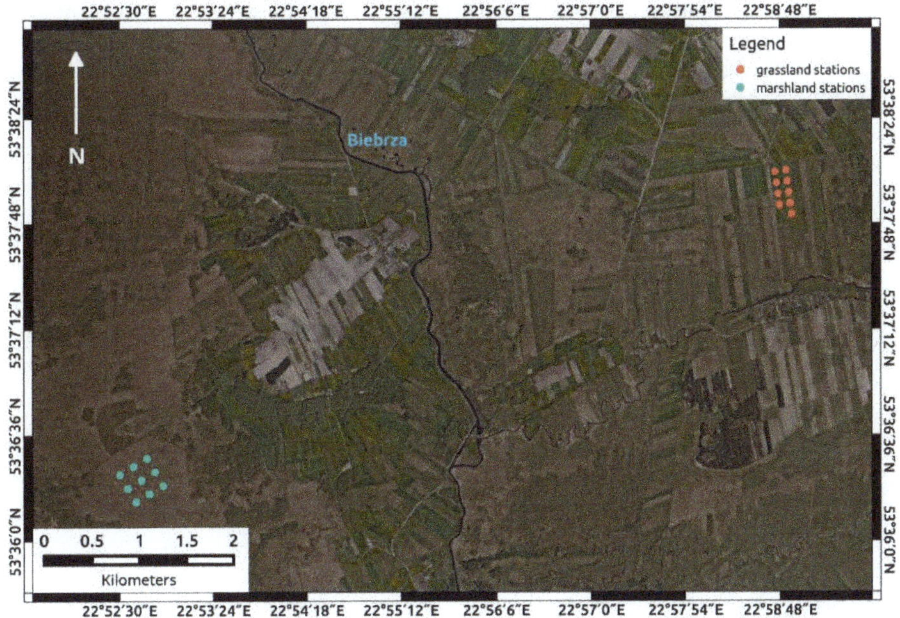

Figure 1. Location of S-1 soil moisture sites at the Biebrza Welands overlapped to the Geoportal maps image (www.geoportal.gov.pl).

Figure 2. Grassland site.

Figure 3. Marshland site.

The marshland site had a regular 500 × 500 m measuring grid composed of nine SM stations equipped with five probes each, measuring at the following depths: 5, 10, 20, and 50 cm. The grassland site had analogous instrumentation, with the stations arranged in two rows (230 × 580 m), one with four SM stations, and the second with five SM stations. In total, 90 Decagons GS3 soil moisture sensors were installed.

The grassland and marshland sites featured different soil moisture values and both sites were flooded during the spring. At the marshland site, the water table was very high; therefore, only the soil layer at 5 cm exhibited noticeable variations in water content. The deeper layers were close to saturation point (80–90 vol. %) through the year. An apparent drop of SM values that occurred in winter was related to the ground freezing. At the grassland site, the water table was lower; thus, only the 50 cm soil layer was permanently close to saturation level. The surface soil layers featured a strong annual cycle with a maximum amplitude of around 60 vol. %. A more in-depth description of the sites is available in [33]. The measurements collected from both sites are available through the International Soil Moisture Network (ISMN) [34].

2.2. In Situ Data

The in situ data were collected during field campaigns carried out in the years 2015–2017, simultaneous to the satellite overpasses. The positions of the measurement plots were determined using GPS (Global Positioning System). This information was essential for preparing the layer of special measurement points that was needed for the reading and processing of satellite data. Soil moisture (volumetric) was measured by 90 Decagons GS3 sensors calibrated to specific soil conditions at four depths: 5, 10, 20, and 50 cm. The GS3 sensor uses an electromagnetic field to measure the dielectric permittivity of the surrounding medium. The dielectric value is then converted to substrate water content by a calibration equation that is specific to the soil conditions. Regarding the observation modes, the SM measurements were performed every 15 min. Additionally, the height of the vegetation (m) and the biomass wet and dry (gm^{-2}) were measured. These data supported the SM analysis with ancillary information about the variables influencing the SAR signal (biomass, vegetation conditions).

During the course of the study, the season of 2015 was extremely dry, whereas conditions in 2017 were extremely wet. In 2016, soil moisture levels were regarded as being moderate.

2.3. Satellite Data

Within the study, the following satellite images were used: Sentinel-1 and Terra MODIS. From the SciHUB (Sentinel Scientific Data Hub), Sentinel-1 Level-1 GRDH (Ground Range Detected at High resolution) products, in IWS (Interferometric Wide Swath) acquisition mode (spatial resolution 10 × 10 m) and in a WGS84 ellipsoid, were downloaded. The S-1 images were acquired in the C-band (5.5 GHz) in dual polarization: VV and VH. The nominal acquisition frequency of a single S-1 satellite over the Biebrza Wetlands during the period of the study was 12 days for a single track. However, the grassland site was covered by four different S-1 tracks (two descending and two ascending orbits), and the marshland site was covered by three different S-1 tracks (one descending and two ascending orbits). Furthermore, the availability of the two Sentinel-1A and Sentinel-1B platforms doubled the revisit time, which on average equaled four days for a single satellite and two–three days for two satellites. Table 1 presents the tracks and local incidence angles at the grassland and marshland test sites for selected S-1 relative orbits.

Table 1. Local incidence angles for selected S-1 orbit passes (A-ascending, D-descending) and tracks.

Pass/Track	Marshland Incidence Angle	Grassland Incidence Angle
A/29	43.49°	43.10°
A/131	35.59°	35.13°
D/80	-	45.65°
D/153	38.57°	38.18°

MODIS images as MOD09Q1 version 6 (V006) products were downloaded from the US Geological Survey website. The MOD09Q1 V006 product provided Bands 1 and 2 (620–670, 841–876, appropriately) at a 250 m resolution in an 8 day gridded level-3 product in the sinusoidal projection. The surface spectral reflectance of Bands 1–2 was corrected for atmospheric conditions such as gasses, aerosols, and Rayleigh scattering. For each pixel, a value was selected from all of the acquisitions within the 8-day composite period, taking into account the cloud coverage and the solar zenith angle [35].

MODIS NDVI 8-day compositions were paired with Sentinel-1 daily satellite images, so that the nearest day of S-1 acquisition to the middle date of 8-day composition of MODIS was taken; therefore, it was assumed that NDVI values could be used to represent the vegetation effect for the modeling of the backscattering coefficients of the S-1. The area of an SM sensors sites is 500×500 m. The soil moisture, σ° and NDVI were taken as the average values for this area.

2.4. Methods

Sentinel-1 products were processed with the Sentinel-1 Toolbox (SNAP S1TBX v5.0.4 software) software provided by the European Space Agency (ESA). The processing included: speckle filtering applying a Lee Sigma speckle filter, radiometric calibration, and data conversion to a backscattering coefficient (σ°) (dB). Then, the scenes were geometrically registered to the local projection PUWG1992, and the σ° S-1 values, which corresponded to the measurement sites, were extracted using ERDAS software (Hexagon Geospatial/Intergraph®, Norcross, GA, USA).

The methodology consists of models that were developed for soil moisture retrieval by applying the following Sentinel-1 data: VH and VV polarizations, VH-VV, VV/VH and the NDVI values from the Terra MODIS data. Soil moisture retrieval was based on simplified Water Cloud Model with application of the Least Squares Method.

2.4.1. Water Cloud Model with the Least Squares Method

The Water Cloud Model represents the total backscatter from the canopy (σ°) as the sum of the contribution of the vegetation σ°_{veg} and of the underlying soil σ°_{soil} [36]:

$$\sigma^\circ = \sigma^\circ_{veg} + \tau^2 \, \sigma^\circ_{soil} \tag{1}$$

where:

$$\sigma^\circ_{veg} = A \, V_1 \cos(\theta) \, (1 - \tau^2) \tag{2}$$

$$\tau^2 = \exp(-2B \, V_2/\cos(\theta)) \tag{3}$$

where: θ—incidence angle, τ^2—two way attenuation through the canopy: V_1 and V_2 are descriptors of the canopy, A and B are fitted parameters of the model that depend on the vegetation descriptor and the radar configuration. As the vegetation descriptors (V_1 and V_2), the NDVI values derived from MODIS data were taken. The B parameter is connected with the density of vegetation and its strength of the attenuation during the growing season. For the specific, homogeneous area, we can assume the fixed value of B and apply linearized nonlinear method to solve the WCM model (instead of nonlinear iterative methods). Figure 4 presents the simulation of the strength of attenuation depending on NDVI values for different values of B parameter.

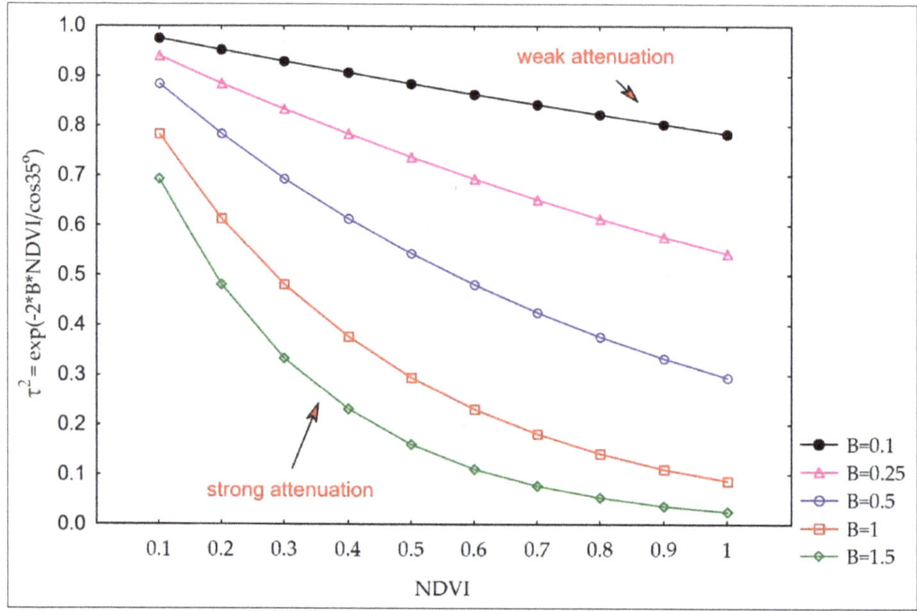

Figure 4. Evolution of attenuation (τ^2) depending on NDVI for different values of B parameter.

For bare soil the response of backscatter to soil moisture (σ°_{soil}) is a linear function. It was assumed that in early spring at the wetlands area the soil has dominated impact on backscatter. Therefore we applied modified WCM, where σ°_{soil} (Equation (1)) was represented by measured SM values. The measurements were conducted during two full years at even time interval, so the relation soil-vegetation can be assumed to be well represented. The following two components of data were designed to describe the effect of the vegetation and the underlying soil on σ° VH value: $\tau^2 * $ SM, and $(1 - \tau^2)* \cos(\theta)*$ NDVI. The first component represents the interaction of the incident radiation between the vegetation and the underlying soil. τ^2 reduces the impact of the soil on backscatter when the vegetation cover is dense. τ^2 takes the value from 0–1 and is inversely proportional to the vegetation index and to the incidence angle. The second component describes the remaining part of the backscatter that depends on the vegetation canopy covering the soil. The parameters of the model with σ° VH as a dependent variable, and $\tau^2 * $ SM and $(1 - \tau^2)* \cos(\theta)*$ NDVI as independent variables, were estimated by applying the Least Squares Method. Data were limited to the vegetation season, i.e., from 60–300 days of each year. The form of modified WCM model is the following:

$$\sigma^{\circ} \text{ VH} = a + b\, \tau^2 \text{ SM} + c(1 - \tau^2) \cos(\theta) \text{ NDVI} \tag{4}$$

where: a, b, c are parameters of regression, that have to be estimated.

2.4.2. Vegetation Descriptors

First, it was assumed that the vegetation index (NDVI) derived from Terra MODIS (described in Section 2.3) could be used as a proxy for the vegetation descriptor of biomass.

Second, the vegetation biomass (expressed by NDVI) was represented by two combinations of sigma VH and sigma VV—the difference and the ratio. This assumption was performed following the approach of using the sigma difference VH−VV as the roughness of the vegetation (in this case, NDVI) following Rao et al. [37]. The σ° VH and σ° VV values were taken from the processed Sentinel-1 data (described in Section 2.3).

The popular NDVI index works as an indicator that describes the greenness or the density, and the health of the vegetation, based on the measurements of absorption and reflectance. The NDVI was calculated from MODIS MOD09Q1 V006 images on the basis of spectral reflectance from the soil-vegetation surface in the visible red (Band 1) and near-infrared (Band 2) spectra of electromagnetic waves according to:

$$NDVI = (R_{NIR} - R_{RED})/(R_{NIR} + R_{RED}), \tag{5}$$

where: R_{RED}—spectral reflectance in the red spectrum, R_{NIR}—spectral reflectance in the near-infrared spectrum. For calculating NDVI all pixels with the spectral reflectance values larger than 0 and lower than 10,000 (16 bit unsigned integer) were taken. Then, from Band 3 (Surface Reflectance 250 m State flags) of MOD09Q1 product the pixels flagged as: water, clouds/cloud shadows, and snow/ice were extracted and applied to NDVI images. The values of spectral reflectance were the ratios of the reflected radiation over the incoming radiation in each spectral channel individually (albedo); hence, the NDVI takes on values between 0–1.

2.4.3. Statistical Analyses

Statistical analyses were completed in STATISTICA software using the following quality measures: Pearson's correlation, Kendall's tau correlation, R (correlation coefficient), R^2 (coefficient of determination), MAPE (Mean Absolute Percentage Error), MPE (Mean Percentage Error), RMSE (Root Mean Square Error), and MBE (Mean Bias Error). The data were checked for the normal distribution and significance prior to all analyses. Validation of the retrieved SM values against the in situ measurements was preformed based on the RMSE error.

3. Results

3.1. Correlation between $\sigma°$ Calculated from S-1 and Soil Moisture Measured at Different Depths

The in situ data and satellite data were used in statistical analyses to develop an inversion approach for the estimation of soil moisture from the Sentinel-1 data over the grassland and marshland sites.

Table 2 presents the results of Pearson's correlation (R values) for the marshland site between the backscattering coefficient ($\sigma°$) in the polarizations VH and VV, as calculated from Sentinel-1 (S-1), and the soil moisture (SM) when measured in situ at three depths: 5, 10, and 20 cm. The values cover the dates of 26 April 2015 to 30 June 2017. Table 3 presents the same values for grassland site.

Table 2. Pearson's correlation (R values) for the marshland site between $\sigma°$ VH and VV from S-1 and soil moisture (GS3), measured in situ at three depths: 5, 10, and 20 cm.

Marshland 2015–2017 Pearson Correlation (R)						
Sentinel-1			Soil Moisture GS3			Number of Observations
Polarization	Track	Orbit Pass	5 cm	10 cm	20 cm	N
	153	D [1]	0.49	0.34	0.40	57
VH	29	A [2]	0.51	0.39	0.49	70
	131	A [2]	0.56	0.46	0.59	66
	153	D [1]	0.47	0.27	0.36	57
VV	29	A [2]	0.40	0.22	0.28	70
	131	A [2]	0.55	0.39	0.52	66

[1] Descending, [2] Ascending.

The highest correlation was noted for the S-1 track 131 (ascending pass, low local incidence angles) and the soil moisture as measured at a 5 cm depth. The values of the correlation coefficient in any case were not higher than 0.59 for the marshland site and 0.72 for the grassland site.

For further analysis, the orbit pass ascending (A), and the depth of the soil moisture measurements at a 5 cm depth were taken into account (the highest correlation was found for these dataset).

Table 3. Pearson's correlation (R values) for the grassland site between σ° VH and VV from S-1, and soil moisture (GS3) measured in situ at three depths: 5, 10, and 20 cm.

	Grassland 2015–2017 Pearson Correlation (R)					
	Sentinel-1		Soil Moisture GS3			Number of Observations
Polarization	Track	Orbit Pass	5 cm	10 cm	20 cm	N
VH	153	D [1]	0.48	0.48	0.48	67
	29	A [2]	0.47	0.49	0.49	79
	80	D [1]	0.28	0.29	0.27	73
	131	A [2]	0.55	0.53	0.47	72
VV	153	D [1]	0.54	0.53	0.46	67
	29	A [2]	0.58	0.58	0.50	79
	80	D [1]	0.39	0.37	0.26	73
	131	A [2]	0.72	0.69	0.55	72

[1] Descending, [2] Ascending.

3.2. Impact of Vegetation on σ° Calculated from S-1 under Different Soil Moisture Conditions

It was noted that there was a different contribution from the vegetation, as represented by the NDVI, when dry conditions (SM < 30 vol. %) or moist conditions (SM > 60 vol. %) occurred. Figures 5 and 6 show the results of the statistical analyses that were performed between the backscattering coefficient (σ°) value as calculated from VH, and the NDVI as calculated from MODIS for the grassland site. Figure 5 presents the relationship between the σ° value and the NDVI for high, i.e., SM > 60 vol. %, soil moisture when measured at a 5 cm depth. In this case, the vegetation played a role in the process of attenuation when the wave penetrated the vegetation to reach the soil. A different situation was observed when the soil was dry, i.e., SM < 30 vol. %, at a 5 cm depth (Figure 6). The impact of vegetation on the σ° VH was stronger than the impact of soil moisture. Higher biomass values were represented by the NDVI, and hence a higher amount of vegetation moisture content dominated the influence of vegetation on the σ° values. Under low SM conditions, an increase in the NDVI values caused an increase in the σ° VH values, as vegetation impact on backscatter dominates. Under high SM conditions, the vegetation plays the role in two way attenuation of the beam (Equation (3)), an increase of NDVI values caused a decrease in the σ° VH values.

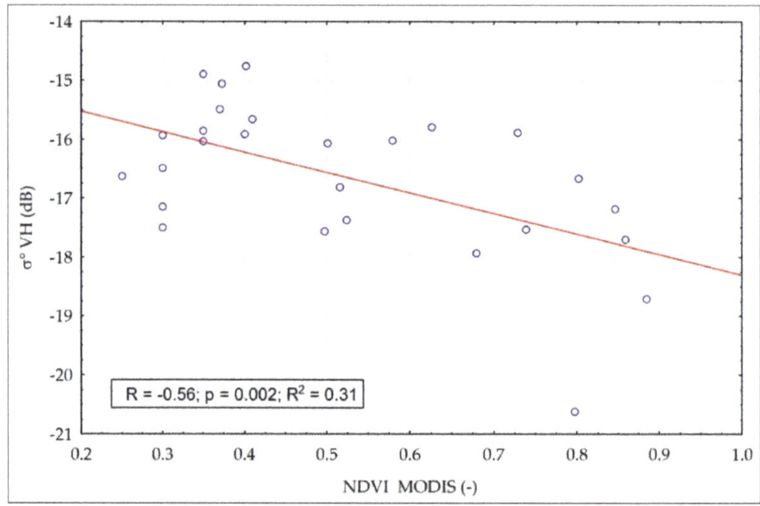

Figure 5. Relationship between the NDVI and σ° VH for the SM values measured at a 5 cm depth > 60 vol. % at the grassland site.

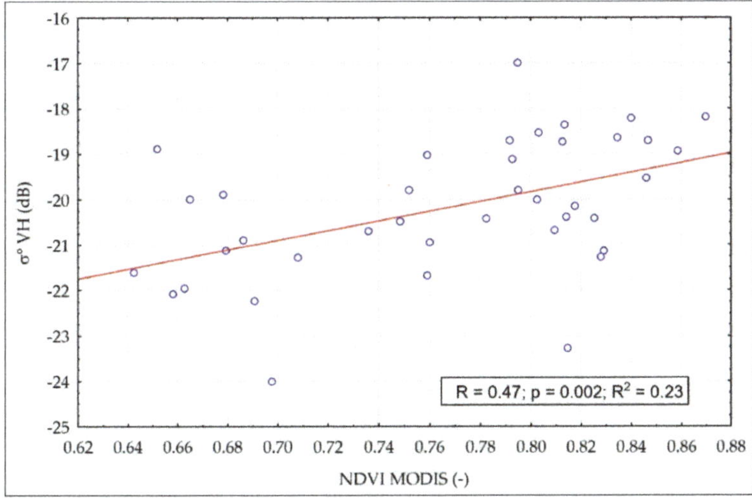

Figure 6. Relationship between the NDVI and σ° VH for the SM values measured at a 5 cm depth < 30 vol. % at the grassland site.

3.3. Impact of Soil Moisture on σ° Calculated from S-1 under a Quasi-Constant NDVI

If the amount of marshland/grassland vegetation biomass represented by the NDVI is constant in time, the variability of σ° S-1 is consistent with the variability of the soil moisture. Experimental data for the NDVI were gathered for each month separately, and the regression equation between the SM that was measured at a 5 cm depth, and σ° S-1 of the growing season (March–October) was estimated. The obtained correlation coefficients between the soil moisture, and σ° VH and VV were high (Table 4). It was assumed that during the month, the vegetation biomass did not vary significantly, which was confirmed by the low standard deviations values (Table 4) for the NDVI for the particular months. Therefore, it can be assumed that the variability of the backscatter responds to the variability of the soil moisture in areas with homogeneous vegetation cover. However the correlation is significant with the best correlation coefficient (R) for April, May, and October. For the rest of the month the correlation is poor but still significant.

Table 4. Correlations between σ° VH and VV and SM at a 5 cm depth for the grassland and marshland sites during the seasons of 2015–2016.

Month	NDVI		SM and σ° VH			SM and σ° VV			N [4]
	Mean	SD [1]	R [2]	*p*-Value	S [3] dB/Vol.%	R [2]	*p*-Value	S [3] dB/Vol.%	
March	0.43	0.08	0.87	0.00	0.24	0.64	0.02	0.14	13
April	0.44	0.10	0.83	0.00	0.14	0.86	0.00	0.12	30
May	0.63	0.16	0.87	0.00	0.10	0.85	0.00	0.11	33
June	0.78	0.08	0.58	0.00	0.04	0.18	0.35	-	29
July	0.77	0.06	0.61	0.00	0.04	0.19	0.27	-	31
August	0.80	0.08	0.53	0.00	0.03	0.24	0.17	-	32
September	0.76	0.07	0.83	0.00	0.06	0.51	0.00	0.03	35
October	0.63	0.10	0.86	0.00	0.07	0.80	0.00	0.06	41

[1] Standard deviations, [2] Correlation coefficient, [3] Sensitivity, [4] Number of observations.

Sensitivity of backscatter to the soil moisture is the measure of the change in σ° with the change in soil moisture. It was defined as the slope of the regression line between them at a given vegetation conditions. The higher values of sensitivity occurred in early spring when vegetation cover was lower than in later part of the growing season (Table 4).

3.4. Compatibility of Seasonal Trends in the Course of the Vegetation Descriptor NDVI, and the $\sigma°$ Difference VH−VV and Ratio VV/VH

The time series of $\sigma°$ indices that were calculated as the difference of polarization VH−VV, or the ratio VV/VH, presented seasonality trends, i.e., variations that were specific to a particular timeframe. There was a systematic increase of $\sigma°$ VH−VV and VV/VH values during the growing season, and a decrease in autumn, similar to the behavior of NDVI. Figure 7 presents the temporal evolution of the NDVI and $\sigma°$ VH−VV values during the vegetation season in 2016 at the grassland test site as an example. Mann-Kendall tau statistics were performed for both sites for the seasons of 2016–2017 separately (two complete growing seasons of observations). It revealed that the compatibility of the seasonal trends of $\sigma°$ VH−VV and VV/VH with the NDVI were statistically significant (Table 5).

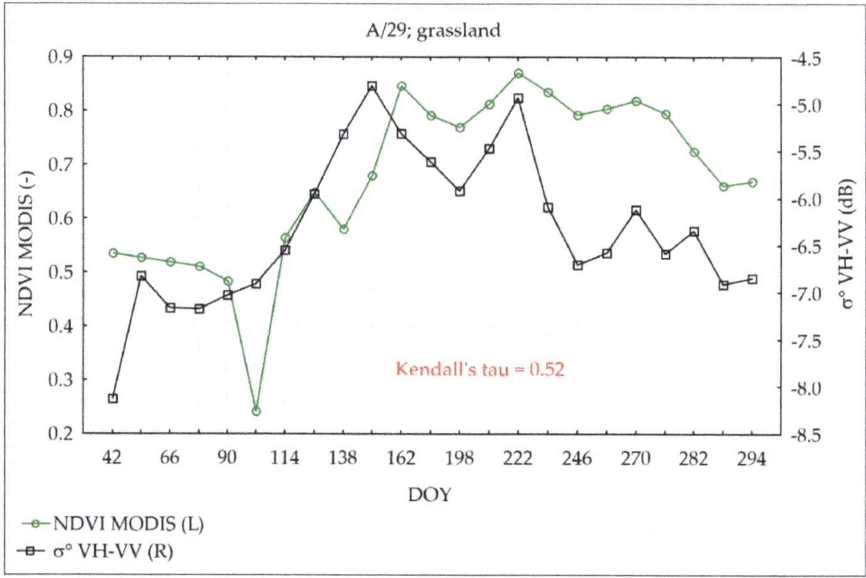

Figure 7. Temporal evolution of the NDVI and $\sigma°$ VH−VV during the vegetation season of 2016 on the grassland site.

Table 5. Kendall's tau statistics between the NDVI and the $\sigma°$ indices VH−VV and VV/VH for the grassland and marshland sites.

Site	Year	S-1 Track	Kendall's Tau for VH−VV	N [1]	Kendall's Tau for VV/VH
grassland	2016	29	0.52	37	0.42
		131	0.39	36	0.46
	2017	29	0.51	27	0.28
		131	0.50	26	0.28
marshland	2016	29	0.35	37	0.37
		131	0.54	36	0.56
	2017	29	0.68	27	0.39
		131	0.74	25	0.32

[1] Number of observations.

Thus, it has been assumed that the influence of vegetation on $\sigma°$ S-1 values could be expressed by indices of the difference between $\sigma°$ VH and VV (VH−VV) and the ratio of $\sigma°$ VV/VH. Analyzing Kendall's tau coefficients for all test sites, tracks, and seasons, it was found that both $\sigma°$ VH−VV and $\sigma°$ VV/VH indices were in monotonic correlation with the NDVI, and that they could replace the NDVI values in soil moisture modeling. In the experiment, the values of $\sigma°$ VV/VH was always positive and less than 1.

By applying the indices calculated using the S-1 data in modeling SM, the independence from the optical data (often overcast conditions) was ensured. Also, it allowed for quick calculations of soil moisture, which often changes rapidly and has to be observed regularly.

The two following approaches are presented in building the model for soil moisture retrieval:

1   Using the NDVI as a vegetation descriptor
2   Substituting the NDVI by the index $\sigma°$ VH−VV and the index $\sigma°$ VV/VH

3.5. Soil Moisture Retrieval Using $\sigma°$ from Sentinel-1 and NDVI from MODIS

Figure 4 shows, that the attenuation of radar signal by vegetation at high moisture conditions of soil was in the range of 3 dB, while the whole range of $\sigma°$ VH variability was 12 dB. Taking the level of attenuation as a middle, the value of B = 0.5 was chosen for further analysis. Thus, it was assumed that radar signal is attenuated by the vegetation in wetland according to:

$$\tau^2 = \exp(-\mathrm{NDVI}/\cos(\theta)) \tag{6}$$

The parameters in (Equation (4)) were estimated as follows:
Model 1a:

$$\sigma° \text{ VH} = -28.3 + 0.2\tau^2 \text{ SM} + 14.7(1 - \tau^2)\cos(\theta)\text{ NDVI} \tag{7}$$

where: R = 0.92; R^2 = 0.85; $p < 0.0000$; N = 147; Std. Err. = 0.79 dB, for ascending orbit.

The partial correlations for the soil and vegetation components were 0.89 and 0.54, respectively, which means that soil moisture influenced $\sigma°$ VH more strongly than the vegetation cover. Figure 8 presents a comparison between the observed values of $\sigma°$ VH (derived from S-1 images) and those that were predicted using Model 1a (Equation (7)).

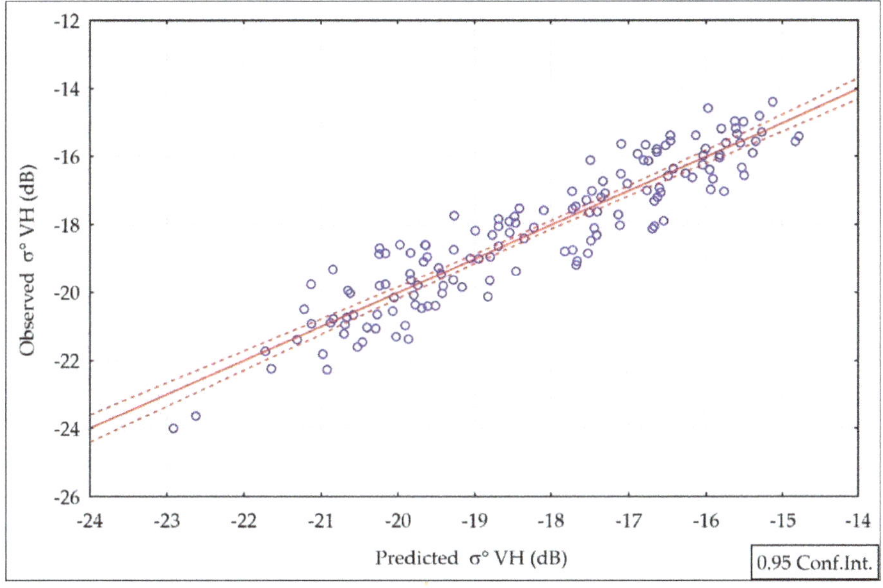

Figure 8. The $\sigma°$ VH values observed and predicted by Model 1a (Equation (7)).

Applying Linear Multiple Regression Model (Equation (4)), three parameters of the WCM model were estimated. Parameter "c" equal to 14.7 in (7) corresponds to "A" in WCM (Equation (2)). The remaining two parameters were interpreted as follows: "b" equal to 0.2 as sensitivity and "a" equal to -28.3 as intercept of SAR backscatter under fixed NDVI = 0 conditions. Intercept is the backscatter value expected for the dry soils. It is mainly a function of surface roughness [38]. For bare soil, where NDVI = 0 (theoretically), what means $\tau^2 = 1$ and $\sigma°_{veg} = 0$, the Equation (7) takes the following form: $\sigma°$ VH = $-28.3 + 0.2*$ SM. For the early spring measurements, when the vegetation has not started yet to grow, estimated equation has the following form: $\sigma°$ VH = $-34.4 + 0.21*$SM, where R = 0.89; N = 34. In both simulated and estimated equations, the regression slope that means sensitivity, is the same. The intercept parameters which are connected with roughness of soil and vegetation cover, differ. This is the measure of the difference between the soil, theoretically bare, according to model (Equation (7)) and our assumption.

Model 1b:

$$\sigma° \text{ VV} = -21.5 + 0.19\tau^2 \text{ SM} + 12.3(1 - \tau^2) \cos(\theta) \text{ NDVI} \tag{8}$$

where: R = 0.91; $R^2 = 0.82$; $p < 0.0000$; N = 170; Std. Err. = 0.84 dB, for ascending orbit.

The partial correlation for the soil and vegetation components were 0.87 and 0.50 respectively, which means that soil moisture influenced $\sigma°$ VV more strongly than the vegetation cover. Figure 9 presents a comparison between the $\sigma°$ VV values observed (derived from satellite images) and predicted by Model 1b according to Equation (8).

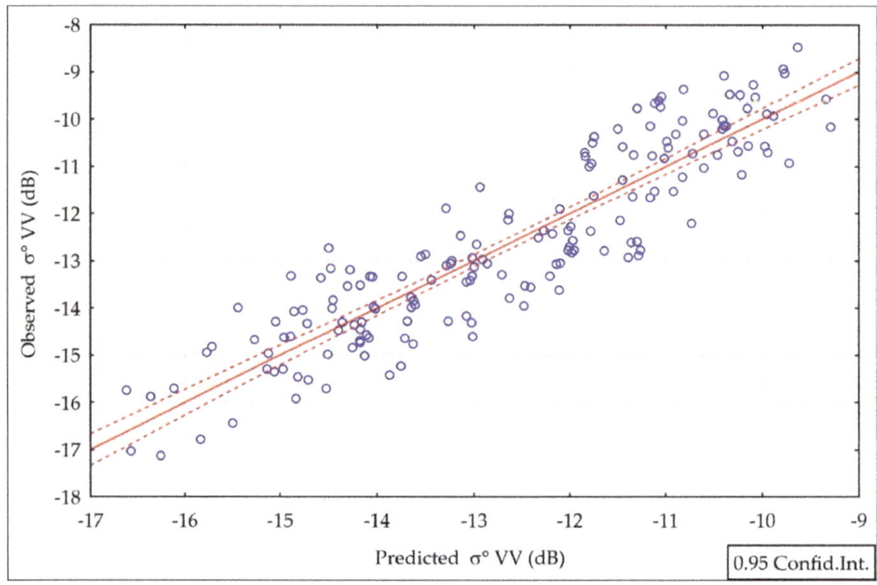

Figure 9. The $\sigma°$ VV values observed and predicted by Model 1b (Equation (8)).

The models 1a–1b present the influence of soil moisture and vegetation cover (expressed by NDVI from MODIS) on the S-1 backscatter. The standard errors of estimation for $\sigma°$ VH and $\sigma°$ VV were 0.79 dB and 0.84 dB, respectively.

Table 6 presents the mean absolute percentage errors (MAPE) of the $\sigma°$ S-1 ascending pass, assessed by Model 1a and Model 1b for the years 2015–2017 for the two sites and the two tracks separately. MAPE1 applies to Model 1a, and MAPE2 applies to Model 1b. The mean percentage error for $\sigma°$ VH estimation was 6.6%, and for $\sigma°$ VV estimation, it was 8.8% for all observations (not only the teaching set). The distribution of the error was well balanced on the sites and the tracks.

Table 6. Mean absolute percentage error (MAPE) errors of σ° VH and VV derived from Model 1a and Model 1b for the years 2015–2017.

Site	Track	MAPE1 [1] (%)	MAPE2 [2] (%)	Number of Observations
Grassland	131	5.7	8.7	62
	29	5.9	8.8	56
Marshland	131	7.6	8.8	45
	29	7.2	8.8	47
All		6.6	8.8	200

[1] Errors applies to Model 1a, [2] Errors applies to Model 1b.

Figures 10 and 11 present the simulation of σ° VH and σ° VV with the increase of the NDVI for various values of soil moisture from the range of 10–90 vol. %. The increase of σ° with the increase of the NDVI was significant with low soil moisture, the attenuation of the signal was small. When the soil moisture was high, the increase of the NDVI influences the decrease of σ°.

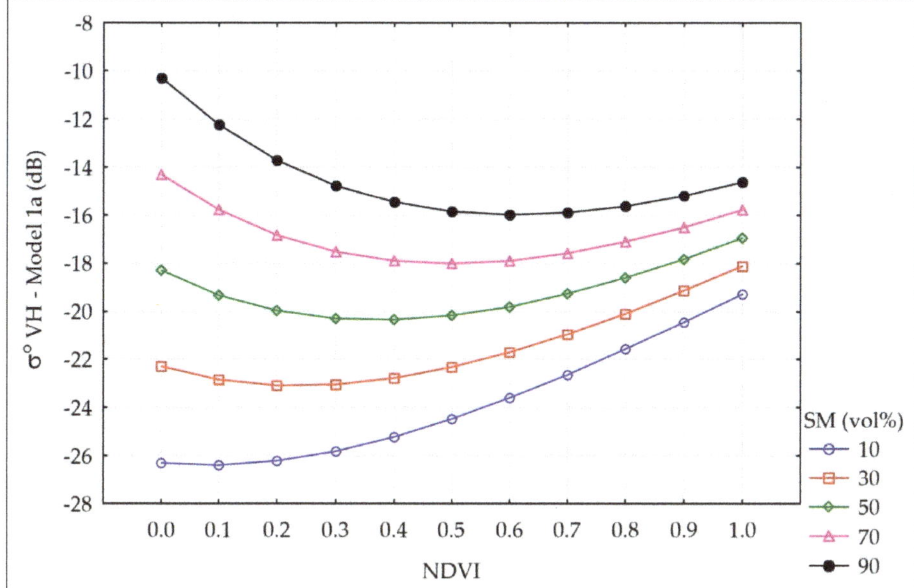

Figure 10. Impact of NDVI on σ° VH under various levels of soil moisture (SM) according to Model 1a.

Figures 10 and 11 present the soil and vegetation impact on σ° VV and σ° VH according to Models 1a–1b. The sensitivity of both polarizations on soil moisture under given vegetation condition (NDVI) was similar at wetland area (parameter b in Equations (7) and (8)). Taking the observed range of NDVI as 0.3–0.8, the sensitivity of σ° VH was calculated. For the satellite track 29 (θ = 43°10′) the obtained highest sensitivity was about 0.088 dB/vol. % and the lowest–0.022 dB/vol. %, while for the satellite track 131 (θ = 35°13′) − 0.095 dB/vol. % and 0.028 dB/vol. %, respectively.

The soil moisture can be retrieved through the inversion of Model 1a (Equation (7)) with an accuracy of 9.8 vol. % (Equation (9)). The errors were similar for two sites.

$$\text{SM} = (\sigma° \text{ VH} + 28.3 - 14.7 * (1 - \tau^2) * \cos(\theta) * \text{NDVI})/(0.2 * \tau^2) \tag{9}$$

Table 7 presents the RMSE errors (vol. %) for selected ranges of soil moisture values (5 cm depth) based on Model 1a. It was noted that for the high SM values (in the range of 80–100 vol %) errors were lower than those of the remaining SM ranges.

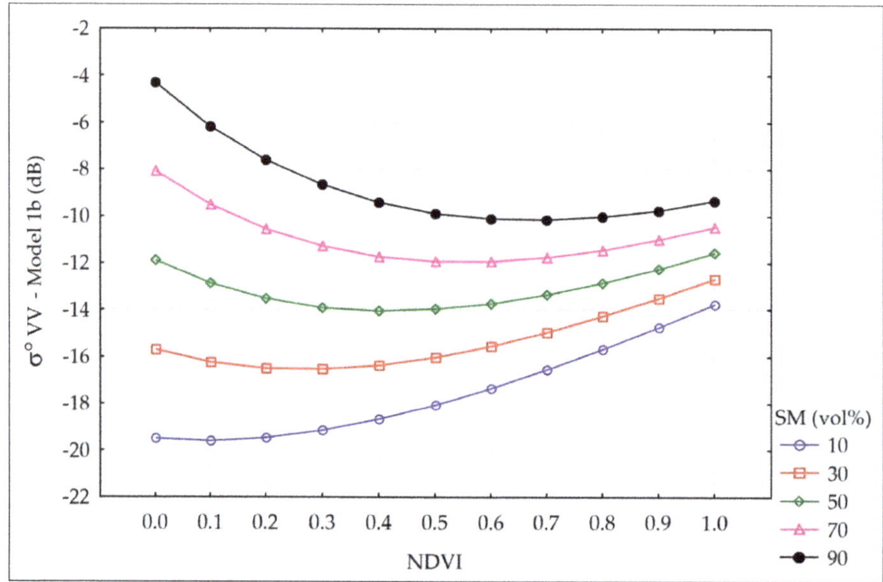

Figure 11. Impact of NDVI on σ° VV under various levels of soil moisture (SM) according to Model 1b.

Table 7. Errors of soil moisture retrieval by Model 1a (ascending pass) 2015–2017.

SM Range	N [1]	RMSE (vol. %)
20–40	39	11.8
40–60	39	9.5
60–80	35	9.9
80–100	34	8.4
All	147	9.8

[1] Number of observations.

Table 8 presents the RMSE errors (vol. %) for selected ranges of the NDVI values based on Model 1a. The RMSE errors were between 7.4–11.5 vol. %. It was clearly visible that the error was higher with denser vegetation cover (higher NDVI values).

Table 8. Errors of soil moisture retrieval by Model 1a for different densities of vegetation.

NDVI Range	N [1]	RMSE (vol. %)
0.2–0.4	24	7.4
0.4–0.6	40	8.5
0.6–0.8	47	10.4
0.8–0.9	36	11.5
All	147	9.8

[1] Number of observations.

3.6. Soil Moisture Retrieval Using the σ° Indices from Sentinel-1

Replacing vegetation index NDVI in Equation (4) by σ° VV/VH values we receive:

$$\tau^2 = \exp(-2(\sigma^\circ \ VV/VH)/\cos(\theta)) \tag{10}$$

where: σ° VV and VH had the only negative values in our study, and σ° VV/VH <1, and B was fixed to 1. The choice of B value was preceded by the same analysis as in the case of Models 1a–1b. Two components were designed to describe the effect of the underlying soil and vegetation on the σ° VH value: τ^2*SM and $(1 - \tau^2)$*cos (θ)*σ°(VH–VV)^2. Then, σ° VH was modeled according to Model 2 applying linearized nonlinear regression method.

Model 2:

$$\sigma^\circ \ VH = -18.9 + 0.33\tau^2 \ SM - 0.14(1 - \tau^2) \ \cos(\theta) \ \sigma^\circ(VH\text{-}VV)^2 \tag{11}$$

where: R = 0.91; R^2 = 0.82; p < 0.000; N = 252; Std. Err. = 0.70 dB, (Figure 12), for ascending orbits.

There is no redundancy of independent components in the multiple regression model. The correlation between them is R^2 = 0.002.

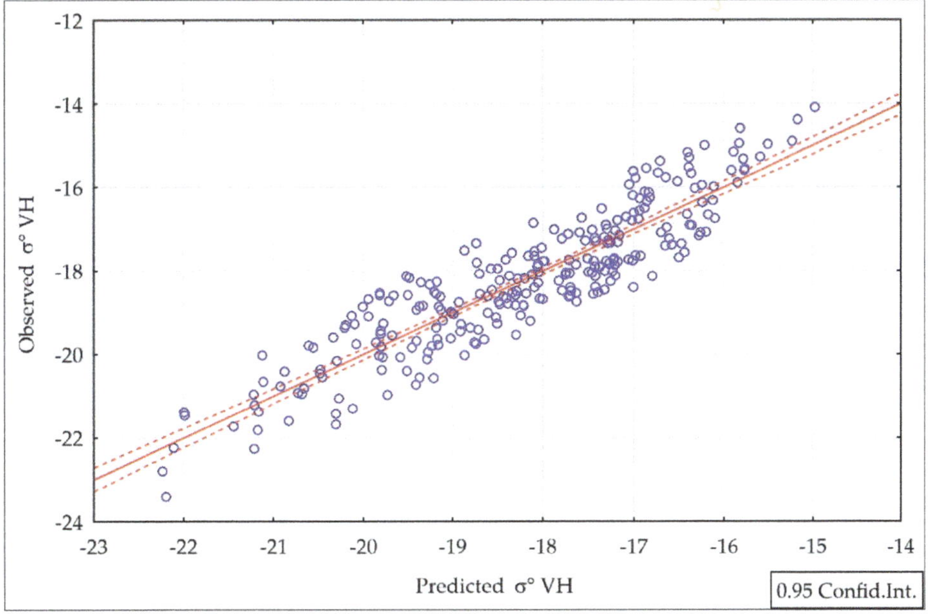

Figure 12. The σ° VH values observed and predicted by Model 2 (Equation (11)).

Three regression parameters could be interpreted as follows: c = 0.14 as vegetation parameter corresponding to A in Equation (2); b = 0.33 as sensitivity of SAR backscatter for τ^2 = 1; constant a = −18.9 is the state of balance between the impact of vegetation and the underlying soil on σ° VH (SM about 50 vol. %, Figure 13). Under σ° VV < 0 the attenuation factor τ^2 (Equation (10)) is always less than 1, so the sensitivity does not reaches the value of 0.33, it is lower. Theoretically, sensitivity of SAR backscatter to soil moisture increases when the ratio σ° VV/VH decreases. Figure 14 shows the periods under low vegetation conditions.

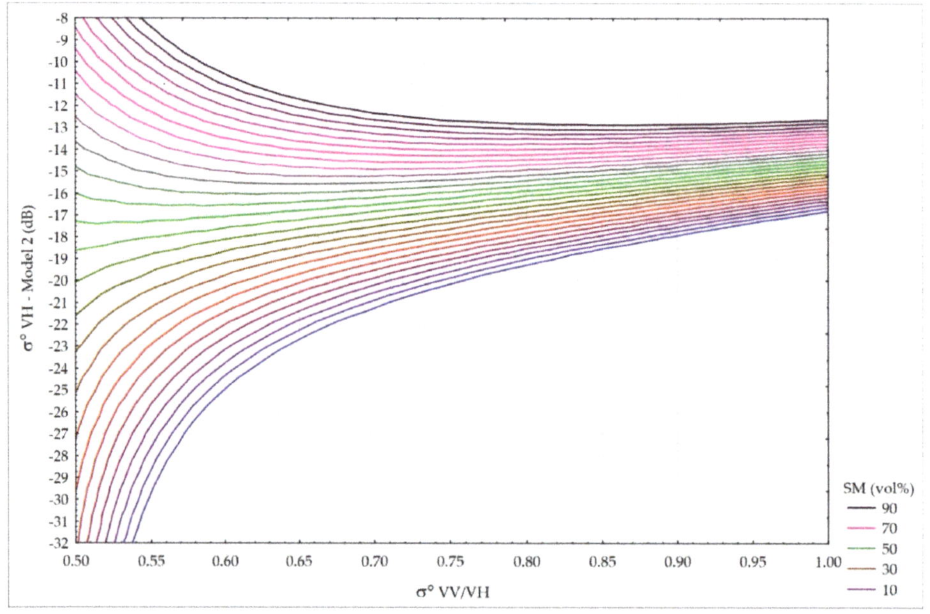

Figure 13. Impact of vegetation described by σ° VV/VH on σ° VH for different SM values according to Model 2.

Figure 13 presents the changes of σ° VH sensitivity during the vegetation development represented by σ° VV/VH. Taking the observed range of σ° VV/VH as 0.5–0.9, and τ^2 for each track separately, the range of sensitivity of σ° VH backscatter was calculated. For the satellite track 29 (θ = 43°10′), the highest sensitivity was 0.084 dB/vol. % and the lowest was 0.029 dB/vol. %, while for the satellite track 131 ((θ = 35°13′) − 0.096 dB/vol. % and 0.036 dB/vol. %, respectively. It is compatible with the results when the NDVI from optical data were used (Figures 10 and 11). For low SM there is the increase of σ° VH. For high values of SM, there is the attenuation of the beam by vegetation. Model 2 can be applied in all weather conditions, independently of sky conditions, on which the acquisition of optical images depends.

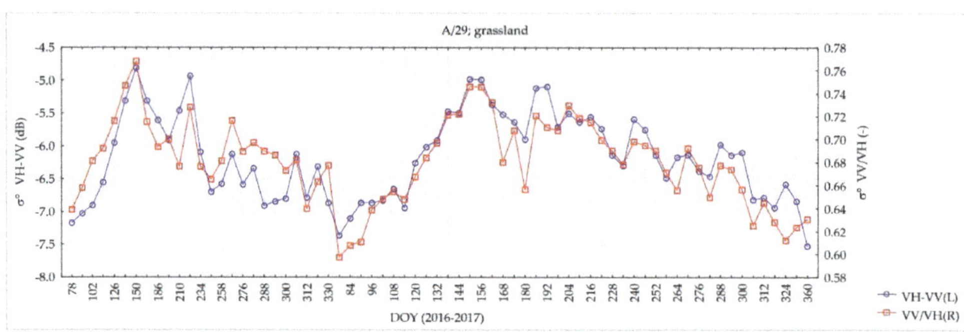

Figure 14. Time series of σ° VH-VV and σ° VV/VH during the years 2016–2017 for the grassland site.

From the inversion of Model 2 (Equation (11)), the soil moisture was calculated as follows:

$$SM = (\sigma° \text{ VH} + 18.9 - 0.14(1 - \tau^2) \cos(\theta)\, \sigma°(\text{VH}-\text{VV})^2/(0.33\, \tau^2) \tag{12}$$

The mean RMSE error of the soil moisture retrieved from Model 2 Equation (12) was 13 vol. % (Tables 9 and 10). Table 9 presents the RMSE errors from data for the whole year when the soil temperature is >278 °K. Table 10 presents the RMSE errors for the data from the vegetation season, i.e., from the DOY (Day Of the Year) 60–300.

Table 9. Errors analysis for different ranges of SM (5 cm depth) as retrieved by Model 2 (whole year).

SM Range (vol. %)	N [1]	RMSE (vol. %)
20–40	51	14.8
40–60	53	11.8
60–80	64	13.8
80–100	40	13.6
All	252	13.0

[1] Number of observations.

Table 10. Errors analysis for different ranges of SM (5 cm depth) as retrieved by Model 2 (growing season).

SM Range (vol. %)	N [1]	RMSE (vol. %)
20–40	60	12.1
40–60	70	12.3
60–80	60	14.7
80–100	62	11.8
All	208	13.5

[1] Number of observations.

The validation of Model 2 was performed for the S-1 data between September 2017–May 2018. The data from December–March were excluded, as the soil temperatures were lower than 278 °K. Table 11 presents the RMSE errors for the data used in the validation procedure.

Table 11. Errors analysis for different ranges of SM (5 cm depth) retrieved by Model 2 for validation data.

SM Range (vol. %)	N [1]	RMSE (vol. %)
40–60	14	11.0
60–80	29	8.5
80–100	35	15.2
All	76	12.6

[1] Number of observations.

For S-1 satellite track 29 ($\theta = 43°10'$), where the incident angle was higher than for track 131 ($\theta = 35°13'$), all of the models gave higher errors of soil moisture estimation. Table 12 presents the mean RMSE errors for both of these tracks separately.

Table 12. Errors of soil moisture estimation from developed models for two satellite tracks.

Orbit/Track	Model 1a	Model 1b	Model 2
	RMSE (vol. %)		
A/29	10.0	10.8	15.2
A/131	9.51	9.9	10.7

Figures 15 and 16 present a comparison between the soil moisture retrieved by the Model 2 inversion according to Equation (12), and the soil moisture measured at a 5 cm depth by the Decagons GS3 sensors at the grassland and marshland sites. As can be seen in the figures, high compatibility occurred between the SM values that were modeled and measured; however, it was higher for the marshland site. The lack of response of the Decagon probes to precipitation during the extreme drought in June to September of 2015 can be explained by the hydrophobic effect of the dry peat [39]. The time of reaction of the soil moisture and the retention of water in the soil to precipitation in peat soils is much slower than in mineral soils. After the precipitation that occurred in July and at the beginning of August 2017, the soil moisture has raised in the middle of August at the grassland site and at the end of August at the marshland site.

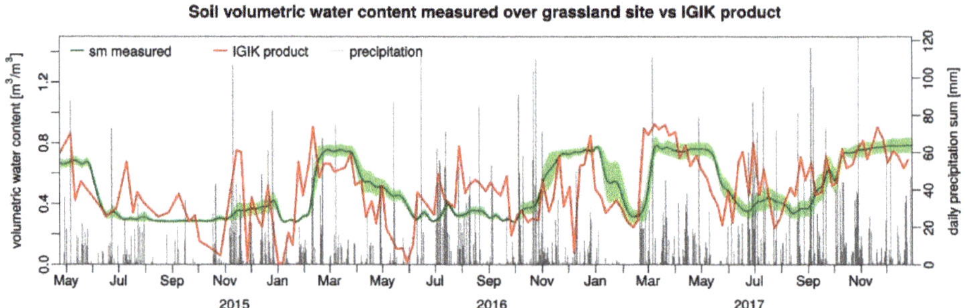

Figure 15. Comparison between the soil moisture retrieved by the inversion of Model 2 according to Equation (12) (IGiK (Institute of Geodesy and Cartography) product) and soil moisture measured at a 5 cm depth (sm) by the Decagons GS3 sensors at the grassland site.

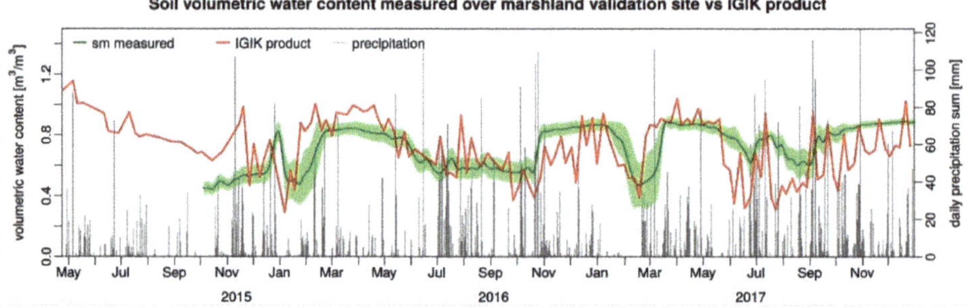

Figure 16. Comparison between the soil moisture retrieved by the inversion of Model 2 according to Equation (12) (IGiK (Institute of Geodesy and Cartography) product), and soil moisture measured at a 5 cm depth (sm) by the Decagons GS3 sensors at the marshland site.

The developed model reacts well on the increase of precipitation due to increase of soil moisture and vegetation moisture.

4. Discussion

Although previous studies have identified relationships between S-1 σ° and the surface soil moisture [16–19,23], this study, for the first time, to our knowledge, in the Biebrza Wetlands, demonstrates the relationships under an extreme range of SM conditions (from dry to wet) i.e., 27–90 vol. %, and different wetland vegetation biomasses (NDVI). The moisture ranges presented, and the diversity of the vegetation biomass, depicts the wetland ecosystems well. The developed models for soil moisture retrieval could be implemented into the system for monitoring areas of wetlands, and in developing decision support and early warning systems.

Two models have been developed based on σ° VH and VV, and the NDVI from MODIS. It is evident in Table 6 that for both sites (grassland and marshland) when considered together, the MAPE errors of σ° as modeled by Model 1a (Equation (7)) and Model 1b (Equation (8)) are comparable; however, for Model 1b, they are slightly higher. Generally, the inversion of the developed σ° models can retrieve the SM with a mean accuracy that is close to 10 vol. %, which is acceptable for the wetland ecosystem authorities and the decision makers. This is especially important for the wetlands areas that are not easily accessible.

The σ° indices as VH−VV and VV/VH, which could replace the vegetation cover as expressed by the NDVI values in soil moisture modeling, have been used to develop Model 2 (Equation (11)). Inversion of Model 2 allows the soil moisture to be retrieved by solely using Sentinel-1 data with a mean accuracy of 13 vol. % (Table 9). Although the accuracy of the soil moisture retrieval using Model 2 was slightly lower than applying Models 1a and 1b, it was still acceptable. Moreover, Model 2 required only microwave data, which is advantageous, especially in areas that are often cloudy.

5. Conclusions

The study has shown that the retrieval of soil moisture based on Sentinel-1 data, which considers wetland ecosystems, can be used effectively and with reasonable accuracy (below 10 vol. %). These developments are valuable for areas where in situ data are not available due to the inaccessibility of the area, and when only satellite data can provide suitable tools for decision makers.

The setup of two dense soil moisture measuring networks located over the wetlands offered unprecedented capabilities for modeling the soil moisture from the Sentinel-1 data. The data collected within the study corresponded to from extremely dry (2015) to extremely wet (2017) conditions, which is favorable for the development and validation of soil moisture retrieval models over the wetlands. Also, the selected grassland and marshland sites feature different soil moisture regimes.

Vegetation has to be considered in the relationship between the backscatter and the soil moisture. The vegetation contribution could be expressed by NDVI, or by VV/VH and VH−VV indices that are calculated from the S-1 data.

It has been noted that there is a different contribution of vegetation that is represented by the NDVI when there are dry conditions (SM < 30 vol. %) and moist conditions (SM > 60 vol. %). It was noticed that the values from 50–60 vol. % of soil moisture are within the threshold for the SM influence on σ° VH and VV.

There are discrepancies between Sentinel-1A and Sentinel-1B data. Ascending orbits are better for soil moisture retrieval because the descending overpasses occur during the night when there is dew. The most significant correlation coefficients between the S-1 backscatter and the soil moisture were found for the ascending tracks and for 5 cm depths. A validation was performed for the period of September 2017 until May 2018. The average error was close to 12.6%. It has to be emphasized that the range of the soil moisture in the wetlands was high, at 27–90 vol. %. Such a moisture extent does not occur in agriculture sites. This could also affect the range of the error.

Developed models could be applied for cloudy conditions for sites other than the European Wetlands.

Further work is needed, especially when HH polarization of S-1 is available, to predict the moisture status in wetland ecosystems. The time of reaction of soil moisture and retention of water on precipitation in peat soil was much slower than the reaction to precipitation of other soils. That is why it will be good to examine the time of reaction of SM to precipitation in peat soil.

Author Contributions: K.D.-Z. conceived and designed the experiments and wrote the paper; J.M. performed the experiments established the soil moisture network over the B.W. and processed microwave satellite data; A.M. performed statistical analyses; M.B. analyzed the data and contributed in writing the paper; R.G. performed the in situ experiments; W.K. processed and analyzed the optical satellite data; M.B. established the in situ measurements; P.G. performed the database analysis.

Funding: The study was funded by ESA EXPRO, contract no. 4000112578/14/NL/MP "Biebrza Sentinel-1 Supersite", and the National project Narodowe Centrum Nauki 2016/23/B/ST10/03155.

Conflicts of Interest: The authors declare no conflict of interest.

References

1. Available online: https://www.biebrza.org.pl/lang,2 (accessed on 6 December 2018).
2. Turbiak, J.; Miatkowski, Z. Assessment of organic mass mineralization rate in deeply drained peat-muck soil based on losses of soil mass and CO_2 emission. *ITP Water Environ. Rural Areas* **2016**, *16*, 73–85.
3. Huang, W.; Hall, S.J. Elevated moisture stimulates carbon loss from mineral soils by releasing protected organic matter. *Nat. Commun.* **2017**, *8*, 1774. [CrossRef] [PubMed]
4. Dembek, W.; Oswit, J.; Rycharski, M. Torfowiska i torfy w Pradolinie Biebrzy. In *Przyroda Biebrzanskiego Parku Narodowego*; Dyrcz, A., Werpachowski, C., Eds.; Biebrzanski Park Narodowy: Osowiec-Twierdza, Poland, 2005; pp. 33–58.
5. Kornelsen, K.C.; Coulibaly, P. Advances in soil moisture retrieval from synthetic aperture radar and hydrological applications. *J. Hydrol.* **2013**, *476*, 460–489. [CrossRef]
6. Allen, C.T.; Ulaby, F.T. Modelling the Polarization Dependence of the Attenuation in Vegetation Canopies. In Proceedings of the Geoscience and Remote Sensing Symposium (IGARSS'84), Strasbourg, France, 27–30 August 1984; pp. 119–124.
7. Le Toan, T.; Lopes, A.; Huet, M. On the relationships between radar backscattering Coefficient and Vegetation Canopy Characteristics. In Proceedings of the Geoscience and Remote Sensing Symposium (IGARSS'84), Strasbourg, France, 27–30 August 1984; pp. 155–160.
8. Attema, E.P.; Ulaby, F.T. Vegetation modeled as a water cloud. *Radio Sci.* **1978**, *13*, 357–364. [CrossRef]
9. Dabrowska–Zielinska, K.; Inoue, Y.; Kowalik, W.; Gruszczynska, M. Inferring the effect of plant and soil variables on C- and L-band SAR backscatter over agricultural fields, based on model analysis. *Adv. Space Res.* **2007**, *39*, 139–148. [CrossRef]
10. Mattia, F.; Satalino, G.; Dente, L.; Pasquariello, G. Using a priori information to improve soil moisture retrieval from ENVISAT ASAR AP data in semiarid regions. *IEEE Trans. Geosci. Remote Sens.* **2006**, *44*, 900–912. [CrossRef]
11. Balenzano, A.; Satalino, G.; Pauvels, V.; Mattia, F. Soil moisture retrieval from dense temporal series of C-band SAR data over agricultural sites. In Proceedings of the 2011 IEEE International Geoscience and Remote Sensing Symposium (IGARSS), Vancouver, BC, Canada, 24–29 July 2011. [CrossRef]
12. Baghdadi, N.; El Hajj, M.; Zribi, M.; Fayad, I. Coupling SAR C-band and optical data for soil moisture and leaf area index retrieval over irrigated grasslands. *IEEE J. Sel. Top. Appl. Earth Obs. Remote Sens.* **2015**, *9*, 1–15. [CrossRef]
13. Baghdadi, N.; Choker, M.; Zribi, M.; El Hajj, M.; Paloscia, S.; Verhoest, N.; Lievens, H.; Baup, F.; Mattia, F. A new empirical model for Radar scattering from bare surfaces. *Remote Sens.* **2016**, *8*, 920. [CrossRef]
14. Choker, M.; Baghdadi, N.; Zribi, M.; El Hajj, M.; Paloscia, S.; Verhoest, N.; Lievens, H.; Mattia, F. Evaluation of the Oh, Dubois and IEM models using large dataset of SAR signal and experimental soil measurements. *Water* **2017**, *9*, 38. [CrossRef]
15. Pasolli, L.; Notarnicola, C.; Bertoldi, G.; Bruzzone, L.; Remelgado, R.; Greifeneder, F.; Niedrist, G.; Della Chiesa, S.; Tappeiner, U.; Zebisch, M. Estimation of soil moisture in mountain areas using SVR technique applied to multiscale active radar images at c-band. *IEEE J. Sel. Top. Appl. Earth Obs. Remote Sens.* **2015**, *8*, 262–283. [CrossRef]
16. Mattia, F.; Satalino, G.; Balenzano, A.; Pauwels, V.; DeLathauwer, E. *GMES Sentinel-1 Soil Moisture Retrieval Algorithm Development*; European Space Agency contract n. 4000101352/10/NL/MP/ef; Final Report; European Space Agency: Paris, France, 2011.
17. Balenzano, A.; Mattia, F.; Satalino, G.; Pauvels, V.; Snoeij, P. SMOSAR algorithm for soil moisture retrieval using Sentinel-1 data. In Proceedings of the IEEE International Geoscience and Remote Sensing Symposium (IGARSS), Munich, Germany, 22–27 July 2012. [CrossRef]
18. Paloscia, S.; Pettinanto, S.; Santi, E.; Notarnicola, C.; Pasolli, L.; Reppucci, A. Soil moisture mapping using Sentinel-1 images: Algorithm and preliminary validation. *Remote Sens. Environ.* **2013**, *134*, 234–248. [CrossRef]
19. Santi, E.; Paloscia, S.; Pettinato, S.; Fontanelli, G. Application of artificial neural networks for the soil moisture retrieval from active and passive microwave spaceborne sensors. *Int. J. Appl. Earth Obs. Geoinf.* **2016**, *48*, 61–73. [CrossRef]

20. Kasischke, E.; Bourgeau-Chavez, L.; Rober, A.; Wyatt, K.; Waddington, J.; Turetsky, M. Effects of soil moisture and water depth on ERS SAR backscatter measurements from an Alaskan wetland complex. *Remote Sens. Environ.* **2009**, *113*, 1869–1873. [CrossRef]

21. Lang, M.W.; Kasischke, E.S.; Prince, S.D.; Pittman, K.W. Assessment of C-band synthetic aperture radar data for mapping and monitoring Coastal Plain forested wetlands in the Mid-Atlantic Region, U.S.A. *Remote Sens. Environ.* **2008**, *112*, 4120–4130. [CrossRef]

22. Gao, Q.; Zribi, M.; Escorihuela, M.J.; Baghdadi, N. Synergetic Use of Sentinel-1 and Sentinel-2 Data for Soil Moisture Mapping at 100 m Resolution. *Sensors* **2017**, *17*, 1966. [CrossRef] [PubMed]

23. El Hajj, M.; Baghdadi, N.; Zribi, M.; Bazzi, H. Synergic Use of Sentinel-1 and Sentinel-2 Images for Operational Soil Moisture Mapping at High Spatial Resolution over Agricultural Areas. *Remote Sens.* **2017**, *9*, 1292. [CrossRef]

24. Baghdadi, N.; El Hajj, M.; Zribi, M.; Bousbih, S. Calibration of the Water Cloud Model at C-Band for Winter Crop Fields and Grasslands. *Remote Sens.* **2017**, *9*, 969. [CrossRef]

25. Tomer, S.K.; Al Bitar, A.; Sekhar, M.; Zribi, M.; Bandyopadhyay, S.; Sreelash, K.; Sharma, A.K.; Corgne, S.; Kerr, Y. Retrieval and Multi-scale Validation of Soil Moisture from Multi-temporal SAR Data in a Semi-Arid Tropical Region. *Remote Sens.* **2015**, *7*, 8128–8153. [CrossRef]

26. Dabrowska-Zielinska, K.; Budzynska, M.; Tomaszewska, M.; Malinska, A.; Gatkowska, M.; Bartold, M.; Malek, I. Assessment of Carbon Flux and Soil Moisture in Wetlands Applying Sentinel-1 Data. *Remote Sens.* **2016**, *8*, 756. [CrossRef]

27. Vreugdenhil, M.; Wagner, W.; Bauer-Marschallinger, B.; Pfeil, I.; Teubner, I.; Rüdiger, C.; Strauss, P. Sensitivity of Sentinel-1 Backscatter to Vegetation Dynamics: An Austrian Case Study. *Remote Sens.* **2018**, *10*, 1396. [CrossRef]

28. Greifeneder, F.; Notarnicola, C.; Hahn, S.; Vreugdenhil, M.; Reimer, C.; Santi, E.; Paloscia, S.; Wagner, W. The Added Value of the VH/VV Polarization-Ratio for Global Soil Moisture Estimations from Scatterometer Data. *IEEE J. Sel. Top. Appl. Earth Obs. Remote Sens.* **2018**, *11*, 3668–3679. [CrossRef]

29. Hosseini, M.; McNairn, H.; Merzouki, A.; Pacheco, A. Estimation of Leaf Area Index (LAI) in corn and soybeans using multi-polarization C- and L-band radar data. *Remote Sens. Environ.* **2015**, *170*, 77–89. [CrossRef]

30. Hosseini, M.; McNairn, H. Using multi-polarization C- and L-band synthetic aperture radar to estimate biomass and soil moisture of wheat fields. *Int. J. Appl. Earth Obs. Geoinf.* **2017**, *58*, 50–64. [CrossRef]

31. Available online: http://www.biebrza.org.pl/index.php (accessed on 6 December 2018).

32. Silakowski, M. Charakterystyka fizyczno-geograficzna-Klimat. In *Biebrzanski Park Narodowy*; Sienko, A., Grygoruk, A., Eds.; Biebrzanski Park Narodowy: Osowiec-Twierdza, Poland, 2003; pp. 17–22.

33. Musial, J.P.; Dabrowska-Zielinska, K.; Kiryla, W.; Oleszczuk, R.; Gnatowski, T.; Jaszczynski, J. Derivation and validation of the high resolution satellite soil moisture products: A case study of the Biebrza Sentinel-1 validation sites. *Geoinf. Issues* **2016**, *8*, 37–53.

34. Dorigo, W.; Wagner, W.; Hohensinn, R.; Hahn, S.; Paulik, C.; Xaver, A.; Gruber, A.; Drusch, M.; Mecklenburg, S.; Oevelen, P.V.; et al. The International Soil Moisture Network: A data hosting facility for global in situ soil moisture measurements. *Hydrol. Earth Syst. Sci.* **2011**, *15*, 1675–1698. [CrossRef]

35. Vermote, E. MOD09Q1 MODIS/Terra Surface Reflectance 8-Day L3 Global 250m SIN Grid V006 [Data set]. *NASA EOSDIS LP DAAC.* **2015**. [CrossRef]

36. Prevot, L.; Champion, I.; Guyot, G. Estimating surface soil moisture and leaf area index of a wheat canopy using a dual—Frequency (C and X bands) scatterometer. *Remote Sens. Environ.* **1993**, *46*, 331–339. [CrossRef]

37. Rao, S.; Sahadevan, D.; Wadodkar, M.; Nagaraju, M.; Chattaraj, S.; Joseph, W.; Rajankar, P.; Sengupta, T.; Venugopalan, M.; Das, S.; et al. Soil Moisture Model with Multi Angle and Multi Polarisation Risat-1 Data. *ISPRS Ann. Photogramm. Remote Sens. Spat. Inf. Sci.* **2014**, *2*, 145. [CrossRef]

38. Narvekar, P.S.; Entekhabi, D.; Seung-Bum, K.; Njoku, E.G. Soil Moisture retrieval using L-band radar observations. *IEEE J. Trans. Geosc. Rem. Sens.* **2015**, *53*, 6. [CrossRef]

39. Available online: http://iopscience.iop.org/article/10.1088/1755-1315/105/1/012083/pdf (accessed on 6 December 2018).

 *remote sensing*

Article

Exploiting Time Series of Sentinel-1 and Sentinel-2 Imagery to Detect Meadow Phenology in Mountain Regions

Laura Stendardi [1,*], Stein Rune Karlsen [2], Georg Niedrist [3], Renato Gerdol [4], Marc Zebisch [5], Mattia Rossi [5] and Claudia Notarnicola [5]

[1] Faculty of Science and Technology, Free University of Bozen-Bolzano, 39100 Bolzano, Italy
[2] NORCE Norwegian Research Centre AS, Measurement Science, P.O. Box 6434, N-9294 Tromsø, Norway; skar@norceresearch.no
[3] EuracResearch, Institute for Alpine Environment, 39100 Bolzano, Italy; georg.niedrist@eurac.edu
[4] Department of Life Sciences and Biotechnology, University of Ferrara, I-44121 Ferrara, Italy; grn@unife.it
[5] EuracResearch, Institute for Earth Observation, 39100 Bolzano, Italy; marc.zebisch@eurac.edu (M.Z.); mattia.rossi@eurac.edu (M.R.); claudia.notarnicola@eurac.edu (C.N.)
* Correspondence: Laura.Stendardi@natec.unibz.it

Received: 1 February 2019; Accepted: 26 February 2019; Published: 6 March 2019

Abstract: A synergic integration of Synthetic Aperture Radar (SAR) and optical time series offers an unprecedented opportunity in vegetation phenology monitoring for mountain agriculture management. In this paper, we performed a correlation analysis of radar signal to vegetation and soil conditions by using a time series of Sentinel-1 C-band dual-polarized (VV and VH) SAR images acquired in the South Tyrol region (Italy) from October 2014 to September 2016. Together with Sentinel-1 images, we exploited corresponding Sentinel-2 images and ground measurements. Results show that Sentinel-1 cross-polarized VH backscattering coefficients have a strong vegetation contribution and are well correlated with the Normalized Difference Vegetation Index (NDVI) values retrieved from optical sensors, thus allowing the extraction of meadow phenological phases. Particularly for the Start Of Season (SOS) at low altitudes, the mean difference in days between Sentinel-1 and ground sensors is compatible with the acquisition time of the SAR sensor. However, the results show a decrease in accuracy with increasing altitude. The same trend is observed for senescence. The main outcomes of our investigations in terms of inter-satellite comparison show that Sentinel-1 is less effective than Sentinel-2 in detecting the SOS. At the same time, Sentinel-1 is as robust as Sentinel-2 in defining mowing events. Our study shows that SAR-Optical data integration is a promising approach for phenology detection in mountain regions.

Keywords: Sentinel-1 and Sentinel-2; time series analysis; start of season, harvest, mountain region

1. Introduction

Agricultural management in European mountain regions is a key strategy for preserving ecosystem stability and regional economies [1,2]. Phenology is defined as "the study of the timing of recurring biological events, the causes of their timing regarding biotic and abiotic forces, and the interrelation among phases of the same or different species" [3].Vegetation phenology is a relevant indicator of crop productivity and health. Phenological stage monitoring is therefore crucial in the decision-making process of the agricultural management [4]. In mountainous regions, agricultural areas are generally of small size and the vegetation is characterized by a heterogeneous distribution. In addition, mountain crops are vulnerable to climate variability [5–7]. Satellite imagery plays a unique and important role in monitoring crop and soil conditions for farm management [8–10]. In the past years, most studies using satellite imagery for crop and natural vegetation monitoring have focused on the use of optical imagery. By exploiting the reflectance of visible and Near Infra-Red (NIR) radiation and the emittance of thermal Infra-Red (IR) radiation, canopy characteristics have been mapped over large areas [11]. The Normalized Difference Vegetation Index (NDVI) [12] has been widely used to detect phenological phases [13–18]. In this case, cloud contamination and topographic effects in mountain regions compromise data significantly in the optical domain [19].

Microwave wavelengths have important advantages over optical remote sensing for agricultural applications, because they pass through the atmosphere and clouds with negligible attenuation [20]. This allows frequent measurements over the short growing season of mountain crops. Conversely, the radar signal can be difficult to interpret as the total radar backscatter is a complex sum of the backscatter from vegetation and soil. The radar beam can penetrate both the canopy and soil to a difficult-to-determine depth, making it complicated to determine if the signal is dominated by either vegetation or soil conditions [21,22].

Reliable ground measurements of crop growth stages and soil moisture throughout the growing season are therefore important to understand the relative influence of these factors on the microwave signal. Additionally, dense time series are necessary to understand the Synthetic Aperture Radar (SAR) signal behavior with regards to crops. From the first attempt to monitor rice crops [23], relevant results were found by combining different SAR sensors [24], incidence angles [25], different polarizations [26–31], and Interferometric SAR technique [32]. A data fusion approach was developed using a dynamical framework based on particle filter (PF). This approach has shown that the incorporation of additional sources to the NDVI time series can improve the phenological monitoring. The inclusion of SAR images in particular increases the sensitivity to crop dynamical development and improves results in the process of estimating specific phenological states [33]. Furthermore, the grassland mowing event detection were explored by applying coherence estimation on interferometric acquisitions [34] and using radar polarimetry [35,36]. Crop structure, dielectric properties of the canopy, soil roughness, and moisture influence the backscattering coefficients. Moreover, the crop structure and plant water content vary depending on phenological stages and crop condition. With multipolarization, it is possible to explore the sensitivity of waves to different orientation, shape and dielectric properties of elements in the scattering field [37]. Both the HH and VV polarizations operating in C-band are sensitive to soil moisture variations, whereas the cross-polarized backscatter is primarily associated with volume scattering of vegetation [38]. The different attenuation of VV and HH polarization is useful for discriminating crop types and the cross-polarized channel with a higher dynamic can improve the crop separability. Moreover, grassland and crop discrimination is achievable by using multitemporal SAR images [39]. Also, for phenology and its parameters, the cross-polarized channel gives a higher contrast between high and low productivities [38,40]. The trends in radar backscatter, measured on different dates, can be correlated with soil moisture content, since the effects of spatial roughness variations are smoothed [41]. To reduce these factors, [42] suggested that the ratio of backscatter measured on two close successive dates might be a simple and effective way to decouple the effect of vegetation and surface roughness from the effect of soil moisture changes, when volumetric scattering by the crop canopy is not dominant.

For robust retrieval methods, the temporal change of backscattering coefficients on mountain ecosystems still needs to be documented. Moreover, an integration of multisensor time series has to be evaluated on meadow phenology detection.

Within the Copernicus programme we now have the possibility to explore different sensors. Both Sentinel-1A and 1B satellites with their SAR sensors provide time series of medium and high resolution of C-band data [43], simultaneously Sentinel-2A and 2B optical sensors acquire 13 spectral bands in the visible, the NIR, and the Short Wave IR (SWIR) [44]. Combining the two Sentinel satellites with a revisiting time of six and five days, respectively, offers an unprecedented opportunity to monitor crop in mountain regions.

In this study, we analyzed time series from the Sentinel-1 (S-1) and Sentinel-2 (S-2) together with proximal sensors to understand their temporal behavior for mountain meadow areas. The main objectives of this paper are: (1) to understand and quantify the impact on multi temporal SAR images of different grassland types and soil conditions in the perspective of data integration; (2) exploit the synergic use of SAR and optical data to retrieve maps of mountain phenology (start of the season and harvesting time).

With respect to the above presented studies, the novelties of this work are:

1. Detection of phenological stages of meadows in mountain ecosystem using multitemporal SAR imagery;
2. Mapping of phenology with SAR data using a statistical approach.

For the first time, S-1 and S-2 are evaluated in synergy in the phenological retrieval process. A multisensor methodology is presented and compared to establish a common and complementary approach for the detection of mountain phenology.

2. Study Area and Datasets

2.1. Study Area

The study area is the South Tyrol region located in northern Italy (Figure 1). South Tyrol has an area of 7400 km^2 and is situated in the center of the Alps with steep elevation gradients stretching from 190–3890 m a.s.l. Typical agricultural land-use types are meadows, pastures, orchards, and vineyards. Around 79% of the region is above 1200 m a.s.l., with small valley floor surfaces and steep slopes. Moreover, around 50% (3228 km^2) of South Tyrol is covered by forest and about 30% is used for agriculture [45,46].

Within the MONALISA project (http://www.monalisa-project.eu/en/home/Pages/default.aspx) several environmental stations were installed in the area with the main aim to monitor vegetation and soil properties. Figure 1 shows the land-cover types, the elevation of the area and the stations used in our study as ground reference. Each name of the station includes information about location, vegetation cover type, and slope of the area. Moreover, the names include the altitude of the stations. For a comprehensive description of the acronyms see Table 1. The list of the stations can be viewed at the following website: http://monalisasos.eurac.edu/sos/static/client/helgoland/index.html#/map.

Table 1. Acronyms of ground stations used in this study.

Ground Station	Acronym
domef 1500	do = Dolomites, me = meadow, f= flat, 1500 = 1500 m a.s.l.
domef 2000	do = Dolomites, me = meadow, f= flat, 2000 = 2000 m a.s.l.
vimef 2000	vi = Vinschgau/Venosta valley, me = meadow, f= flat, 2000 = 2000 m a.s.l.
vimes 1500	vi = Vinschgau/Venosta valley, me = meadow, s= steep, 2000 = 2000 m a.s.l.

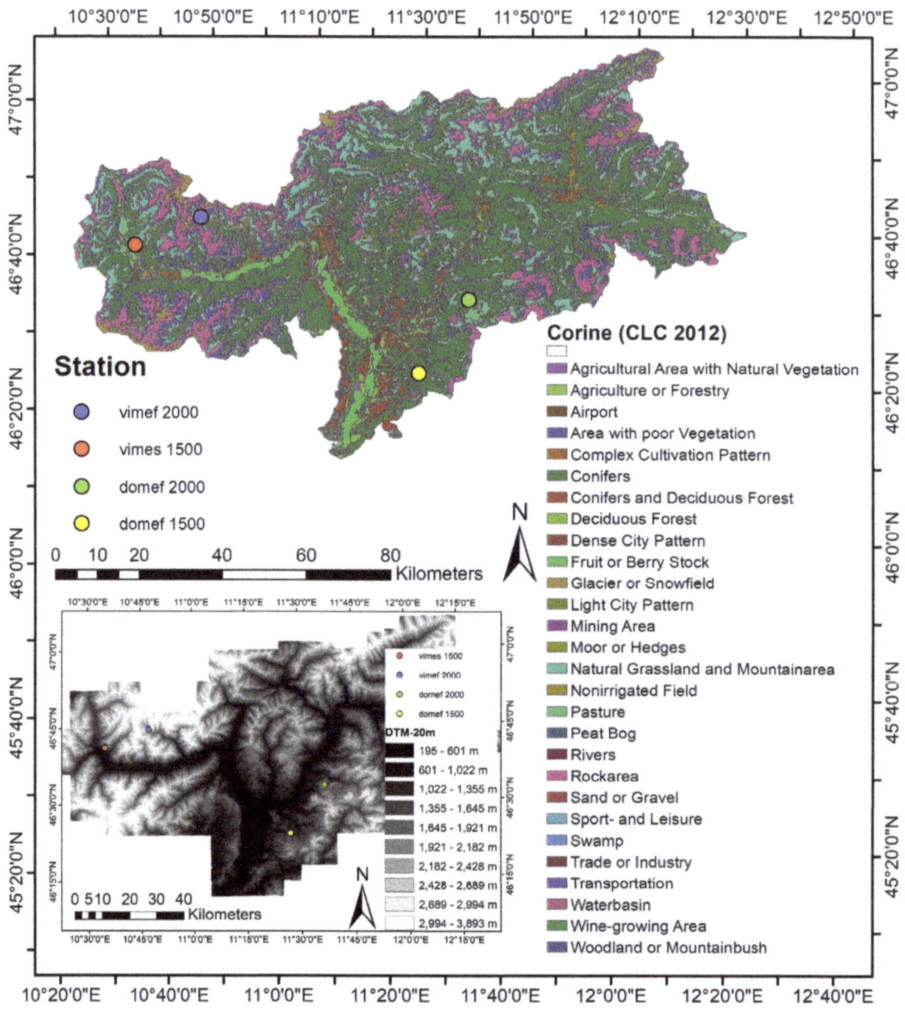

Figure 1. Corine Land-cover map (CLC 2012) of South Tyrol and Digital Terrain Model (DTM 20 m). On each map we overlaid four ground stations.

2.2. Datasets

The analyzed data sets are composed of 59 S-1A images acquired from October 2014 to September 2016. The data belongs to track 168, which covers the region almost entirely, every 12 days. The 78 S-2A images span from June 2015 to November 2016, with a temporal coverage of 10 days. Although the number of S-2 images is higher than S-1, the presence of clouds reduced the data availability in our time series. Clouds generated missing data in the optical images especially during the winter period and at high altitudes. Out of 78 S-2 scenes we were able to use an average of 39 scenes in the areas at 1500 m and 31 images in the areas at 2000 m a.s.l. On the contrary, in the microwave domain the images were consistent during the seasons, except for the missed acquisitions caused by onboard satellite problems. S-1A instrument was unavailable during 2016, between 8 June and 14 July [47]. Together with satellites images, ground data were available for meadow and pasture areas. Most of the stations are equipped with sensors providing information on Soil Water Content (SWC), Soil Temperature (ST), and NDVI-PRI from the Spectral Reflectance Sensor (SRS, Decagon Devices Inc., Pullman, WA, USA). The data have been recorded since 2015 with a time step of 15 min. SWC is available at different depths from 2 to 20 cm (Soil Water Content Reflectometer, Campbell Scientific, Edmonton, AB, Canada). A few of these are further equipped with PhenoCams acquiring both an RGB and a combined RGB + IR image in 30-min intervals with 1296 × 960-pixel resolution. The cameras used were StarDot Hybrid IP 1.3 Megapixel Netcams (StarDot Technology, Buena Park, CA, USA) mounted atop the MONALISA stations in 2015.

In this study, we selected four representative stations, fully equipped with SWC, ST at 2, 5, 20 cm, NDVI & PRI and PhenoCam and located at 1500 and 2000 m a.s.l. For the detailed description of the four selected ground stations see Table 2.

Table 2. List of the stations and related measurements available: SWC (Soil Water Content), ST (SoiL Temperature), PAR (Photosynthetic Active Radiation), NDVI (Normalized Difference Vegetation Index), PRI (Photochemical Reflectance Index).

Ground Station	Parameters	Latitude	Longitude	Altitude
domef 1500	SWC & ST 2, 5, 20 cm, PAR, NDVI & PRI, PhenoCam	46.401002	11.454211	1500 m a.s.l.
domef 2000	SWC & ST 2, 5, 20 cm, PAR, NDVI & PRI, PhenoCam	46.556687	11.614836	2000 m a.s.l
vimef 2000	SWC & ST 2, 5, 20 cm, PAR, NDVI & PRI, PhenoCam	46.745151	10.788845	2000 m a.s.l
vimes 1500	SWC & ST 2, 5, 20 cm, NDVI & PRI, PhenoCam	46.686163	10.579881	1500 m a.s.l

3. Methodology

The overall scheme of the proposed methodology is illustrated in Figure 2. The central aim of the procedure is to derive from time series of S-1 and S-2 images the main phenological features. The whole procedure is divided into four main steps. After a preprocessing of the S-1 and S-2 images, they are co-registered as to refer to the same area of interest. Then the S-1 and S-2 time series are extracted over the selected areas where also ground data are available. Subsequently, the backscatter from S-1 and the NDVI from S-2 and ground sensors are modeled to extract the main features of the phenocycle such as start of the season and mowing event. Finally, maps are produced, and the validation is carried out. In the following, each step of the procedure is described in detail.

3.1. Preprocessing of S-1, S-2 Images and Ground Observation

The S-1 data preprocessing encompasses several standard steps to derive geocoded intensity images starting from the Ground Range Detected (GRD) data.

These operations were performed using the tools provided by SNAP (Sentinel Application Platform) and custom algorithms developed in Python. Beside the standard operations, a spatial and temporal speckle filter was used [38]. The S-2 images were preprocessed using the Sen2Cor processor (v.2.3) without cirrus or topographic correction. All non-vegetated areas were masked using the CORINE 2012 (CLC 2012) land-cover information. Both S-1 and S-2 data were corrected to eliminate layover/shadow zones and to reduce the contamination due to cloud presence based on the Sen2Cor scene classification, respectively. The values recorded by the ground sensors were averaged according to intervals of time, averaging four values each interval. The SRS sensor has wavebands centered at 650 nm (Red) and 810 nm (NIR) [48]. To calculate the NDVI, we used the formula [12]:

$$NDVI = \frac{NIR - Red}{NIR + Red} \tag{1}$$

Subsequently, we calculated average of measurements around 10:00 a.m. (Sentinel-2 acquisition time, UTC time zone). To define the start, maximum and the end of the growing season at the four stations we did a visual analysis of the images from the PhenoCams. We used a common protocol as in [49], defining the Start Of the growing Season (SOS) as 50% green leaves developed, time of maximum (MAX) as full-size leaves and the End Of the growing Season (EOS) as 50% yellow leaves. In addition, we recorded the harvesting time.

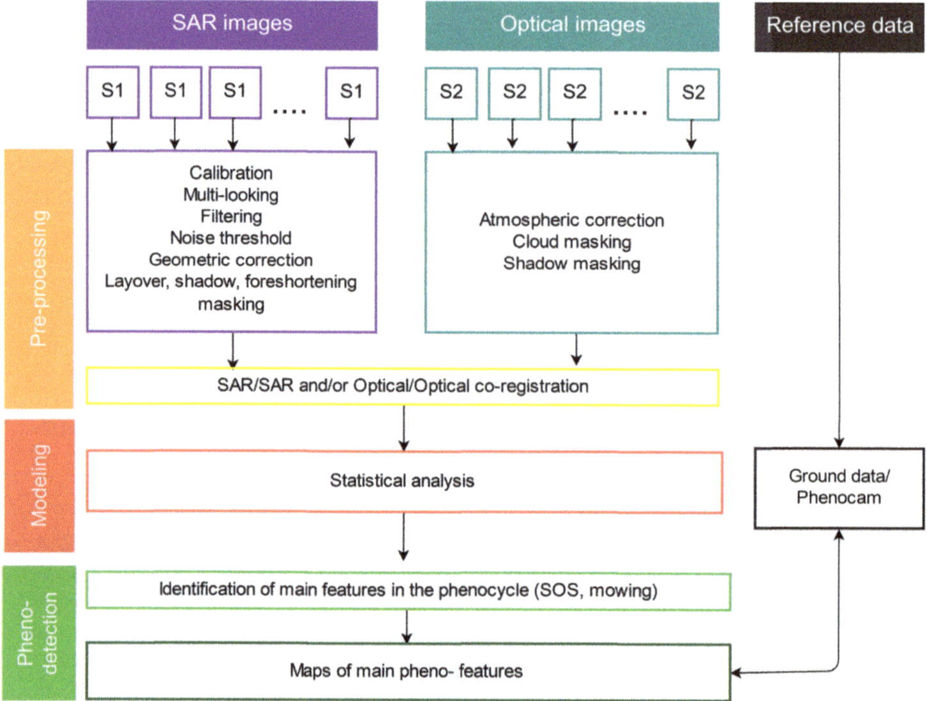

Figure 2. Flow chart of the proposed approach to detect the main features of the phenocycle for the selected areas.

3.2. Statistical and Electromagnetic Modeling

The correlation and modeling analyses are a preparatory phase to detect the phenocycle features from S-1 time series. The aim is to understand the temporal behavior of S-1 imagery and correlate this trend to soil and vegetation parameters. Even though the current study focuses on meadows phenology, as a first step, S-1 backscattering coefficients in VV and VH polarization were extracted over different crops and land-cover types to understand the radar signal dynamics to different vegetation types. A stack of images has been created and, considering a land-cover map of the area (CORINE Land Cover 2012), several regions of interest (around 10 for each land-cover type) were extracted for the following land-cover classes: meadow, pasture, orchard, vineyard, and cereal. Coniferous and deciduous forest have also been considered for the sake of comparison with the other classes. The size of each area of interest (around 200 m^2) was selected as a compromise between homogeneity of the area and number of pixels. Trends in VH and VV were analyzed for the different land-cover classes.

Subsequently, time series of S-1, in areas of interest corresponding to the ground stations, were compared with other sources of information. In detail, they are compared with time series of NDVI derived from S-2 and SRS, and with soil parameters such as ST and SWC. This analysis is carried out for meadows, which are the target land-cover type observed by the ground stations. The analyses were done for the season 2016, considering the data availability of the different sensors. For comparing the backscatter with point data, the areas of interest were extracted close to the station, in homogeneous areas. After obtaining time series of values from the selected areas, a -correlation between both optical and SAR sensors was performed and noise-reduction filters were applied depending on the investigation purpose. The cross-correlation analysis was performed by using the t-series v0.1-2 package [50] on R (v. 3.4.3.). After with analyzed the SAR signal for different land-cover types, as a last step, a simulation with electromagnetic models were carried out, specifically for meadows. The simulation was performed to understand in a quantitative way the impact of the different parameters (soil and vegetation) on the SAR signal.

The total scattering from vegetated soils was simulated by using the Water Cloud Model (WCM) [22]:

$$\sigma^0_{pq,tot} = \frac{A\cos(\theta)}{2b\text{NDVI}}(1 - \exp(-2b\text{NDVI}\sec(\theta))) + \sigma^0_{pq,b}\exp(-2b\text{NDVI}\sec(\theta)) \tag{2}$$

where in Equation (1) the dependence on the vegetation is expressed through NDVI as a proxy of vegetation water content; θ is the local incidence angle, σ^0_{pq} is the scattering from bare soil that for the VH polarization was simulated with the Oh model [51]. A and b are parameters for crop type and were fitted against ground data. A range of crop parameters were tested to determine the best-fitting combination. The simulated backscatter was analyzed through linear regressions. Subsequently, the influence of the signal components were examined through a dominant factor analysis [52].

3.3. Phenological Phases Extraction

A Best Index Slope Extraction (BISE) [53] filter was applied on optical and SAR time series to extract phenological parameters; BISE filter was chosen to remove the noisy points affected by the mowing that could interfere with the detection of annual vegetation cycle. Then, four filter techniques were tested: Savitzky Golay, Double-logistic, Linear Filter and Fast Fourier Transform (FFT). For each of the modeled time series we extracted the SOS and EOS, with a threshold of 50% and the maximum (MAX) of the curve [15]. The analyses were performed using the Phenex package [54] on R (v. 3.4.3.). Then, the results were validated by using the phenological information derived by PhenoCams and SRS sensors. Subsequently, a phenological map of the SOS was created as a final product for the season 2016, using a Linear Filter both for SAR and optical images, on codes developed in Python (v. 2.7). To detect the SOS we applied a pixel-by-pixel-based method used in [55] and discussed in [56]. The optical map was created as a reference for the SAR one.

Conversely, to identify the harvest time, a linear filter was applied. This filter allowed to preserve unaltered seasonal trend of a given time series. Based on the maximum (MAX), obtained from the previous analysis, a minimum between intervals of time was used to detect the first and, eventually, the second mowing. Next, S-1 and S-2 maps of the harvest time were produced and compared to PhenoCam images through a visual interpretation, to detect the date of mowing.

4. Results

First, the trends of VV and VH polarization are obtained over different crop types. Next, the capability of the WCM to reproduce vegetation characteristics is evaluated and discussed. Then, in the selected areas, the phenological phases extraction is performed both for S-1 and S-2 time series and compared with the observations of the fixed stations. Finally, phenological maps of SOS and harvesting time are produced on large scale.

4.1. Statistical and Electromagnetic Modeling

Both VH and VV polarization signals show a strong correlation in C-band for all different land-use types, as a result of the high sensitivity to vegetation biomass with respect to SWC. The highest σ^0 values in both polarizations were associated with the period in which the crop green biomass generally reaches its maximum. The trends belonging to vineyards, orchards, and deciduous forest show a higher level of the signal, ranging from -11.5 dB to -7.5 dB and from -18.5 dB to -13.5 dB for VV and VH polarization, respectively. Forests of conifer show a lower signature in both polarizations. Cereals present a similar trend in the VV and VH channel, while meadows and pastures exhibit seasonal changes in the dynamic range, depending on the polarization. In the VV channel, pastures show a high dynamic range, from −14 dB to −6.5 dB, with a peak in November and lower values during the summer due to the scarcity of water, which was particularly strong in 2015 (http://weather.provinz.bz. it/historical-data.asp). Conversely, in the VH polarization the σ^0 values demonstrate less sensitivity to seasonal dynamics, with value ranging from −17 dB to −21 dB. The signature of meadows, which are strongly managed in terms of fertilization, irrigation, and mowing in this area, varies between −14.5 dB and −8.5 dB and from −24.5 dB and −15.5 dB for VV and VH polarization, respectively. In the VH polarization, the trend of meadows shows the highest dynamic and is clearly distinguishable from pastures and from the other classes. Figure 3 illustrates an example of the S-1 backscatter trend for different land-use types. The smoothing lines show a local polynomial regression fitting (loess), done using neighboring values, weighted by their distance to the point [57].

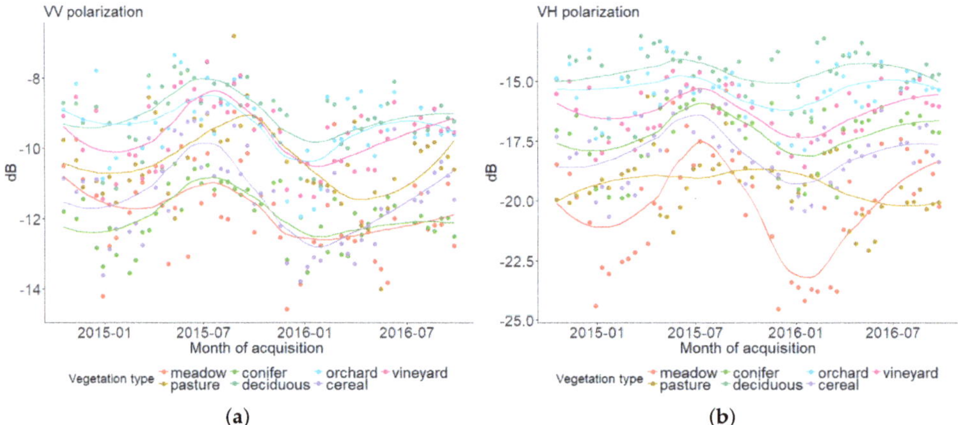

Figure 3. Trends of S-1 backscatter for different land-use types: orchards, meadows, crops, deciduous forest, coniferous forest, and pastures. (**a**) VV polarization over the area of interest. (**b**) VH polarization over the area of interest. The smoothing lines are obtained with a local polynomial regression fitting.

The relationship between Sentinel-2 NDVI and backscattering coefficients was compared for four meadow areas with NDVI measured at the ground (SRS sensor). The strongest correlation between S-1 and S-2 in semi-natural habitats was found in the VH channel ($R^2 = 0.52$).

For illustration purposes, the results are presented for a single area, while Tables 3 and 4 summarizes the statistics. Figure 4a,b illustrates time series of sensors in the area of vimef 2000, where a similar trend is visible during the entire acquisition period with an increase of VH signal of around 4–6 dB in the summer period as the NDVI advances from 0.5 to 0.8 after the snow melting. The NDVI from Sentinel-2 shows a higher value during all the temporal profile compared to the ground sensors, with values around 0.8–0.9, during the summer peak. Figure 4c illustrates a similar trend between σ_{VH} and temperature, while Figure 4d demonstrates a shift in the lag between SWC and scattering coefficients.

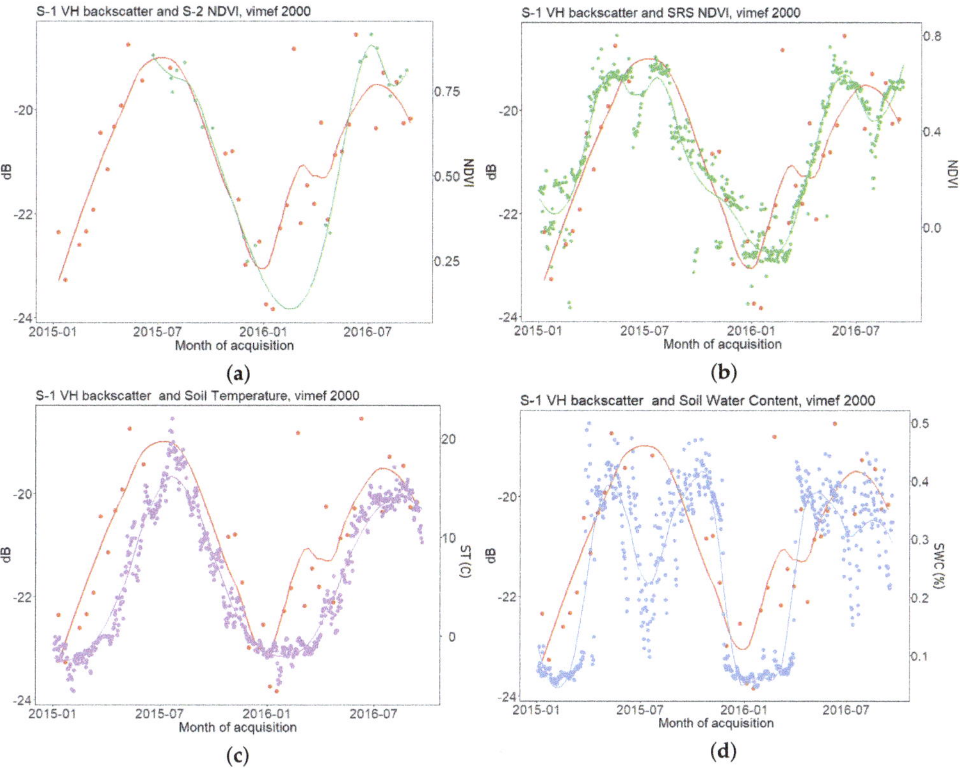

Figure 4. Temporal evolution of the mean backscattering coefficient VH (red)—Sentinel-2 NDVI (dark green) (**a**), VH (red)—NDVI from the SRS ground sensor (green) (**b**), VH (red)—soil temperature at 2 cm (violet) (**c**) and VH (red)—SWC (blue) (**d**), for the area domef 1500, during the acquisition period 2015–2016.

To understand the evolution of the VH signal in respect of the NDVI of both S-2 and SRS sensors, a temporal analysis of the season 2016 was computed. First, we analyzed the interaction between these three series of data through a cross-correlation function. The correlation was positive between sensors in all the areas. As shown in Table 3, for the areas of domef 2000 and vimes 1500 there is a minor shift in lags (h) for all the cross correlations. Conversely, the most dominant cross correlations in the area domef 1500 occur between lags −10 and −12 among S-1 and S-2, and lags 19 and 22 among S-1 and SRS. Furthermore, due to a high cloud cover contamination in the S-2 time series, the vimef 2000 area shows a shift in the lag both for the S-2/SRS and the S-1/S-2 correlation. The most dominant cross correlations occurred instead between lags −1 and 1 for S-1/SRS.

Additionally, Table 3 demonstrates a high Pearson's product-moment correlation between each sensor. There is a positive relationship, in the range 0.69 to 0.84, between Sentinel-1 σ_{VH} and Sentinel-2 NDVI, and a strong positive relationship between both the NDVI (from 0.54 to 0.69) as well as between SRS and Sentinel-1 (from 0.45 to 0.88). Due to the cloud contamination at high altitude in the optical domain, the Pearson's product-moment produces higher values between SAR and ground sensor time series in the areas domef 2000 and vimef 2000.

Table 3. Pearson's product-moment correlation and the most dominant cross correlations at lags = h, between S-1 VH backscatter, S-2 NDVI and SRS NDVI of the areas of interest.

Ground Station	S-2/SRS (Person Correlation and Lags)	S-1/S-2 (Person Correlation and Lags)	S1/SRS (Person Correlation and Lags)
domef 1500	0.54, h = [−2, 0]	0.70, h = [−10, −12]	0.51, h = [19, 22]
domef 2000	0.69, h = [−5, −3]	0.80, h = [−6, −4]	0.88, h = [−3, 0]
vimes 1500	0.66, h = [−1, 1]	0.84, h = [−2, 0]	0.45, h =[−2, 0]
vimef 2000	0.57, h = [13, 15]	0.69, h = [6, 8]	0.71, h = [−1, 1]

Based on the output of our analysis, σ^0 VH time series follow the trend of the NDVI during all seasons in the selected areas thus indicating the possibility of phenological phases extraction. To complete the analysis, we investigated the impact of soil and vegetation on this signal for meadows, through simulations with the WCM. The crop parameters have been fitted to achieve the best match between backscatter coefficients and WCM. After testing a permutation of A and b variables (100 samples each) in Equation (1), $A = 0.001$ and $b = 0.002$ were the combination that best fitted the WCM for meadows. The simulated backscatter from vegetated soil follows the trend of the measured σ_{VH} in domef 1500 (Figure 5a) and vimes 2000 (Figure 5b). Conversely, the poor result of vimes 1500 (Figure 5c) is due to problems in the ground data acquisition. For the season 2015 the NDVI reaches a saturation on July 20 and remains with values around 0.98 until December. Therefore, this station cannot be involved in the analysis of the results. Moreover, the station domef 2000 starts to acquire the ground NDVI only from May 2015. This reduces the output of the simulation (Figure 5d).

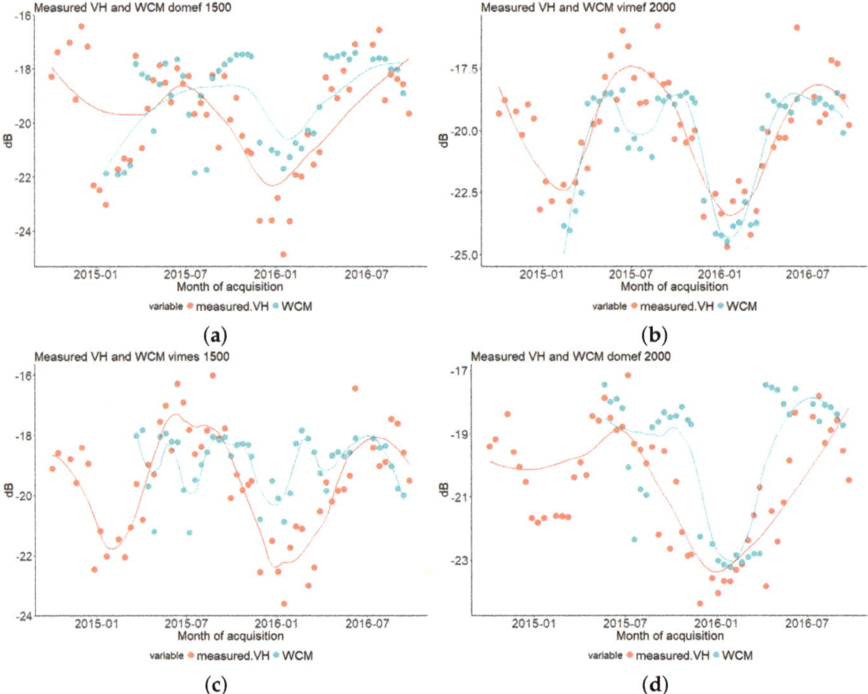

Figure 5. Measured VH backscatter and simulation through the WCM: the four panels show the stations domef 1500 (**a**), vimef 2000 (**b**), vimes 1500 (**c**) and domef 2000 (**d**). The smoothing lines are obtained with a local polynomial regression fitting.

Table 4 summarizes the results of model fitting between σ_{VH} and different parameters (WCM, NDVI and SWC) for the four areas. The linear regression between simulation and VH signal has an adjusted R^2 of 0.52 and 0.55 for the stations domef 1500 and vimes 2000, respectively. In addition, the Root-Mean-Square Error (RMSE) corresponds to 1.40 dB and 1.64 dB. The relation between the model and the σ_{VH} drops to an R^2 of 0.23 in domef 2000 station, while the RMSE rises to 2.40 dB. The ground NDVI shows on average a higher adjusted R^2 (R^2 mean of 0.6) in the linear model with the VH backscatter, than the SWC (R^2 mean of 0.37). By comparing the resulted linear relations, through the method of dominant factors, the influence of vegetation expressed in terms of NDVI on the WCM is greater than the soil component.

Table 4. Summaries of the results of linear model fitting between parameters.

Linear Model	Ground Station	Adjusted R^2	p-Value (<)	RMSE (dB)
VH~WCM	domef 1500	0.52	1.369×10^{-8}	1.40
VH~WCM	domef 2000	0.23	0.001	2.40
VH~WCM	vimes 1500	0.04	0.096	1.90
VH~WCM	vimef 2000	0.55	3.39×10^{-9}	1.64
VH~NDVI	domef 1500	0.32	2.20×10^{-5}	
VH~NDVI	domef 2000	0.68	1.15×10^{-10}	
VH~NDVI	vimes 1500	0.05	0.071	
VH~NDVI	vimef 2000	0.80	2.2×10^{-16}	
VH~SWC	domef 1500	0.47	2.82×10^{-8}	
VH~SWC	domef 2000	0.17	0.006	
VH~SWC	vimes 1500	0.04	0.096	
VH~SWC	vimef 2000	0.47	9.24×10^{-9}	

4.2. Phenological Phases Extraction

4.2.1. SOS

The retrieved phenological phases from the four areas are compared with the information extracted from the PhenoCams in the field. Figure 6 shows an example of the SAR modeled time series with the phases of SOS (red line), MAX (yellow line) and EOS (blue line). Meanwhile Table 5 summarizes the results, expressed in Day Of Year (DOY), of sensors and PhenoCams.

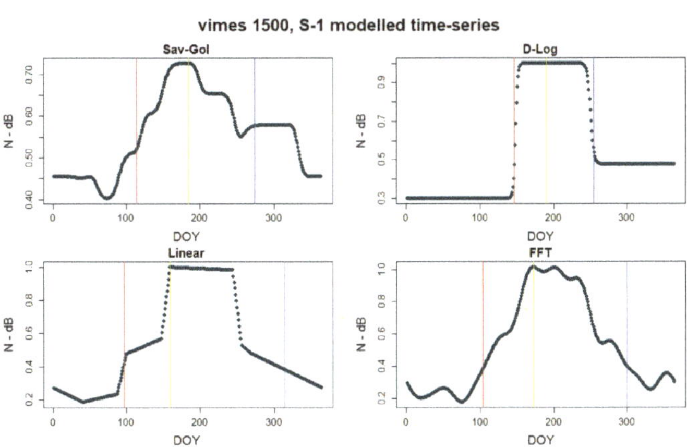

Figure 6. Example of modeled time series using Savitzky Golay, Double-logistic, Linear, and FFT filters. The figure shows the phenocycle phases extraction in vimes 1500 area using the normalized S-1 VH backscatter (N-dB).

Table 5. Day Of Year (DOY) of phenological phases for all the filtered time series. In bold the DOY extracted from PhenoCams.

Area	Phenophase	Filter	Phenocam (DOY)	S-1 (DOY)	S-2 (DOY)	SRS (DOY)
domef 1500	SOS		**102**			
	SOS	Sav-Gol		66	106	106
	SOS	D-Log		151	101	101
	SOS	Linear		90	76	105
	SOS	FFT		94	82	106
	MAX		**185**			
	MAX	Sav-Gol		191	246	241
	MAX	D-Log		186	198	111
	MAX	Linear		208	245	115
	MAX	FFT		290	247	127
	EOS		**301**			
	EOS	Sav-Gol		246	307	301
	EOS	D-Log		265	317	312
	EOS	Linear		259	306	218
	EOS	FFT		253	309	236
domef 2000	SOS		**147**			
	SOS	Sav-Gol		96	38	149
	SOS	D-Log		64	90	151
	SOS	Linear		67	92	23
	SOS	FFT		75	82	79
	MAX		**205**			
	MAX	Sav-Gol		230	124	242
	MAX	D-Log		229	124	201
	MAX	Linear		208	189	166
	MAX	FFT		196	145	272
	EOS		**282**			
	EOS	Sav-Gol		298	230	284
	EOS	D-Log		260	230	288
	EOS	Linear		318	270	284
	EOS	FFT		318	252	289
vimes 1500	SOS		**99**			
	SOS	Sav-Gol		127	95	96
	SOS	D-Log		147	96	96
	SOS	Linear		89	97	97
	SOS	FFT		81	104	97
	MAX		**180**			
	MAX	Sav-Gol		179	210	137
	MAX	D-Log		164	176	261
	MAX	Linear		160	219	175
	MAX	FFT		181	229	139
	EOS		**306**			
	EOS	Sav-Gol		265	309	313
	EOS	D-Log		306	300	330
	EOS	Linear		309	295	327
	EOS	FFT		303	282	290
vimef 2000	SOS		**140**			
	SOS	Sav-Gol		122	154	128
	SOS	D-Log		154	162	131
	SOS	Linear		110	156	131
	SOS	FFT		75	149	139
	MAX		**199**			
	MAX	Sav-Gol		145	209	210
	MAX	D-Log		162	235	192
	MAX	Linear		159	188	169
	MAX	FFT		163	191	217
	EOS		**284**			
	EOS	Sav-Gol		217	273	285
	EOS	D-Log		258	269	285
	EOS	Linear		250	266	286
	EOS	FFT		249	271	274

To better explain the results, Table 6 shows the averages of the days of difference between the sensors and PhenoCams. The first three items correspond to the total average between the areas, while the last four items are divided based on the altitude.

Table 6. Average number of days of difference between sensors and PhenoCams.

Sensor	Areas	SOS (day)	MAX (day)	EOS (day)
S-1	all	10	10	20
S-2	all	4	10	8
RSR	all	1.5	18.5	2
S-1	1500	9	1	18.5
S-1	2000	14	19.5	21
S-2	1500	1.5	8.5	4
S-2	2000	9	12	11.5

Moreover, Figure 7 shows the DOY of the SOS extracted from each sensor (a) and percent error (b) on the SOS date estimation for all the areas. In Figure 7a, S-1 (light blue bars) detects the SOS before S-2 in stations at 1500 m, while it is delayed at 2000 m a.s.l. On the other hand, when comparing satellites with ground sensors, S-1 is less effective than S-2 in almost all stations, as illustrated in in Figure 7b. Only in the domef 2000 area, S-1 has a better result than S-2.

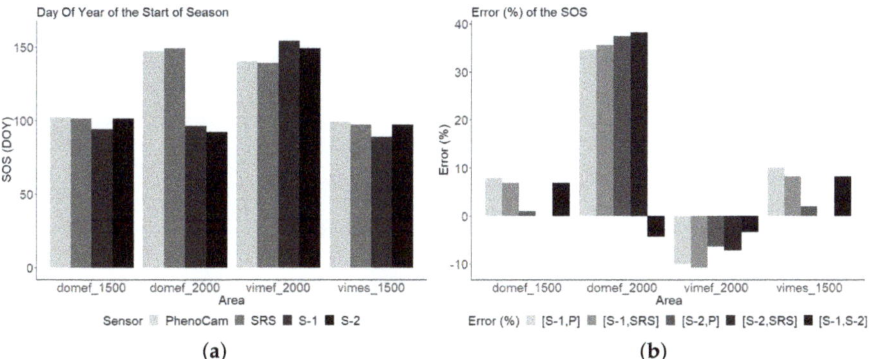

Figure 7. DOY detected and percent errors between sensors in the analysis of the SOS: in (**a**) the different color bars represent the results for each sensor; in red are shown in (**b**) the percent difference between S-1-PhenoCams ([S-1,P]) and S-1-SRS ([S-1,SRS]), in yellow between S-2-PhenoCams ([S-2,P]) and S-2-SRS ([S-2,SRS]), while in orange the percent difference between S-1-S-2 ([S-1,S-2]).

Finally, we mapped the SOS with S-1 and S-2 time series, by using a linear filter. The first map is obtained from S-1 σ_{VH} time series with a backscatter threshold of 0.9 (Figure 8a), and the second one from S-2 NDVI time series using a NDVI threshold of 0.7 (Figure 8b). Each class corresponds to a different SOS interval of time for a total of 10 classes, from DOY 61 to DOY 210. All non-vegetated areas are masked using CORINE 2012 land-cover information. Above a certain elevation though, as indicated in the statistical analysis, the σ_{VH} loses the sensitivity to vegetation and gave unreasonable results. For this reason, the SAR map (Figure 8a,c) is masked above 2100 m of altitude. A detail of the map is shown in Figure 8c,d, where the station domef 1500 is located.

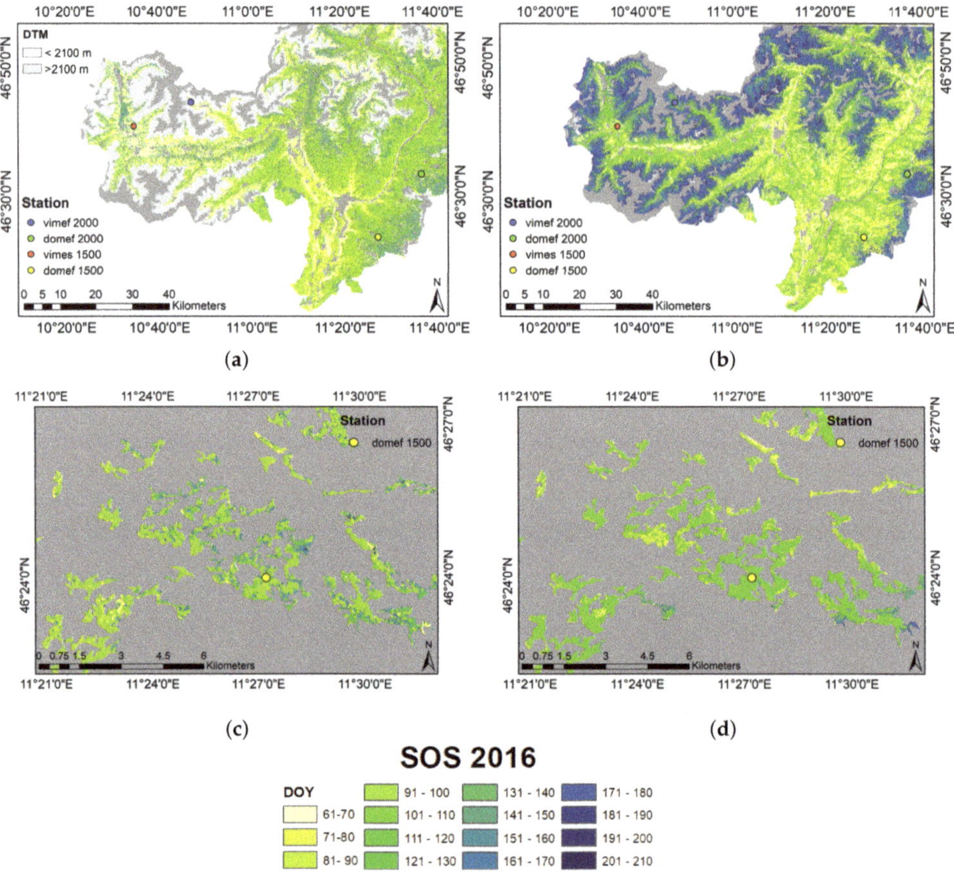

Figure 8. Time of start of the growing season in the South Tyrol region, using S-1 σ_{VH} (**a**) and S-2 NDVI (**b**) time series. The SAR map is masked above 2100 m of altitude while only a non-vegetated area mask is applied on the optical one. In the bottom panel a detail of the maps of meadows areas around domef 1500 station, obtained with S-1 (**c**) and S-2 (**d**).

4.2.2. Harvest

In our study, first we automatically detected the harvest time in all the sensors, we compared them visually with images from PhenoCams, and finally we retrieve S-1 and S-2 harvest time maps. Figure 9 shows the detected timings of mowing for each sensor. In domef 1500 and vimes 1500 areas (Figure 9a,c) both optical sensors catch two mowing events; conversely, S-1 VH backscatter recognizes an individual event in between. In Figure 9d, S-2 detects two events instead of one due to a heavy snowfall corresponding to the day of acquisition of the optical satellite.

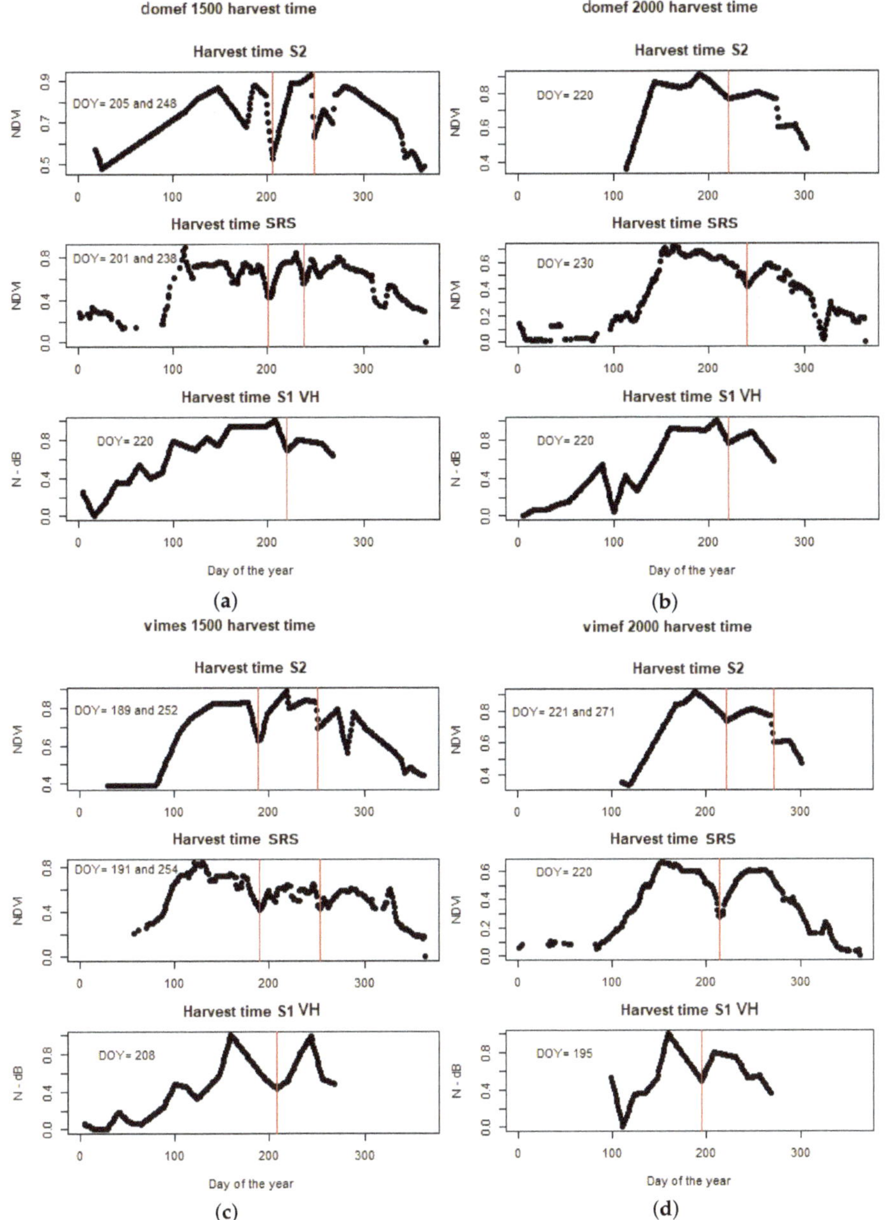

Figure 9. Detection of the harvest time (red line) using S-2, SRS, and S-1 time series. In the upper panel domef 1500 (**a**) and domef 2000 (**b**), in the bottom panel vimes 1500 (**c**) and vimef 2000 (**d**).

Table 7 illustrates the results from the S-1, S-2, and SRS compared with the harvest time detected from PhenoCams. S-1, as shown in Figure 10a, compared to both ground sensors and S-2, is delayed in the definition of the first mowing event, in low altitude stations. Conversely, S-1 is in advance in areas at 2000 m a.s.l., except for domef 2000, where S-1 and S-2 give the same result. In terms of percent error (Figure 10b), the results of S-1 are less accurate compared to S-2, except for the area domef 2000.

Table 7. Harvest time retrieved by PhenoCams, SAR and optical sensors.

Area	Phenocam (DOY)	S-2 (DOY)	SRS (DOY)	S-1 (DOY)
domef 1500	198 and 252	205 and 248	201 and 238	220
domef 2000	230	220	230	220
vimes 1500	188 and 250	189 and 252	191 and 254	208 and 268
vimef 2000	210	221 and 271	220	195

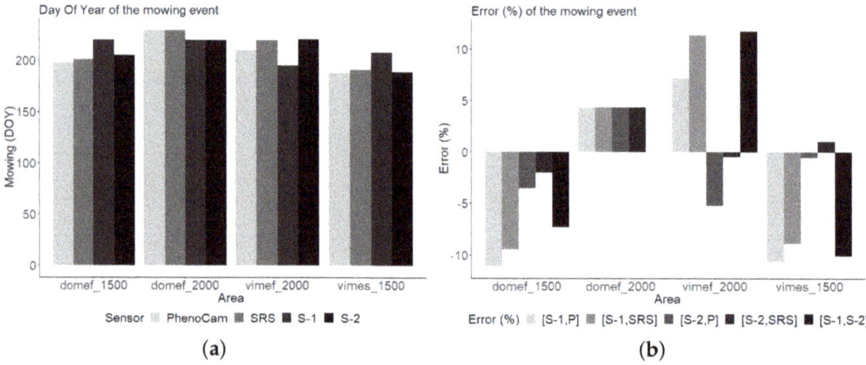

Figure 10. First mowing event detected by the different sensors: in (**a**) the different color bars represent the results of each sensor; in light blue are shown in (**b**) the percent error between S-1/PhenoCams ([S-1,P]) and S-1/SRS ([S-1,P]), in orange between S-2/PhenoCams ([S-2,P]) and S-2/SRS ([S-2,SRS]), while in purple the difference between S-1/S-2 ([S-1,S-2]).

Figure 11a,b illustrates the harvest time between DOY 180 and DOY 221 of meadow areas close to the station vimes 1500. The first map (a) is obtained from Sentinel-1 VH time series, while the second map (b) is generated from Sentinel-2 NDVI. The SAR map shows a harvesting time between DOY 201 and 220 for most of the meadow areas (green and yellow color); conversely, the optical map presents an earlier harvest among DOY 180 and 200 (pink and violet color). Except for the area where the mowing is beyond DOY 220, the two maps give a result that do not correspond, with a shift in time of the SAR result. SAR and optical time series give a corresponding result from DOY 221, as we can see in orange on the maps.

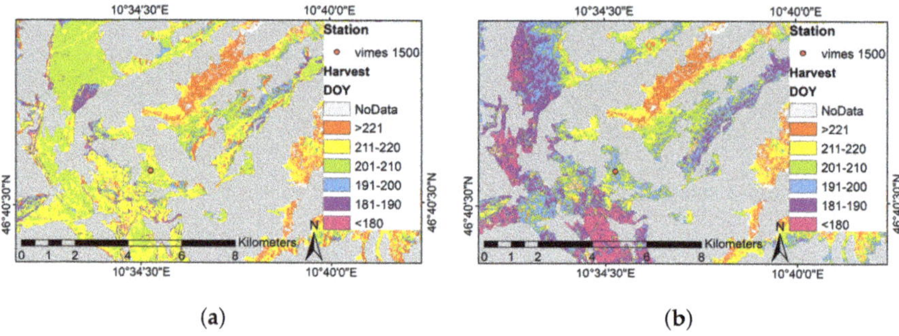

Figure 11. Harvest maps of the first mowing event generated from S-1 and S-2 (summer 2016). The maps show a detail of the surrounding vimes 1500 area: on the left panel (**a**) the map obtained from σ_{VH} backscatter; on the right panel (**b**) the map obtained from Sentinel-2 NDVI.

5. Discussion

Our investigation revealed that time series of Sentinel-1 C-band allows phenological dynamics detection in different vegetation land-cover types. As shown in previous studies [58,59], VV and VH polarizations have a clear seasonal dynamic, with a peak that might correspond to the maximum of biomass production. It seems also possible to discriminate the signal of different vegetation classes. Moreover, the results achieved in the current study suggest that the backscatter of mountain meadows has a high dynamic range.

In pastures class, backscattering profiles are stable in VH polarization, while changing trend was observed in the VV channel. It might be possible that the contribution of bare soil and vegetation structure (short and thin leaves) for this class increases the sensitivity to variations in the water content of soil, instead of vegetation cover, as demonstrated in [60]. Furthermore, at high altitudes, there are limitations on SAR σ^0 VH sensitivity: the presence of low biomass with narrow leaves might increase the absorption effect, causing a flat trend in the backscatter coefficients [61].

Conversely, the VH channel better describes meadow phenology. As previously demonstrated [62], the correlation between σ^0 coefficients and NDVI is stronger in the VH channel in meadows areas, due to the volume scattering of vegetation. Similar results were found in [63], where σ^0 VH sharply raises during the phase of green-up, it is stable during the vegetation reproduction, and decreases rapidly due to the harvest. The cross-correlation between S-1 σ^0 VH backscatter and S-2 NDVI for the season 2016, shows a positive correlation for the selected areas. Moreover, the Pearson's product-moment correlation between S-1 and the SRS ground sensor reveals that in areas where the cloud cover limits S-2 data acquisition, backscattering coefficients can support the phenological phase detection. Our analyses were limited to one year, 2016. Since, time series contain a combination of seasonal, gradual and abrupt changes [64], a decomposition analysis should be applied to longer time series. A seasonal-trend analysis could therefore be useful in further studies when multiple years of Sentinel data will be available.

To derive useful quantitative information regarding the contribution of the vegetation to the SAR backscatter, we used the WCM. This semi-empirical model represents the power backscattered by the whole canopy as the incoherent sum of the contribution of the vegetation and soil [65]. Including NDVI in the model allows understanding the vegetation contribution to the VH channel. As emphasized by [66], in the VH channel, the vegetation contribution to the backscattering coefficient is higher than the soil component, when the vegetation is well developed [67].

The results of the comparison between the values predicted by the WCM and σ^0 VH time series show that the model is adequate to describe vegetation in mountainous areas. The statistical results are in line with previous studies for both Adjusted R^2 and Root-Mean-Square Error [68,69]. Moreover, the altitude does not seem to interfere with the simulation, giving a RMSE of 1.63 dB in an area at 2000 m a.s.l. Furthermore, through the analysis of R^2, $sigma_{VH}$ is more influenced by the vegetation growth than SWC. Hence, our results confirm that VH C-band SAR data combined with optical data may be applicable to estimate the vegetation phenology in mountain meadows.

To obtain the best mapping results, we evaluated different filter techniques, based on previous studies [70,71]. It is important to underline that in the validation phase, there are significant limitations in comparing satellite sensors and ground observation [72]. Whereas the NDVI is a direct measure of radiation absorption by the canopy [19], PhenoCam visual analysis, has different sources of uncertainties, especially to track when the first leaves appear from the surrounding vegetation and the mixture of senescence leaf colors [49,73,74]. For this reason, we evaluated the accuracy of our results through both the NDVI ground sensor (SRS) and PhenoCam images. The two results were in good agreement, with a mean of 1.5 day of difference for the SOS and 2 days for the senescence. Phenocams resulted essential to detect the harvesting time, by directly observing the mowing operations.

All four filters clearly describe the trend of the growing season in each area and none show better performance compared to the others. As expected, each filter applied to SRS NDVI time series approximates well the seasonal phenology, even though, despite BISE noise-reduction techniques, the mowing events interfere with the detection of the EOS.

The days of difference of S-1 with respect to the dates extracted from the PhenoCams and SRS sensor increase with the altitude of the areas. For the SOS, at 1500 m a.s.l., the distance between the field data and the SAR data is compatible with the time of acquisition of the satellite. Conversely, in the areas at 2000 m, the distance in days exceeds the temporal resolution of the SAR satellite. The same tendency is repeated for the EOS, where, however, the difference increases with respect to the ground data. The optical data follows the trend of the SAR, with fewer days of difference. In this context, the percent error ranges between −10% in the worst scenario and 8% in the best one for S-1 and ground sensors; −7% and 2% of error respectively, for S-2 and ground sensors. In this analysis we do not consider the SOS extracted from both S-1 and S-2 time series in the area domef 2000; this exception is determined by the fact that in this area:

- a heavy snowfall in April, corresponding to the day of S-1 acquisition, caused a signal drop and consequently errors in filter modeling;
- in the optical domain, during the period January–October 2016, only 13 images were cloud free in this area.

Although the results of S-1 are in most cases less accurate than those of S-2, we expect that applying our detection method on flat areas and/or with different vegetation cover and leaves structure, we could have consistent results among SAR and optical sensors. Furthermore, we think that by increasing the temporal resolution, with the S-1B and S-2B acquisitions, the accuracy in the phenology estimation process would increase for both sensors.

In the mapping process, since from our comparative test the filters perform equally well, to have an identical approach in the optical and microwaves domains, and for simplicity reason, we applied a Linear Filter to both the time series. In both maps the growing season follows an altitude-based gradient, with an early start of vegetation growth at valley floors, anticipated in the wider areas, which is gradually delayed at high altitudes and especially in narrow valleys. For the vegetation that is covered by snow several months during the year, i.e., above 2500 m a.s.l., the green-up starts between the end of May (DOY 147) and the start of August (DOY 210). The map obtained from the σ_{VH} shows less sensitivities at high altitude, where the vegetation decreases in height and biomass. Furthermore, the presence of bare soil strongly influences the SAR signal. The optical map shows an earlier start at the bottom of the valleys (around DOY 60–70), compared to the SAR detection (around DOY 80–90) and emphasizes the green-up gradient going from low to high altitude. When we zoomed in the map, the S-1 backscatter gave a delay in the SOS of around 10 days in some areas. However, the S-1 backscatter seems to be more sensitive than the S-2 NDVI, diversifying more SOS periods. The comparison between the SOS maps of South Tyrol, obtained from S-1 and S-2, illustrates that SAR data can be used to detect the onset of the growing season in meadow areas. However, as demonstrated in the vegetation type analyses, the same procedure is not applicable to pasture areas. In this class the σ_{VH} time series, with a flat trend, does not allow the phenology detection. A sensitivity analysis of VV channel and the ratio VV/VH needs to be further investigated to understand a possible contribution to phenology detection of pasture class. In addition, both maps should be validated at different altitude and on different vegetation cover types (i.e., forest classes).

In mountain regions, there is a transition from fertilized to unfertilized meadows and pastures. Grasslands located at low altitudes or in the valley are usually mowed several times during the growing season. With increasing elevation, agriculture is less intensive, and the mountain meadows are mown once a year and mostly grazed in autumn [75]. Our areas of interest located at 1500 m a.s.l. are usually mowed twice a year, while those at 2000 m a.s.l. only once. Starting from the assumption that optical sensors well describe the radiation changes related to physiological conditions of plants, but they do not explain modification of the vegetation geometry [33], we expected to obtain a better harvest time detection with SAR time series. However, the S-1 GRD products were missing for the season 2016 because of an onboard anomaly recorded between 8 June and 14 July [47]. This led to errors in the definition of the first mowing event. In terms of percent error, the results of S-1 are less accurate compared to S-2, with a range of error between −11% to 11% in the areas at lower altitude. Conversely, when there is only one mowing, in August at high altitudes, S-1 data give promising results (percent error between −8% to 4%) as well as S-2 (percent error between −5% to 4%). In these areas the mowing maps follow, indeed, the same trend. This demonstrates that S-1A instrument unavailability caused errors in the first mowing detection. Concurrently, the result suggests that in the presence of time series without missing data, S-1 gives results similar to S-2, allowing to overcome the problems of cloud cover in optical images. Having consistent data is indeed decisive in the definition of a mowing event. Furthermore, the advances/delays in the harvesting time detection are derived by the averaging of the selected areas which include different time of mowing. An example is shown in Figure 12 where in (a) at the top left we can see the start of mowing operations on DOY 220 and in (b) the end of them on DOY 235. Therefore, even in the case of mowing detection, using images from S-1B and S-2B, would improve our results.

Optical remote sensing provides a powerful tool to monitor phenology in mountain ecosystems and, our investigation has shown that SAR data might be effective in meadows phenology detection as well as complementary to the optical information. However, to test the applicability of the method on different vegetation classes more validation points are needed as well as a threshold's optimization. We cannot expect to obtain the same results in microwave and optical domains, due to different physical mechanisms: the first based on the structure, roughness, dielectric constant, and slope/orientation of scattering surfaces [22,76], and the second one on the reflectance properties of leaves, illumination angle, leaf orientation, and background [77]. In this context, our approach aims at understanding the behavior of the backscattering coefficients in meadow areas to complement the optical data with SAR images to reduce missing information caused by clouds contamination and atmospheric effects in the optical domain.

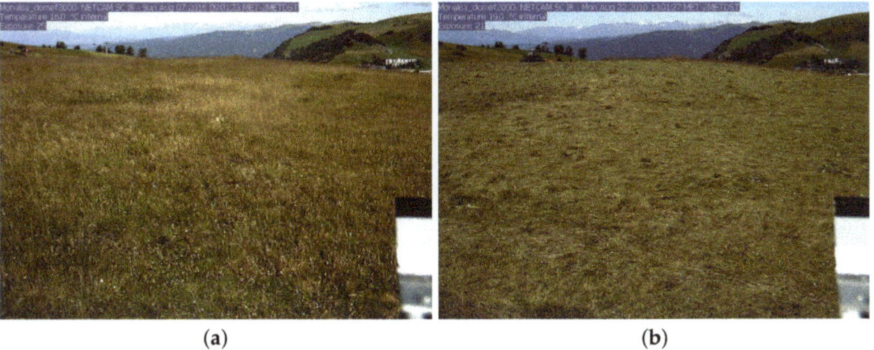

(a)   (b)

Figure 12. Images from the PhenoCam at the domef 2000 station. In (**a**) the DOY 220: at the top left the start of the mowing. In (**b**) DOY 235 the mowing right in front of the PhenoCam.

6. Conclusions

The paper describes multitemporal Sentinel-1 C-band and Sentinel-2 NDVI application on mountain meadows monitoring. The main aim was to test the feasibility of phenocycle phases retrieval from SAR time series and compare the results with the optical sensors, in the perspective of data integration.

From our analysis:

- The statistical analysis of σ^0 time series showed that the SAR signal can detect phenological cycles in different vegetation cover types.

- The significant correlation, with a negligible shift in lags, between σ_{VH} and the NDVI from optical sensors, allowed the extraction of the phases of start, maximum and EOS, in addition to the mowing period.
- SAR data can be used to detect the phenological phases in meadows areas, with an accuracy compatible with the temporal resolution of S-1 until 1500 m a.s.l.

This result appears promising in the SAR-Optical data integration process for phenology detection. However, it needs to be confirmed for different altitudes and vegetation types. The data unavailability during the mowing period led to errors in the definition of the first harvest time. For this reason, future studies should be considered Sentinel-1B and Sentinel-2B acquisitions to increase the data consistency.

Author Contributions: All authors contributed extensively to the work presented in this paper. Project conceived and designed by C.N., L.S., G.N., S.R.K., R.G. and M.Z.; preprocessing of S-2 data by M.R.; Fieldwork designed and data acquired by G.N.; data analyses carried out by L.S. with support from S.R.K. and C.N.; Manuscript preparation lead by L.S., with support (substantial critical feedback, revisions and additions to text) from all co-authors; Final version of the manuscript read and approved by all co-authors.

Funding: This research received no external funding.

Acknowledgments: The station network is supported by the Autonomous Province of Bolzano/Bozen-South Tyrol within the frame of the projects MONALISA. We are grateful to Eurac research (Alessandro Zandonai and Stefano Della Chiesa) for field maintenance and Norut for the for logistical support and data preprocessing. We will also thank Carlo Marin for S-1 time series preprocessing. Finally, a grateful remark to Piyush Kumar for the language critical reading support.

Conflicts of Interest: The authors declare no conflict of interest.

References

1. MacDonald, D.; Crabtree, J.R.; Wiesinger, G.; Dax, T.; Stamou, N.; Fleury, P.; Lazpita, J.G.; Gibon, A. Agricultural abandonment in mountain areas of Europe: environmental consequences and policy response. *J. Environ. Manag.* **2000**, *59*, 47–69. [CrossRef]

2. Flury, C.; Huber, R.; Tasser, E. Future of mountain agriculture in the Alps. In *The Future of Mountain Agriculture*; Mann, S., Ed.; Springer: Berlin/Heidelberg, Germany, 2013; pp. 105–126, ISBN 978-3-642-33584-6.

3. Lieth, H. (Ed.) Purposes of a Phenology Book. In *Phenology and Seasonality Modeling*; Springer: Berlin/Heidelberg, Germany, 1974; pp. 3–19, ISBN 978-3-642-51863-8.

4. Chmielewski, F.M.; Müller, A.; Bruns, E. Climate changes and trends in phenology of fruit trees and field crops in Germany, 1961–2000. *Agric. For. Meteorol.* **2004**, *121*, 69–78. [CrossRef]

5. Tasser, E.; Tappeiner, U. Impact of land use changes on mountain vegetation. *Appl. Veg. Sci.* **2002**, *5*, 173–184. [CrossRef]

6. Beniston, M. Climatic change in mountain regions: A review of possible impacts. *Clim. Chang.* **2003**, *59*, 5–31.:1024458411589. [CrossRef]

7. Theurillat, J.-P.; Guisan, A. Potential impact of climate change on vegetation in the European Alps: A review. *Clim. Chang.* **2001**, *50*, 77–109. [CrossRef]

8. Jackson, R.D. Remote sensing of vegetation characteristics for farm management. *Remote Sens. Crit. Rev. Technol.* **1984**, *475*, 81–98. [CrossRef]

9. Bauer, M.E. Spectral inputs to crop identification and condition assessment. *Proc. IEEE* **1985**, *73*, 1071–1085. [CrossRef]

10. Moran, M.S.; Inoue, Y.; Barnes, E.M. Opportunities and limitations for image-based remote sensing in precision crop management. *Remote Sens. Environ.* **1997**, *61*, 319–346. [CrossRef]

11. Pinter, P.J., Jr.; Hatfield, J.L.; Schepers, J.S.; Barnes, E.M.; Moran, M.S.; Daughtry, C.S.T.; Upchurch, D.R. Remote sensing for crop management. *Photogramm. Eng. Remote Sens.* **2003**, *69*, 647–664. [CrossRef]

12. Rouse, J.W., Jr.; Haas, R.H.; Schell, J.A.; Deering, D.W. *Monitoring Vegetation Systems in the Great Plains with ERTS*; NASA. Goddard Space Flight Center 3d ERTS-1 Symp.; SEE 19740022592, Sect. A; NASA: Washington, DC, USA, 1974; Volume 1, pp. 309–317.

13. Lloyd, D.A. phenological classification of terrestrial vegetation cover using shortwave vegetation index imagery. *Int. J. Remote Sens.* **1990**, *11*, 2269–2279. [CrossRef]

14. Reed, B.C.; Brown, J.F.; VanderZee, D.; Loveland, T.R.; Merchant, J.W.; Ohlen, D.O. Measuring phenological variability from satellite imagery. *J. Veg. Sci.* **1994**, *5*, 703–714. [CrossRef]
15. White, M.A.; Thornton, P.E.; Running, S.W. A continental phenology model for monitoring vegetation responses to interannual climatic variability. *Glob. Biogeochem. Cycles* **1997**, *11*, 217–234.
16. Jönsson, P.; Eklundh, L. TIMESAT—A program for analyzing time-series of satellite sensor data. *Comput. Geosci.* **2004**, *30*, 833–845. [CrossRef]
17. Xin, J.; Yu, Z.; van Leeuwen, L.; Driessen, P. Mapping crop key phenological stages in the North China Plain using NOAA time series images. *Int. J. Appl. Earth Obs. Geoinform.* **2002**, *4*, 109–117. [CrossRef]
18. Pan, Z.; Huang, J.; Zhou, Q.; Wang, L.; Cheng, Y.; Zhang, H.; Blackburn, G.A.; Yan, J.; Liu, J. Mapping crop phenology using NDVI time-series derived from HJ-1 A/B data. *Int. J. Appl. Earth Obs. Geoinform.* **2015**, *34*, 188–197. [CrossRef]
19. Myneni, R.B.; Maggion, S.; Iaquinta, J.; Privette, J.L.; Gobron, N.; Pinty, B.; Kimes, D.S.; Verstraete, M.M.; Williams, D.L. Optical remote sensing of vegetation: Modeling, caveats, and algorithms. *Remote Sens. Environ.* **1995**, *51*, 169–188. [CrossRef]
20. Richards, M.A. Introduction to Radar Systems. In *Fundamentals of Radar Signal Processing*; Tata McGraw-Hill Education: New York City, NY, USA, 2005; pp. 45–47, ISBN 978-0071444743.
21. Ulaby, F. Radar response to vegetation. *IEEE Trans. Antennas Propag.* **1975**, *23*, 36–45. [CrossRef]
22. Attema, E.P.W.; Ulaby, F.T. Vegetation modeled as a water cloud. *Radio Sci.* **1978**, *13*, 357–364. [CrossRef]
23. Kurosu, T.; Fujita, M.; Chiba, K. Monitoring of rice crop growth from space using the ERS-1 C-band SAR. *IEEE Trans. Geosci. Remote Sens* **1995**, *33*, 1092–1096. [CrossRef]
24. Choudhury, I.; Chakraborty, M.; Parihar, J.S. Estimation of rice growth parameter and crop phenology with conjunctive use of RADARSAT and ENVISAT. In *Proccedings of the ENVISAT Symposium*, Montreux, Switzerland, 23–27 April 2007.
25. Lopez-Sanchez, J.M.; Vicente-Guijalba, F.; Ballester-Berman, J.D.; Cloude, S.R. Retrieving rice phenology with X-band co-polar data: A multi-incidence multi-year experiment. In Proceedings of the 2015 IEEE International Geoscience and Remote Sensing Symposium (IGARSS), Milan, Italy, 26–31 July 2015; pp. 3977–3980. [CrossRef]
26. Lopez-Sanchez, J.M.; Cloude, S.R.; Ballester-Berman, J. David Rice phenology monitoring by means of SAR polarimetry at X-band. *IEEE Trans. Geosci. Remote Sens.* **2012**, *50*, 2695–2709. [CrossRef]
27. Lopez-Sanchez, J.M.; Vicente-Guijalba, F.; Ballester-Berman, J.D.; Cloude, S.R. Polarimetric response of rice fields at C-band: Analysis and phenology retrieval. *IEEE Trans. Geosci. Remote Sens.* **2014**, *52*, 2695–2709. [CrossRef]
28. Vicente-Guijalba, F.; Martinez-Marin, T.; Lopez-Sanchez, J.M. Crop phenology estimation using a multitemporal model and a Kalman filtering strategy. *IEEE Geosci. Remote Sens. Lett.* **2014**, *11*, 1081–1085. [CrossRef]
29. Yang, Z.; Li, K.; Liu, L.; Shao, Y.; Brisco, B.; Li, W. Rice growth monitoring using simulated compact polarimetric C band SAR. *Radio Sci.* **2014**, *49*, 1300–1315. [CrossRef]
30. Shang, J.; Jiao, X.; McNairn, H.; Kovacs, J.; Walters, D.; Ma, B.; Geng, X. Tracking crop phenological development of spring wheat using Synthetic Aperture Radar (SAR) in northern Ontario, Canada. In Proceedings of the 2013 Second International Conference onAgro-Geoinformatics (Agro-Geoinformatics), Fairfax, VA, USA, 12–16 August 2013; pp. 517–521. [CrossRef]
31. Küçük, Ç.; Taşkın, G.; Erten, E. Paddy-rice phenology classification based on machine-learning methods using multitemporal co-polar X-band SAR images. *IEEE J. Sel. Top. Appl. Earth Obs. Remote Sens.* **2016**, *9*, 2509–2519. [CrossRef]
32. Rossi, C.; Erten, E. Paddy-rice monitoring using TanDEM-X. *IEEE Trans. Geosci. Remote Sens.* **2015**, *53*, 900–910. [CrossRef]
33. De Bernardis, C.; Vicente-Guijalba, F.; Martinez-Marin, T.; Lopez-Sanchez, J.M. Contribution to real-time estimation of crop phenological states in a dynamical framework based on NDVI time series: Data fusion with SAR and temperature. *IEEE J. Sel. Top. Appl. Earth Obs. Remote Sens.* **2016**, *9*, 3512–3523. [CrossRef]
34. Tamm, T.; Alite, K.; Voormansik, K.; Talgre, L. Relating Sentinel-1 interferometric coherence to mowing events on grasslands. *Remote Sens.* **2016**, *8*, 802. [CrossRef]
35. Voormansik, K.; Jagdhuber, T.; Olesk, A.; Hajnsek, I.; Papathanassiou, K.P. Towards a detection of grassland cutting practices with dual polarimetric TerraSAR-X data. *Int. J. Remote Sens.* **2013**, *34*, 8081–8103. [CrossRef]

36. Zhao, L.; Yang, J.; Li, P.; Zhang, L. Characteristics analysis and classification of crop harvest patterns by exploiting high-frequency multipolarization SAR data. *IEEE J. Sel. Top. Appl. Earth Obs. Remote Sens.* **2014**, *7*, 3773–3783. [CrossRef]

37. Lopez-Sanchez, J.M.; Ballester-Berman, J.D. Potentials of polarimetric SAR interferometry for agriculture monitoring. *Radio Sci.* **2009**, *44*, 1–20. [CrossRef]

38. McNairn, H.; Brisco, B. The application of C-band polarimetric SAR for agriculture: A review. *Can. J. Remote Sens.* **2004**, *30*, 525–542. [CrossRef]

39. Dusseux, P.; Corpetti, T.; Hubert-Moy, L.; Corgne, S. Combined use of multi-temporal optical and radar satellite images for grassland monitoring. *Remote Sens.* **2014**, *7*, 6163–6182. [CrossRef]

40. Baghdadi, N.; Boyer, N.; Todoroff, P.; El Hajj, M.; Bégué, A. Potential of SAR sensors TerraSAR-X, ASAR/ENVISAT and PALSAR/ALOS for monitoring sugarcane crops on Reunion Island. *Remote Sens. Environ.* **2009**, *113*, 1724–1738. [CrossRef]

41. Paloscia, S.A. summary of experimental results to assess the contribution of SAR for mapping vegetation biomass and soil moisture. *Can. J. Remote Sens.* **2002**, *28*, 246–261. [CrossRef]

42. Balenzano, A.; Mattia, F.; Satalino, G.; Davidson, M.W.J. Dense temporal series of C-and L-band SAR data for soil moisture retrieval over agricultural crops. *IEEE J. Sel. Top. Appl. Earth Obs. Remote Sens.* **2011**, *4*, 439–450. [CrossRef]

43. Torres, R.; Snoeij, P.; Geudtner, D.; Bibby, D.; Davidson, M.; Attema, E.; Potin, P.; Rommen, B.; Floury, N.; Brown, M.; et al. GMES Sentinel-1 mission. *Remote Sens. Environ.* **2012**, *120*, 9–24. [CrossRef]

44. Drusch, M.; Del Bello, U.; Carlier, S.; Colin, O.; Fernandez, V.; Gascon, F.; Hoersch, B.; Isola, C.; Laberinti, P.; Martimort, P.; et al. Sentinel-2: ESA's optical high-resolution mission for GMES operational services. *Remote Sens. Environ.* **2012**, *120*, 25–36. [CrossRef]

45. Istituto Provinciale di Statistica, Bolzano (ASTAT) 6° Censimento Generale Dell'Agricoltura. Provincia Autonoma Di Bolzano–Alto Adige. 2010. Available online: http://astat.provincia.bz.it/downloads/JB2017_K2.pdf (accessed on 9 September 2018).

46. Schirpke, U.; Tasser, E.; Tappeiner, U. Mapping Ecosystem Services supply in mountain regions: A case study from South Tyrol (Italy). *Ann. Bot.* **2014**, *4*, 35–43. [CrossRef]

47. Meadows, P.; Hajduch, G. S-1A & S-1B Annual Performance Report for 2016. MPC-0366 DI-MPC-APR; CLS ESA, Issue 1.1, Appendix E; 4 April 2017. Available online: https://sentinel.esa.int/documents/247904/2370914/Sentinel-1-Annual-Performance-Report-2016 (accessed on 12 November 2018).

48. Spectral Reflectance Sensor, Operator's Manual. METER Group, Inc. USA. Available online: http://manuals.decagon.com/Manuals/14597_SRS_Web.pdf (accessed on 27 November 2018).

49. Klosterman, S.; Hufkens, K.; Gray, J.M.; Melaas, E.; Sonnentag, O.; Lavine, I.; Mitchell, L.; Norman, R.; Friedl, M.A.; Richardson, A. Evaluating remote sensing of deciduous forest phenology at multiple spatial scales using PhenoCam imagery. *Biogeosciences* **2014**, *16*, 4305–4320. [CrossRef]

50. Adrian T.; Kurt, H. Tseries: Time Series Analysis and Computational Finance Note = R Package Version 0.10-46. 2018. Available online: https://CRAN.R-project.org/package=tseries (accessed on 11 January 2019).

51. Oh, Y. Quantitative retrieval of soil moisture content and surface roughness from multipolarized radar observations of bare soil surfaces. *IEEE Trans. Geosci. Remote Sens.* **2004**, *42*, 596–601. [CrossRef]

52. Budescu, D.V. Dominance analysis: A new approach to the problem of relative importance of predictors in multiple regression. *Psychol. Bull.* **1993**, *114*, 542–551. [CrossRef]

53. Viovy, N.; Arino, O.; Belward, A.S. The Best Index Slope Extraction (BISE): A method for reducing noise in NDVI time-series. *Int. J. Remote Sens.* **1992**, *13*, 1585–1590. [CrossRef]

54. Maximilian Lange and Daniel Doktor Phenex: Auxiliary Functions for Phenological Data Analysis. Note = R Package Version 1.4-5, 2017. Available online: https://CRAN.R-project.org/package=phenex (accessed on 13 September 2018).

55. Hogda, K.A.; Karlsen, S.R.; Solheim, I. Climatic change impact on growing season in Fennoscandia studied by a time series of NOAA AVHRR NDVI data. In Proceedings of the IEEE 2001 International Geoscience and Remote Sensing Symposium (IGARSS'01), Sydney, Australia, 9–13 July 2001; Volume 3, pp. 1338–1340. [CrossRef]

56. Karlsen, S.R.; Elvebakk, A.; Høgda, K.A.; Johansen, B. Satellite-based mapping of the growing season and bioclimatic zones in Fennoscandia. *Glob. Ecol. Biogeogr.* **2006**, *3*, 416–430. [CrossRef]

57. Cleveland, W.S.; Grosse, E.; Shyu, W.M. Local regression models. In *Statistical Models in S. Wadsworth & Brooks/Cole Advanced Books & Software Pacific Grove, CA*; CRC Press: Boca Raton, FL, USA, 1992; Volume 8. [CrossRef]

58. Inoue, Y.; Kurosu, T.; Maeno, H.; Uratsuka, S.; Kozu, T.; Dabrowska-Zielinska, K.; Qi, J. Season-long daily measurements of multifrequency (Ka, Ku, X, C, and L) and full-polarization backscatter signatures over paddy rice field and their relationship with biological variables. *Remote Sens. Environ.* **2002**, *81*, 194–204. [CrossRef]

59. Inoue, Y.; Sakaiya, E.; Wang, C. Capability of C-band backscattering coefficients from high-resolution satellite SAR sensors to assess biophysical variables in paddy rice. *Remote Sens. Environ.* **2014**, *140*, 257–266. [CrossRef]

60. Moran, M.S.E.; Vidal, A.E.; Troufleau, D.E.; Qi, J.E.; Clarke, T.R.E.; Pinter, P.J.E., Jr.; Mitchell, T.A.E.; Inoue, Y.E.; Neale, C.M.U. Combining multifrequency microwave and optical data for crop management. *Remote Sens. Environ.* **1997**, *61*, 96–109. [CrossRef]

61. Macelloni, G.; Paloscia, S.; Pampaloni, P.; Marliani, F.; Gai, M. The relationship between the backscattering coefficient and the biomass of narrow and broad leaf crops. *IEEE Trans. Geosci. Remote Sens.* **2001**, *39*, 873–884. [CrossRef]

62. Ali, I.; Schuster, C.; Zebisch, M.; Förster, M.; Kleinschmit, B.; Notarnicola, C. First results of monitoring nature conservation sites in alpine region by using very high resolution (VHR) X-band SAR data. *IEEE J. Sel. Top. Appl. Earth Obs. Remote Sens.* **2013**, *6*, 2265–2274. [CrossRef]

63. Moran, M.S.; Alonso, L.; Moreno, J.F.; Cendrero Mateo, M.P.; De La Cruz, D.F.; Montoro, A. A RADARSAT-2 quad-polarized time series for monitoring crop and soil conditions in Barrax, Spain. *IEEE Trans. Geosci. Remote Sens.* **2012**, *50*, 1057–1070. [CrossRef]

64. Verbesselt, J.; Hyndman, R.; Newnham, G.; Culvenor, D. Detecting trend and seasonal changes in satellite image time series. *Remote Sens. Environ.* **2010**, *114*, 106–115. [CrossRef]

65. Prevot, L.; Champion, I.; Guyot, G. Estimating surface soil moisture and leaf area index of a wheat canopy using a dual-frequency (C and X bands) scatterometer. *Remote Sens. Environ.* **1993**, *46*, 331–339. [CrossRef]

66. Baghdadi, N.; El Hajj, M.; Zribi, M.; Bousbih, S. Calibration of the water cloud model at C-Band for winter crop fields and grasslands. *Remote Sens.* **2017**, *9*, 969. [CrossRef]

67. El Hajj, M.; Baghdadi, N.; Zribi, M.; Belaud, G.; Cheviron, B.; Courault, D.; Charron, F. Soil moisture retrieval over irrigated grassland using X-band SAR data. *Remote Sens. Environ.* **2016**, *176*, 202–218. [CrossRef]

68. Notarnicola, C.; Solorza, R. Integration of remotely sensed images and electromagnetic models into a bayesian approach for soil moisture content retrieval: Methodology and effect of prior information. In *Dynamic Programming and Bayesian Inference, Concepts and Applications*; InTech: Horwich, UK, 2014; pp. 48–52. [CrossRef]

69. Dabrowska-Zielinska, K.; Inoue, Y.; Kowalik, W.; Gruszczynska, M. Inferring the effect of plant and soil variables on C-and L-band SAR backscatter over agricultural fields, based on model analysis. *Adv. Space Res.* **2007**, *30*, 139–148. [CrossRef]

70. Beck, P.S.A.; Atzberger, C.; Høgda, K.A.; Johansen, B.; Skidmore, A.K. Improved monitoring of vegetation dynamics at very high latitudes: A new method using MODIS NDVI. *Remote Sens. Environ.* **2006**, *100*, 321–334. [CrossRef]

71. Chen, J.; Jönsson, P.; Tamura, M.; Gu, Z.; Matsushita, B.; Eklundh, L. A simple method for reconstructing a high-quality NDVI time-series data set based on the Savitzky–Golay filter. *Remote Sens. Environ.* **2004**, *91*, 332–344. [CrossRef]

72. Rodriguez-Galiano, V.F.; Dash, J.; Atkinson, P.M. Intercomparison of satellite sensor land surface phenology and ground phenology in Europe. *Geophys. Res. Lett.* **2015**, *42*, 2253–2260. [CrossRef]

73. Kosmala, M.; Crall, A.; Cheng, R.; Hufkens, K.; Henderson, S.; Richardson, A.D. Season Spotter: Using citizen science to validate and scale plant phenology from near-surface remote sensing. *Remote Sens.* **2016**, *8*, 726. [CrossRef]

74. Richardson, A.D.; Hufkens, K.; Milliman, T.; Frolking, S. Intercomparison of phenological transition dates derived from the PhenoCam Dataset V1. 0 and MODIS satellite remote sensing. *Sci. Rep.* **2018**, *8*, 5679. [CrossRef] [PubMed]

75. Lüth, C.; Tasser, E.; Niedrist, G.; Dalla Via, J.; Tappeiner, U. Plant communities of mountain grasslands in a broad cross-section of the Eastern Alps. *Flora-Morphol. Distrib. Funct. Ecol. Plants* **2011**, *206*, 433–443. [CrossRef]
76. Shi, J.; Du, Y.; Du, J.; Jiang, L.; Chai, L.; Mao, K.; Xu, P.; Ni, W.; Xiong, C.; Liu, Q.; et al. Progresses on microwave remote sensing of land surface parameters. *Sci. China Earth Sci.* **2012**, *55*, 1052–1078. [CrossRef]
77. Knipling, E.B. Physical and physiological basis for the reflectance of visible and near-infrared radiation from vegetation. *Remote Sens. Environ.* **1970**, *1*, 155–159. [CrossRef]

MDPI
St. Alban-Anlage 66
4052 Basel
Switzerland
Tel. +41 61 683 77 34
Fax +41 61 302 89 18
www.mdpi.com

Remote Sensing Editorial Office
E-mail: remotesensing@mdpi.com
www.mdpi.com/journal/remotesensing